SIMILARITY SOLUTIONS FOR THE BOUNDARY LAYER FLOW AND HEAT TRANSFER OF VISCOUS FLUIDS, NANOFLUIDS, POROUS MEDIA, AND MICROPOLAR FLUIDS

SIMILARITY SOLUTIONS FOR THE BOUNDARY LAYER FLOW AND HEAT TRANSFER OF VISCOUS FLUIDS, NANOFLUIDS, POROUS MEDIA, AND MICROPOLAR FLUIDS

JOHN H. MERKIN

Department of Applied Mathematics, University of Leeds, Leeds, United Kingdom

IOAN POP

Department of Mathematics, Faculty of Mathematics and Computer Science, Babeş-Bolyai University, Cluj-Napoca, Romania

YIAN YIAN LOK

Mathematics Section, School of Distance Education, Universiti Sains Malaysia, Pulau Pinang, Malaysia

TEODOR GROSAN

Department of Mathematics, Faculty of Mathematics and Computer Science, Babeş-Bolyai University, Cluj-Napoca, Romania

ELSEVIER

ACADEMIC PRESS

An imprint of Elsevier

Academic Press is an imprint of Elsevier
125 London Wall, London EC2Y 5AS, United Kingdom
525 B Street, Suite 1650, San Diego, CA 92101, United States
50 Hampshire Street, 5th Floor, Cambridge, MA 02139, United States
The Boulevard, Langford Lane, Kidlington, Oxford OX5 1GB, United Kingdom

Library of Congress Cataloging-in-Publication Data
A catalog record for this book is available from the Library of Congress

British Library Cataloguing-in-Publication Data
A catalogue record for this book is available from the British Library

ISBN: 978-0-12-821188-5

For information on all Academic Press publications visit our website at https://www.elsevier.com/books-and-journals

Publisher: Matthew Deans
Acquisitions Editor: Brian Guerin
Editorial Project Manager: Hilary Carr
Production Project Manager: Sojan P. Pazhayattil
Cover designer: Victoria Pearson

Typeset by TNQ Technologies

Contents

Preface

Fluid mechanics is a major part of applied mathematics, of physics, and in many branches of engineering, particularly civil, mechanical, chemical, and aeronautical engineering. It also plays an important role in naval architecture, geophysics, as well as in astrophysics, biological and physiological fluid dynamics. Significant contributions to the theory of airfoils came early in the 20th century, and during the whole of the first-half of this century applied aerodynamics was the major incentive, but also dealing with questions that were important in both mechanical and civil engineering.

Partial differential equations are the basis for nearly all technical processes in heat transfer and fluid mechanics. In our research activity, we became aware of the fact that a lot of researchers focus increasingly on numerical methods for solving the partial differential equations involved in their problems. Analytical methods, taught in fluid mechanics and heat transfer courses, are often quickly discarded, because of the common belief that almost all problems appearing in real applications can be solved easily by numerical methods. The basic idea of this book is to show some selected analytical and numerical methods and to explain their application to more complicated problems, which are technically relevant. Of course, this means that some of the standard analytical methods might not be discussed and present in this book (for example, integral transforms, Lie groups, etc.). On the other hand, the analytical methods discussed here are applicable to a range of interesting problems and to applied mathematicians. Also, PhD students or engineers could learn how to solve useful technical problems analytically. There is no doubt that the knowledge of numerical solution methods is very important, and there is an opportunity in using both numerical and analytic tools to gain useful insights into flow physics and heat transfer characteristics. However, numerical methods are always dependent on grid quality and grid size and on a lot of implementation features.

Perhaps the most important development in fluid mechanics during the 20th century was the concept of boundary layer flow introduced by Prandtl in 1904. A boundary layer is a layer of fluid which forms on a surface bounding the moving fluid. Every time a fluid moves along a surface a boundary layer forms on the surface. Therefore, boundary layers exist in the interior of water pipes, in sewer pipes, in irrigation channels, near the earth's surface, and around buildings due to winds, near aeroplane wings, around a moving car, at the river bottom, inside the blood vessels, and so on. The boundary layer concept is, therefore, a very popular field in fluid mechanics for engineers, physicists, and mathematicians. Hundreds of papers are published each year in this field.

The two-dimensional and axisymmetric flows resulting from a continuously moving uniform solid surface through a fluid at rest (Sakiadis flat plate flow problem, 1961) is the converse to Blasius boundary layer flow (1908) in the sense that, in the Blasius flow, the fluid moves and the plate stays at rest. In Sakiadis flow, in comparison with the Blasius flow, the viscous drag is larger and the other physical parameters of boundary layer flow such as the momentum thickness and the displacement thickness also differ from Blasius flow. Because of its fundamental importance, this type of moving-surface flow has become the subject of numerous studies on boundary layer flow, stability, transition, and turbulence. There exists another interesting class of viscous flows, commonly known as stretching sheet flows (Crane, 1970) and shrinking sheet flows (Miklavcic and Wang, 2006), where the wall velocity is in the opposite direction to other applied effects, for example, in the opposite direction to an outer flow or to buoyancy-driven convection. The self-similar boundary layer solutions to these stretching/shrinking and moving surfaces have yet to be explored completely. These interactive flows involve at least one-dimensionless parameter that can result in critical values. This feature has an important effect in both restricting the range of the similarity solution and in the temporal stability of these solutions (Goldstein, 1965).

Very little is known about the stretching/shrinking surfaces in flows in porous media, micropolar fluids, nanofluids, and hybrid nanofluids. Therefore, in comparison with the classical viscous flow over stretching/shrinking surfaces, the non-Newtonian flows due to continuously moving surfaces have not been fully explored and so require more detailed consideration. This will be an important aspect of this book. Similarity solutions will be derived and then solved by various efficient numerical methods such as routine bvp4c from Matlab, Keller-box method, singularity methods, etc. Numerical solutions and asymptotic results for limiting cases will be presented to investigate how the flow develops at the leading edge and its end behavior. Further, several examples of the problems of boundary layer flow and heat transfer in viscous fluid, porous media, micropolar fluids, nanofluids, and hybrid nanofluids due to continuously moving, stretching/shrinking surfaces will be considered to show the adaptability of different nanofluids to various industrial and commercial applications.

The past three decades are the clear evidence for the significant growth in research on nanofluids and hybrid nanofluids. In this book, several nanofluid and hybrid nanofluids published papers are considered to show the adaptability of different nanofluids to various industrial and commercial applications. It is evident that there is an intense interest among the scientific community on the nanofluids research. In the year 2019 alone about 1960 papers were published on nanofluids whereas about 140 articles were published on hybrid nanofluids. In addition, both effective analytic and numerical solutions to MHD flow and heat transfer past a permeable stretching/shrinking sheet in a hybrid nanofluid with suction/injection and convective boundary conditions are presented. Nanoparticles of alumina (Al_2O_3), and copper (Cu), were added to the base fluid (water) to mix as hybrid nanofluid. Proper similarity variables transformations were used to transform the system of partial differential equations into a system of ordinary (similarity) differential equations. The governing boundary-layer equations were exactly solved for both stretching and shrinking sheets. The present book is an attempt to put down some of this experience. Nonlinear problems pose a challenge that is often difficult to resist; each new exact solution is a thing of joy. This book has been written for graduate students, PhD students, mechanical and chemical engineers, researchers as well as an advanced graduated level textbook in the fields of applied mathematics, fluid mechanics, and heat transfer.

In the writing of this book, we have been much helped by Professor Anuar Ishak, Professor Roslinda Nazar, Professor Norihan Mohd Arifin, Professor Mikhail Sheremet, Professor Mustafa Turkylmazoglu, Professor Hang Xu, Professor Emad H. Aly, Dr. Norfifah Bachok, Dr. Fadzilah Ali, Dr. Syakila Ahmad, Dr. Hafidzuddin E. Hafidz, Dr. Mohammad Ghalambaz, Dr. Azizah Rohni, Dr. Kohilavani Naganthran, Dr. Anuar Jamaludin, and PhD students Waini Iskandar and Khashi'ie N. Safwa. We are particularly indebted to our colleagues Dr. Radu Trimbitas, Dr. Natalia C. Rosca, Dr. Alin V. Rosca, and Dr. Cornelia Revnic for their valuable help in the preparation of this book.

Finally, we are very grateful for the very good cooperation during the preparation of this book with Mrs. Hilary Carr from Elsevier for her considerable patience and care.

Our families provided us invaluable support, care, and comfort.

February 2021

CHAPTER 1

Basic equations and mathematical methods

1.1 Basic equations

1.1.1 Viscous fluids and heat transfer

Fluid mechanics is a part of applied mathematics, of physics, of many branches of engineering, certainly civil, mechanical, chemical, and aeronautical engineering, and of naval architecture and geophysics, with astrophysics and biological and physiological fluid dynamics to be added. Significant contributions to the theory of airfoils came early in the century, and during the whole of the first-half of the century applied aerodynamics was to be probably the major incentive, dealing with questions that were important also in mechanical and civil engineering; but geophysical questions, certainly not without charm and fascination, received much attention.

For a classical viscous and incompressible fluids, the continuity, momentum (Navier–Stokes), and energy equations in vectorial form are (Bejan, 2013; Pop & Ingham, 2001; Schlichting & Gersten, 2000)

$$\nabla \cdot \boldsymbol{v} = 0 \tag{1.1}$$

$$\frac{\partial \boldsymbol{v}}{\partial t} + (\boldsymbol{v} \cdot \nabla)\boldsymbol{v} = -\frac{1}{\rho}\nabla p + \nabla(\nu \nabla \boldsymbol{v}) + \frac{\sigma}{\rho}[(\boldsymbol{v} \times \boldsymbol{B}) \times \boldsymbol{B}] + \beta(T - T_0)\boldsymbol{g} \tag{1.2}$$

$$\frac{\partial T}{\partial t} + (\boldsymbol{v} \cdot \nabla)T = \nabla(\alpha \nabla T) + \frac{\mu}{\rho C_p}\Phi + \frac{\sigma\,(\boldsymbol{j} \times \boldsymbol{B})^2}{\rho\left(\rho C_p\right)} + q''' \tag{1.3}$$

$$\nabla \times \boldsymbol{B} = \mu_0 \boldsymbol{j}, \quad \nabla \times \boldsymbol{E} = -\frac{\partial \boldsymbol{B}}{\partial t}, \quad \nabla \cdot \boldsymbol{B} = 0 \tag{1.4}$$

where Φ is the viscous dissipation which can be expressed in the Cartesian coordinates (x, y) as (Bejan, 2013)

$$\Phi = \left[\left(\frac{\partial u}{\partial x}\right)^2 + \left(\frac{\partial v}{\partial y}\right)^2\right] + \left(\frac{\partial u}{\partial y} + \frac{\partial v}{\partial x}\right)^2 \tag{1.5}$$

Similarity Solutions for the Boundary Layer Flow and Heat Transfer of Viscous Fluids, Nanofluids, Porous Media, and Micropolar Fluids
ISBN 978-0-12-821188-5
https://doi.org/10.1016/B978-0-12-821188-5.00002-3

and $\boldsymbol{v}$ is the velocity vector, p is the pressure, and the physical meaning of the other quantities is mentioned in the Nomenclature.

1.1.2 Porous medium

The interest about the porous media has been motivated by the fact that thermally driven flows in porous media have considerable applications in mechanical, chemical, and civil engineering. Applications include food processing and storage, fibrous insulation, thermal insulation of buildings, geophysical systems, electro chemistry, metallurgy, the design of pebble bed nuclear reactors, underground disposal of nuclear or nonnuclear waste, solar power collectors, nuclear reactors, geothermal applications, etc. Comprehensive reviews of heat transfer mechanisms in geothermal systems have been first presented by Cheng (1979, 1985) and Bejan (1987). Since then, many practical applications, both industrial and environmental, have caused a rapid extension of the research, and a substantial number of articles, which relate the boundary-layer flow past surfaces of various flow configuration models embedded in porous media, have been published. Ingham and Pop (1998, 2002, 2005), Pop and Ingham (2001), Vafai (2010, 2015), Bejan et al. (2004), Vadasz (2008), Nield and Bejan (2017), and Bear (2018) identified many applications that highlight the directions where further theoretical and experimental developments and investigations in porous media are required. For the classical porous media, the continuity, momentum (non-Darcy) and energy equations in vectorial form are

$$\nabla \cdot \mathbf{v} = 0 \tag{1.6}$$

$$\nabla p = -\mu/K\,\mathbf{v} + \rho[1 - \beta(T - T_\infty)]\mathbf{g} + \surd(|\mathbf{v}|) \tag{1.7}$$

$$\frac{\partial T}{\partial t} + (\nabla \cdot \mathbf{v})T = \alpha \nabla^2 T \tag{1.8}$$

The physical meaning of the variables and quantities is mentioned in the Nomenclature.

1.1.3 Nanofluids

Convective flows in viscous fluids, nanofluids, and hybrid nanofluids have extensive engineering applications in aerodynamics, geothermal systems, crude oil extractions, groundwater pollution, thermal insulation, heat exchanger, storage of nuclear waste, bulging construction, agriculture, etc. Ultrahigh performance cooling is one of the most vital needs of many industrial technologies. Nanofluids are a relatively new class of fluids which consist of a base fluid with nano-sized particles (1–100 nm) suspended within it. It was introduced by Choi in 1995 at Argonne National

Laboratory, USA. It is well known that conventional fluids do not allow to essentially intensify fluid flow and heat transfer. At the same time, an inclusion in a traditional working fluid nanoparticles of metal or metal oxide with low concentration can lead to an intensification of flow and heat transfer due to a variation of thermal conductivity and viscosity of the suspension (Minkowycz et al., 2013; Shenoy et al., 2016; Tiwari & Das, 2007). However, the study of nanofluids is still not at the end, and it seems very difficult to have a precise idea on the way the use of nanoparticles acts in fluid flow and heat transfer, and complementary works are needed to understand the heat transfer characteristics of nanofluids and identify new and unique applications for these fluids. Usual base fluids are as follows: water, ethylene glycol, oil, engine oil, etc., while the added nanoparticles can be metals (Cu, Al, Ti, Zn, Ni, Fe), carbides (SiC), nitrides (AlN, SiN), nonmetals (graphite, carbon nanotubes, diamond), oxides (CuO, Al_2O_3, TiO_2, ZnO), etc.

Compared to conventional solid–liquid suspensions for heat transfer intensifications, properly engineered thermal nanofluids possess the following advantages: high specific surface area and therefore more heat transfer surface between particles and fluids; high dispersion stability with predominant Brownian motion of particles; reduced pumping power as compared to pure liquid to achieve equivalent heat transfer intensification; reduced particle clogging as compared to conventional slurries, thus promoting system miniaturization; and adjustable properties, including thermal conductivity and surface wettability, by varying particle concentrations to suit different applications.

Nanofluids have a wide range of applications in heat transfer applications and industrial cooling. Due to the lack of abundant sources of clean energy and widespread dissemination of battery operated devices, such as cell phones and laptop, it is essential to use nanofluids as a smart fluid. Nanofluids can be used as a smart material working as a heat valve to control the heat flow (Das et al., 2008; Minkowycz et al., 2013; Shenoy et al., 2016). In nuclear power plant system, the limiting process in the generation of steam is critical heat flux between the fuel rods and the water-when vapor bubbles that end up covering the surface of the fuel rods conduct very little heat as opposed to liquid water. Using nanofluids instead of water, the fuel rods become coated with nanoparticles such as alumina, which actually pushes newly formed bubbles away, preventing the formation of a layer of vapor around the rod and subsequently increasing the critical heat flux significantly (Das et al., 2008; Minkowycz et al., 2013; Shenoy et al., 2016). Nanofluids can be used for liquid cooling of computer processors due to their high thermal conductivity. It is predicted that the next generation of computer chips will produce localized heat flux over 10 MW/m, with the total power exceeding 300 W. In combination with thin film evaporation, the nanofluid oscillating heat pipe cooling system will be able to remove heat flux server 10 MW/m and serve as the next-generation cooling device that will be able to handle the heat dissipation coming from new technology.

Buongiorno (2006) conducted a study of convective transport in nanofluids with a focus on explaining the heat transfer enhancements which were observed during convective situations. Buongiorno (2006) discounts suspension, particle rotation, dispersion, and turbulence as being significant agents which cause heat transfer enhancements, and a new model was proposed by him, which is based on the mechanics of nanoparticles/base-fluid relative velocity. He stated that the nanoparticle absolute velocity can be viewed as the sum of the base fluid velocity and a relative (slip) velocity. He proposed seven slip mechanisms including inertia, Brownian diffusion, thermophoresis, diffusiophoresis, Magnus effect, fluid drainage, and gravity. Buongiorno (2006) concluded that in the absence of turbulent effects, Brownian motion and thermophoresis will become important as slip mechanisms. Then, he developed a nonhomogeneous two-component four-equation model taking into account these two mechanisms. He concluded that Brownian diffusion and thermophoresis dominate when the turbulent effects are absent. Assuming that the nanofluid is incompressible and laminar, and using the nanofluid model as proposed by Buongiorno (2006), the governing equations in vectorial form are

$$\nabla \cdot \boldsymbol{v} = 0 \tag{1.9}$$

$$\frac{\partial \boldsymbol{v}}{\partial t} + (\boldsymbol{v} \cdot \nabla)\boldsymbol{v} = -\frac{1}{\rho}\nabla p + \nabla(\nu \nabla \boldsymbol{v}) + \beta(T - T_0)\boldsymbol{g} \tag{1.10}$$

$$\frac{\partial T}{\partial t} + (\boldsymbol{v} \cdot \nabla)T = \nabla(\alpha \nabla T) + \omega\left(D_B \nabla C \cdot \nabla T + \frac{D_T}{T_0}\nabla T \cdot \nabla T\right) \tag{1.11}$$

$$\frac{\partial C}{\partial t} + (\boldsymbol{v} \cdot \nabla)C = D_B \nabla^2 C + \frac{D_T}{T_0}\nabla^2 T \tag{1.12}$$

The mathematical nanofluid model proposed by Tiwari and Das (2007) is given as

$$\nabla \cdot \boldsymbol{v} = 0 \tag{1.13}$$

$$\frac{\partial v}{\partial t} + (\boldsymbol{v} \cdot \nabla)\boldsymbol{v} = -\frac{1}{\rho_{nf}}\nabla p + \frac{\mu_{nf}}{\rho_{nf}}\nabla^2 \boldsymbol{v} + \beta_{nf}(T - T_0)\boldsymbol{g} \tag{1.14}$$

$$\frac{\partial T}{\partial t} + (\boldsymbol{v} \cdot \nabla)T = \frac{k_{nf}}{\left(\rho C_p\right)_{nf}}\nabla^2 T \tag{1.15}$$

Here the thermophysical properties of a typical nanofluid, $\mu_{nf}, \rho_{nf}, k_{nf}$, and $\left(\rho C_p\right)_{nf}$, are given in Table 1.1, and the physical meaning of these quantities is mentioned in the Nomenclature. The modified physical properties of the nanofluid can be calculated as follows, see Tiwari and Das (2007):

$$\left.\begin{aligned}
&\text{Dynamic viscosity}: \mu_{nf} = \frac{\mu_f}{(1-\varphi)^{2.5}}\\
&\text{Density}: \rho_{nf} = (1-\varphi)\rho_f + \varphi\rho_{s1}\\
&\text{Dynamic viscosity}: \mu_{nf} = \frac{\mu_f}{(1-\varphi)^{2.5}}\\
&\text{Thermal conductivity } k_{nf} = \frac{\left(k_{s1} + 2k_f - 2\varphi\left(k_f - k_{s1}\right)\right)}{\left(k_{s1} + 2k_f + \varphi\left(k_f - k_{s1}\right)\right)} \times \left(k_f\right)\\
&\text{Thermal expansion } (\rho\beta)_{nf} = (1-\varphi)(\rho\beta)_f + \varphi(\rho\beta)_{s1}\\
&\text{Heat capacity } \left(\rho C_p\right)_{nf} = (1-\varphi)\left(\rho C_p\right)_f + \varphi\left(\rho C_p\right)_{s1}\\
&\text{Electrical conductivity } \frac{\sigma_{nf}}{\sigma_f} = 1 + \frac{3\varphi\left(\frac{\sigma_{s1}}{\sigma_f} - 1\right)}{2 + \frac{\sigma 1}{\sigma_f} - \varphi\left(\frac{\sigma_{s1}}{\sigma_f} - 1\right)}
\end{aligned}\right\} \tag{1.16}$$

Convective flow and heat transfer in a porous medium saturated by nanofluids have received much attention because of their possible

Table 1.1 Thermophysical properties of fluid and nanoparticles (Oztop & Abu-Nada, 2008).

Physical properties	Fluid phase (water)	Cu	Al_2O_3	TiO_2
C_p(J/kg K)	4179	385	765	686.2
ρ (kg/m^3)	997.1	8933	3970	4250
k (W/m K)	0.613	400	40	8.9538
$\alpha \times 10^{-7}$ (m^2/s)	1.47	11,163.1	131.7	30.7
$\beta \times 10^{-5}$ (1/K)	21	1.67	0.85	0.9

applications in many branches of science and technology. Practical importance of such applications is, for example, in geothermal systems, nuclear reactors, thermal insulations, drying processes, food processing and storage, fibrous insulation, thermal insulation of buildings, electrochemistry, metallurgy, the design of pebble bed nuclear reactors, underground disposal of nuclear or nonnuclear waste, solar power collectors, nuclear reactors, etc. Comprehensive reviews on this topic can be found in review papers by Barletta et al. (2015) and Khanafer and Vafai (2019). A mathematical nanofluid model for Darcy porous media proposed by Kuznetsov and Nield (2013) is given below:

$$\nabla \cdot \langle \mathbf{v} \rangle = 0 \tag{1.17}$$

$$\frac{\mu}{K}\langle \boldsymbol{v} \rangle = -\nabla p + \left\{ C\rho_p + (1-C)\left[\rho_f(1-\beta(T-T_0))\right]\right\}\boldsymbol{g} \tag{1.18}$$

$$(\boldsymbol{v} \cdot \nabla)T = \alpha \nabla^2 T + \omega\left(D_B \nabla C \cdot \nabla T + \frac{D_T}{T_0}\nabla T \cdot \nabla T\right) \tag{1.19}$$

$$\frac{1}{\varepsilon}(\mathbf{v} \cdot \nabla)C = D_B \nabla^2 C + \frac{D_T}{T_0}\nabla^2 T \tag{1.20}$$

where ε is the porosity of porous media, K is the permeability of the porous medium, and the physical meaning of the other variables is mentioned in the Nomenclature.

To further improve the characteristics of nanofluids, that could possess a number of favorable characteristics, researchers developed a new-generation heat transfer fluid called hybrid nanofluid. These hybrid nanofluids are prepared either by dispersing dissimilar nanoparticles as individual constituents or by dispersing nanocomposite particles in the base fluid. Hybrid nanofluids may possess better thermal network and rheological properties due to synergistic effect.

In the absence of body forces, the governing equations of the hybrid nanofluids are described in vectorial form by Devi and Devi (2016a, 2016b) as

$$\nabla \cdot \boldsymbol{v} = 0 \tag{1.21}$$

$$\frac{\partial \boldsymbol{v}}{\partial t} + (\boldsymbol{v} \cdot \nabla)\boldsymbol{v} = -\frac{1}{\rho_{hnf}}\nabla p + \frac{\mu_{hnf}}{\rho_{hnf}}\nabla^2 \boldsymbol{v} + \frac{\mu_{hnf}}{\rho_{hnf}}\Phi + \frac{\sigma_{hnf}}{\rho_{hnf}}[(\boldsymbol{v} \times B) \times B] + \beta_{hnf}(T - T_0)\boldsymbol{g} \tag{1.22}$$

$$\frac{\partial T}{\partial t}+(\boldsymbol{v}\cdot\nabla)T=\frac{k_{hnf}}{(\rho C_p)_{hnf}}\nabla^2 T+\frac{\mu_{hnf}}{(\rho C_p)_{hnf}}\Phi+\frac{\sigma_{hnf}}{\rho_{hnf}(\rho C_p)_{hnf}}(\boldsymbol{j}\times\boldsymbol{B})^2 \tag{1.23}$$

where Φ is the viscous dissipation, which can be expressed in the Cartesian coordinates (x, y) as

The thermophysical properties of fluid at 25°C and particles are provided in Table 1.1 (Oztop & Abu-Nada, 2008). Following Devi and Devi (2016a, 2016b) and Devi and Devi (2017), a special form of thermophysical properties is introduced for the boundary layer equations of the hybrid nanofluid, which is considered by taking the mixture of Cu nanoparticles into 0.1 volume of Al_2O_3/water to form the required hybrid nanofluid. These thermophysical properties of the hybrid nanofluid are given in Eq. (1.24), where spherical nanoparticles are considered. In this model, initially the nanoparticle of $Al_2O_3(\varphi_1$)is added to the base fluid with 0.1 volume solid volume fraction (i.e., $\varphi_1 = 0.1$), which is fixed throughout the problem hereafter and consequently Cu (φ_2) is added with various solid volume fractions to form the hybrid nanofluid namely $Cu{-}Al_2O_3$/water.

$$\left.\begin{array}{l}
\text{Density:}\quad \rho_{hnf}=(1-\varphi_2)\left[(1-\varphi_1)\rho_f+\varphi_1\rho_{s1}\right]+\varphi_2\rho_{s2}\\
\text{Dynamic viscosity:}\quad \mu_{nf}=\dfrac{\mu_f}{(1-\varphi_1)^{2.5}(1-\varphi_2)^{2.5}}\\
\text{Thermal conductivity}\quad k_{hnf}=\dfrac{\left(k_{s2}+2k_{nf}-2\varphi_2\left(k_{nf}-k_{s2}\right)\right)}{\left(k_{s2}+2k_{nf}+\varphi_2\left(k_{nf}-k_{s2}\right)\right)}\times\left(k_{nf}\right)\quad \text{where } k_{nf}=\dfrac{\left(k_{s1}+2k_f-2\varphi_1\left(k_f-k_{s1}\right)\right)}{\left(k_{s1}+2k_f+\varphi_1\left(k_f-k_{s1}\right)\right)}\times\left(k_f\right)\\
\text{Thermal expansion}\quad \beta_{hnf}=\varphi_1\beta_{s1}+\varphi_2\beta_{s2}+\left(1-\varphi_{hnf}\right)\beta_f\quad \text{where } \varphi_{hnf}=\varphi_1+\varphi_2\\
\text{Heat capacity}\quad (\rho C_p)_{hnf}=(1-\varphi_2)\left[(1-\varphi_1)(\rho C_p)_f+\varphi_1(\rho C_p)_{s1}\right]+\varphi_2(\rho C_p)_{s2}\\
\text{Electrical conductivity}\quad \dfrac{\sigma_{hnf}}{\sigma_f}=\dfrac{\sigma_{s2}+2\sigma_{bf}+2\sigma_2\left(\sigma_{s2}-\sigma_f\right)}{\sigma_{s2}+2\sigma_{bf}-2\sigma_2\left(\sigma_{s2}-\sigma_f\right)},\text{ where } \dfrac{\sigma_{bf}}{\sigma_f}=\dfrac{\sigma_{s1}+2\sigma_f+2\sigma_1\left(\sigma_{s1}-\sigma_f\right)}{\sigma_{s1}+2\sigma_f-2\sigma_1\left(\sigma_{s1}-\sigma_f\right)}
\end{array}\right\} \tag{1.24}$$

The physical meaning of μ_{hnf}, k_{hnf}, ρ_{hnf}, $(\rho C_p)_{hnf}$, and σ_{hnf} is mentioned in the Nomenclature; φ refers to the nanoparticle volume fraction ($\varphi_1 = \varphi_2 = 0$ correspond to a regular fluid).

It is worth mentioning that there are two main approaches to simulate the nanofluids flows, named single-phase (Khanafer et al., 2003, Tiwari & Das, 2007 and Devi & Devi, 2017) and two-phase approaches (Buongiorno, 2006). Single-phase approach supposes nanofluids as a homogenous fluid including base fluid and nanoparticles. In this way of study, effective parameters have been estimated by models which take into account the base fluids' and nanoparticles' thermophysical properties as well as nanoparticle volume concentration. These models have developed regardless of interactions between nanoparticles and base fluid molecules, but usually give acceptable results in low volume concentration (below10%) (Saghir et al., 2016). The two-phase approach looks at nanofluids and hybrid nanofluids as a nonhomogenous mixture with two distinct parts: base fluid and nanoparticles. These models usually take into consideration many forces like drag, gravity, buoyancy, thermophoresis, Brownian motions, etc. These models demonstrate more accurate results but require more processing time and power (Buongiorno et al., 2009).

1.1.4 Micropolar fluids

Micropolar fluids are non-Newtonian fluids with rigid, randomly oriented, and nondeformable particles. It belongs to the class of fluids where the stress tensor is nonsymmetric (Łukaszewicz, 1999). The theory of micropolar fluids was firstly introduced by Eringen in 1966 (Eringen, 1966) to model fluids with microstructure which cannot be adequately described by the classical Navier–Stokes theory. In this theory, the microscopic effects from the micromotions that arise from the microstructure and the intrinsic motion of the fluid elements are treated. The essence of the theory lies in the extension of the constitutive equations for Newtonian flow. In the mathematical modeling of micropolar fluids, the usual transport equations for the conservation of mass, momentum, and energy will have an additional constitutive parameter, on top of that, an additional transport equation representing the conservation of angular momentum needed to be solved simultaneously with the continuity, momentum, and energy equations. In 1972, Eringen (1972) extended the theory of micropolar fluids to thermomicropolar fluids which take into account the thermal effects. Some new effects that were not discussed in classical field theories are presented such as the occurrence of a thermodynamic pressure tensor, the coupling of the temperature gradient with the stress–moment constitutive equations, and the occurrence of the microgyrations in the heat-conduction equation (Eringen, 1972). Since the publication of Eringen (1966), Eringen (1972) micropolar fluid theory, many authors have investigated various flow and heat transfer problems with different geometries, effects, and boundary conditions. A thorough review of micropolar fluid was provided by Ariman et al. (1973) and Ariman et al. (1974), while the books by Łukaszewicz (1999) and (Eringen, 2001) give a comprehensive discussion on the theory and applications of micropolar fluids.

Examples of micropolar fluids include polymeric suspensions, exotic lubricants, colloidal fluids, animal blood, liquid crystal, etc. The presence of dust or smoke in a gas may also be modeled using micropolar fluid dynamics. Migun (1981) and Kolpashchikov et al. (1983) proposed an experimental method to determine the viscosity coefficients that characterize the microstructure of micropolar liquids. For example, the nematic liquid crystals are expected to have smaller viscosity coefficient. The applications of micropolar fluids can be found in the industrially relevant processes such as the extrusion of polymer melts in the production of sheeting materials, the solidification of liquid crystals in material processing, and cooling of a metallic plate in a bath.

The basic equations for micropolar fluids are (see Ahmad et al., 2012)

$$\nabla \cdot \boldsymbol{v} = 0 \tag{1.25}$$

$$\rho\left[\frac{\partial \boldsymbol{v}}{\partial t} + (\boldsymbol{v} \cdot \nabla)\boldsymbol{v}\right] = -\nabla p + (\delta + 2\mu + \kappa)\nabla(\nabla \cdot \boldsymbol{v}) - (\mu + \kappa)\nabla \times (\nabla \times \boldsymbol{v}) - \kappa(\nabla \times \boldsymbol{N}) \tag{1.26}$$

$$\rho j\left[\frac{\partial \boldsymbol{N}}{\partial t} + (\boldsymbol{v} \cdot \nabla)\boldsymbol{N}\right] = (\alpha + \beta + \gamma)\nabla(\nabla \cdot \boldsymbol{N}) - \gamma\nabla \times (\nabla \times \boldsymbol{N}) - \kappa(\nabla \cdot \boldsymbol{v}) - 2\kappa \boldsymbol{N} \tag{1.27}$$

$$\frac{\partial T}{\partial t} + (\boldsymbol{v} \cdot \nabla)T = \alpha \nabla^2 T \tag{1.28}$$

where κ is the vortex viscosity, j is the microinertia density, γ is the spin-gradient viscosity, and the physical meaning of other variables is mentioned in the Nomenclature. These fluid properties are constants, while the viscous dissipation and body couple are neglected.

1.2 Similarity solutions

The deduction of the boundary layer equations from Navier–Stokes' equations has been one of the most important advances in fluid dynamics. Using the order of magnitude analysis, the well-known governing Navier–Stokes' equations can be greatly simplified within the boundary layer. By taking the boundary layer approximations, the flow is divided into an inviscid portion that is easier to be solved by a number of methods. But, the boundary layer problem, which is governed by a set of highly nonlinear partial differential equations, is not easily solvable except for a few precise flows. The boundary layer equations are the set of nonlinear partial differential equations, which are very complicated for the sake of finding the closed-form solutions. However, with the help of similarity techniques, they are transformed to a set of mathematically tractable nonlinear ordinary differential equations (ODEs). The similarity techniques needed for the purpose essentially consist in converting the partial differential equations into ODEs. The reduction of the system of partial differential equations to a set of ODEs is achieved by way of transforming the dependent and independent variables to suitable nondimensional dependent and independent variables with the aid of similarity transformations. These transformed dependent and independent variables are called similarity variables. The subject of similarity boundary layer equations has been studied by many scientists, engineers, geophysicists, etc., over the years because of its numerous applications in the fields of engineering and technology, such as fluid mechanics, biomechanics, agriculture engineering, petroleum engineering, chemical engineering, soil mechanics, solar physics, atmospheric physics, aerodynamics, astrophysics, plasma physics, etc. (see Bejan, 2013; Gebhart et al., 1988; Pop & Ingham, 2001; Schlichting & Gersten, 2000; Rosenhead, 1963, etc.)

In a boundary layer, the flow varies quickly with the normal coordinate y and slowly with the tangential coordinate x. If the core flow is suitable, simplification may result from replacing y with a similarity coordinate $\eta = \frac{y}{\delta(x)}$ for some $\delta(x)$, the boundary layer thickness. Similarity solutions are appealing as they reduce the number of independent variables, which in some cases results in the unknowns satisfying ODEs. This is a much simpler matter to solve compared to the original PDE system (Ludlow, 2000).

It seems that Prandtl (1904) was the first, who has reduced the Navier–Stokes equations to a third-order nonlinear ODE in a similarity variable. Blasius (1908) considered the development of uniform flow over a flat plate. This solution was later generalized to wedge flow, leading to the family of Falkner–Skan solutions (1931). This family has additional three-dimensional solutions. However, many investigations are still required to be carried out on the boundary layer flows to exploit the topic of boundary layer flow.

On the other hand, it should be mentioned that it has been known for some time that the theory of continuous transformation groups gives promise of providing a very general method of analysis (Birkhoff, 1950, 1955; Hansen, 1964; Na & Hansen, 1967). In fact, the basic ideas date back to the last century and are found in the work of the mathematician Sophus Lie (2015). Moreover, the theory of groups has been applied quite extensively in recent times by investigators in the field of similarity analysis. Nevertheless, there still exists a need for a truly in-depth study of the application of Lie's theory to the similarity analysis of problems other than the reduction of the number of variables in a partial differential equation. It therefore seemed reasonable to find how far Lie's ideas might be pursued in formulating a very general approach to similarity analyses. The theory of continuous groups was first applied to the solution of partial differential equations by Birkhoff (1955), who used a one-parameter group. Morgan (1958) proved a theorem which established the condition under which the number of variables can be reduced by one.

1.3 Some numerical methods

1.3.1 Matlab program bvp4c

According to the description in Matlab documentation, bvp4c is a finite difference code that implements the three-stage Lobatto IIIa formula. First, we input the initial mesh of points and initial approximation of solutions. The collocation technique uses a mesh of points to determine the subinterval of the interval of integration. The differential equations along with the boundary conditions are then solved by the bvp solver. Numerical solutions are obtained and the resulting errors are estimated. If the solution does not satisfy the criterion of tolerance, then the solver adapts the mesh based on the residual of solutions and repeats the process of calculation. The obtained numerical solution is fourth-order accurate.

If a problem admits more than one solution, a "good" initial guess is needed to obtain the desired solutions. During the computation, the "infinity" in boundary conditions is replaced with a finite value. This finite value represents "sufficiently large" boundary layer thickness η_∞. The relative error of tolerance ε is set as the convergence criterion. The converged solution is identified when the solutions fulfill the initial and boundary conditions with the residuals less than ε. It is found that by considering the same initial guess with different values of boundary layer thickness will result in different set of solutions. On the other hand, the same boundary layer thickness with different initial guess will give different set of solutions too. If it is hard to provide a good initial guess of solutions, hence a tactic called continuation will be implemented. We start with a set of values for which the problem is easy to solve. The results obtained can then be used as an initial guess for solution of the problem with small variation of the variables.

1.3.2 Keller-Box method

The Keller-Box method is an implicit finite difference method that has been introduced by Keller (1970) in 1970. This numerical scheme with second-order accuracy has been devised for solving parabolic differential equations efficiently. The implicit nature of the Keller-Box method has generated a tridiagonal matrix, like other implicit method; however, the specialty of the Keller-Box method is the entries in the matrices are expressed in blocks rather than scalars. The detail of the procedures of this method can be found in the papers by Keller (1970), Keller and Cebeci (1972) and (Keller, 1978) as well as in the books by Cebeci (2002) and Cebeci and Bradshaw (2013). The Keller-Box method involves the following four steps:

1. Reduce the higher-order ODEs to a system of first-order ODEs by introducing new independent variables.
2. Express the first-order system in the form of finite difference using central differences derivatives and averages at the midpoints so that the continuous variables are replaced by discrete variables or points.
3. Linearize the resulting equations using Newton's method and write them in a matrix-vector form.
4. Solve the linear system along with their boundary conditions by the block-tridiagonal elimination method.

Finite Difference Scheme

Consider the net rectangle, for example, in the $\tau - \eta$ plane as shown in Fig. 1.1.

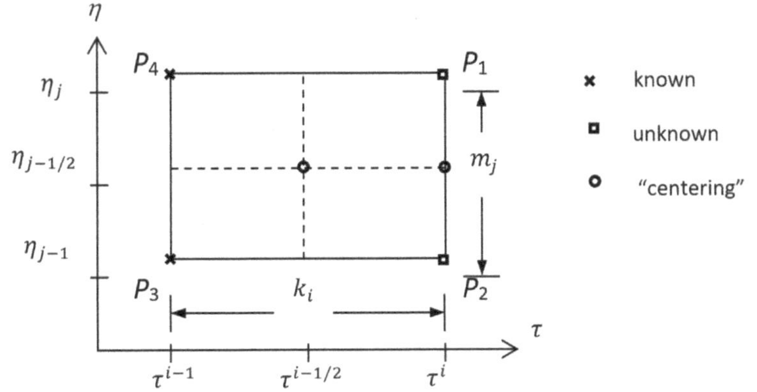

Figure 1.1 Net rectangle for difference approximation.

The net points are defined as Eq. (1.29)

$$\begin{cases} \tau_0 = 0, \quad \tau^i = \tau^{i-1} + k_i, \quad i = 1, 2, \ldots, I, \\ \eta_0 = 0, \quad \eta_j = \eta_{j-1} + m_j, \ j = 1, 2, \ldots, J, \quad \eta_J \equiv \eta_\infty \end{cases} \tag{1.29}$$

where k_i is the $\Delta\tau$ −spacing and m_j is the $\Delta\eta$ −spacing.

The derivatives in the $\tau-$ direction are replaced by

$$\frac{\partial()}{\partial\tau} = \frac{()^i - ()^{i-1}}{k_i} \tag{1.30}$$

while the derivatives in the $\eta-$ direction are replaced by

$$\frac{\partial()}{\partial\eta} = \frac{()_j - ()_{j-1}}{m_j} \tag{1.31}$$

The "centering" about the midpoint $\left(\tau^i, \eta_{j-1/2}\right)$ of the segment P_1P_2 in Fig. 1.1 is given by

$$()^i_{j-1/2} = \frac{1}{2}\left[()^i_j + ()^i_{j-1}\right] \tag{1.32}$$

Substituting the difference scheme Eqs. (1.30)−(1.32) into the first-order system yields the discretized difference equations.

Newton's Linearization

If we assume $()^{i-1}_j$ to be known for $0 \le j \le J$, then we have to obtain the solution of the unknowns $()^i_j$ for $0 \le j \le J$. To linearize the first-order system using Newton's method, we introduce the iterate,

$$()^{(n+1)}_j = ()^{(n)}_j + \delta()^{(n)}_j \tag{1.33}$$

Note that the superscript i is dropped for simplicity of notation, and (n) refers to the number of iterations. A linear diagonal system is obtained after ignoring those quadratic and higher-order terms in $\delta()^{(n)}_j$.

Block-Tridiagonal Elimination Method

The linearized system has a block-tridiagonal structure, which can be written in vector-matrix form

$$A\delta = r \tag{1.34}$$

where

$$A = \begin{bmatrix} [A_1] & [C_1] & & & & \\ [B_2] & [A_2] & [C_2] & & & \\ & \ddots & \ddots & \ddots & & \\ & & \ddots & \ddots & \ddots & \\ & & & \ddots & \ddots & \ddots \\ & & & [B_{J-1}] & [A_{J-1}] & [C_{J-1}] \\ & & & & [B_J] & [A_J] \end{bmatrix}, \quad \delta = \begin{bmatrix} [\delta_1] \\ [\delta_2] \\ \vdots \\ \vdots \\ \vdots \\ [\delta_{J-1}] \\ [\delta_J] \end{bmatrix}, \quad r = \begin{bmatrix} [r1] \\ [r_2] \\ \vdots \\ \vdots \\ \vdots \\ [r_{J-1}] \\ [r_J] \end{bmatrix} \tag{1.35}$$

The coefficient matrix A is known as a tridiagonal matrix, where all elements are zero except those three along the diagonal. Commonly, the block-tridiagonal structure consists of variables or constants, but here, an interesting feature that can be observed is, for the Keller-Box method, it consists of block matrices. The linear system Eqs. (1.34) is solved for δ by using LU decomposition method which involves forward sweep and backward sweep. Once the values of δ are found, then the solution can be obtained from Eqs. (1.33), and the steps can be proceeded to find the solution at the next iterations.

1.3.3 Runge—Kutta method

One of the most known and efficient numerical methods used to solve initial values problems (IVPs) is the Runge—Kutta method. This class of one-step methods, based on the Taylor series, was developed by Runge that proofed in 1885 that it is possible to use intermediary time steps in approximating derivatives with a high order of accuracy and implemented later, in 1901, by Kutta.

In order to keep a desired error on an integration step, variable length steps are required. The most effective method to estimate this error is to couple two Runge—Kutta methods, see Butcher (2008). Well-known pairs of Runge—Kutta methods such as Bogacki—Shampine, Runge—Kutta—Fehlberg, and Dormand—Prince are the basis of the well-known ode23 and ode45 functions from Matlab (Moler, 2014).

The Runge—Kutta method along with shooting technique used to be widely used in numerical simulation related to boundary layer problems (Na, 1979). However, during the last years, the mathematical models coming from the boundary layer approximation (boundary values systems of differential equations) became more complicated and, therefore, more

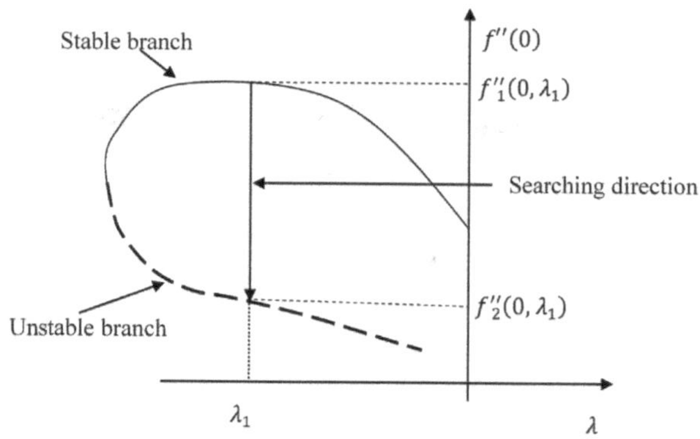

Figure 1.2 Searching for the initial (starting) solution for the unstable branch.

specialized methods are needed and many researchers migrated toward the bvp4c routine from Matlab. However, the Runge–Kutta method can be still used to approximate the initial solution guess for specialized boundary values problems (BVPs) solvers, especially for the cases when multiple solutions are expected.

A typical situation is presented in Fig. 1.2. Sometime it is very difficult to find a good initial solution for the unstable branch for a BVP numerical method. Therefore, starting from the stable branch (full line), in a chosen point λ_1, the boundary condition at infinity of the BVP governing system is replaced with an initially prescribed value of $f_1''(0, \lambda_1)$, the boundary value problem being transformed in an IVP. Next, the IVP is solved using a Runge–Kutta method, the value of $f''(0, \lambda_1)$ being successively decreased with a suitable small step. When the boundary condition at the infinity is fulfilled again, for $f_2''(0, \lambda_1\ )$ say, then the profile of the unstable solution is obtained. Using this profile and the continuation technique (see, Section 1.3.1), it is possible to obtain the unstable branch.

1.4 Analytical solution methods

In the analysis of time-dependent IVP, an essential part is a full understanding of the possible large time solutions, steady states. The ODEs governing these steady states usually involve one or more dimensionless parameters. A full appreciation of how the solutions to the steady states vary with these parameters and, in particular, where solutions do not exist through the identification of critical values is a necessary requirement for assessing how the solution to the IVP will evolve with time. Results obtained from studying the steady states can often suggest cases to examine further for the IVP. Thus, there is a direct interaction between the solutions to the steady states and solutions to the IVP. To calculate the critical points, a linear perturbation of the original steady-state equation is made. This linear equation together with the original equation forms an eigenvalue problem for the identified dimensionless parameter λ say. The value of the critical value λ_{crit} of λ has usually to be determined numerically. If the system of steady-state equations involves any further dimensionless parameters, the values of λ_{crit} can then be determined in terms of these additional parameters and the regions of existence of a solution mapped out in parameter space. For general values of the parameters, the steady-state equations have normally to be solved numerically using one of a range of two-point boundary value solvers freely available. These numerical integrations describe the solution as a distinguished parameter λ (say) is varied, for example, giving the wall stress τ_w or wall heat transfer q_w (say) in terms of λ.

The numerical solutions can indicate that the solution continues to large, positive and/or negative values of λ. To complete this solution, an asymptotic expansion valid for λ large can be constructed. In this a transformation of variables, both dependent and independent, is normally made. A solution of the transformed equations is then sought through a series. A λ asymptotic series can act as a check on any numerical procedure and give results for parameter ranges where it might be difficult to do so. They can also provide insights into the nature of the general problem. When the distinguished parameter λ has critical value at λ_{crit}, there is a saddle-node bifurcation (real eigenvalue passing through zero) at $\lambda = \lambda_{crit}$. This bifurcation gives rise to two solution branches in $\lambda < \lambda_{crit}$ say and also represents a change in the temporal stability of solutions to the original IVP. To calculate the two solution branches, a value is assigned to the wall stress τ_w, for example, and the parameter is then calculated numerically using a numerical procedure capable of doing so. An alternative way of looking at this is that τ_w is a multivalued function of λ, whereas λ is a single-valued function of τ_w.

A further situation that can occur in the numerical integrations is that a singularity appears to be developing in the solution as $\lambda \rightarrow \lambda_0$ say. This manifested by the wall stress or heat transfer, for example, becoming increasingly larger as the singularity is approached in the numerical solutions. To determine the nature of this singularity, an expansion based on the assumption that $[\lambda - \lambda_0]$ is small is taken into account. A transformation of variables is usually required and a solution sought as a series in powers of $[\lambda - \lambda_0]$. Two situations can then arise, either the leading-order term or higher-order terms satisfy all the prescribed boundary conditions. This is then a regular perturbation. Alternatively, the leading-order term satisfies only the wall conditions, giving rise to a singular perturbation. This is then an inner solution and a solution in an outer region of a larger lateral extent is also required and through which the outer conditions are satisfied. The matching of the inner and outer solutions fully completes the solution near the singularity. The identification of nonexistence in the solutions to the steady-state equations for certain parameter values indicates that the solution to the IVP will evolve only to a finite time t_s say. A finite-time singularity develops as $t \rightarrow t_s$ with, for example, the wall stress and/or heat transfer becoming large. A solution based on $(t_s - t)$ being small can be sought with this normally being a singular perturbation. This solution usually involves a narrow region next to the wall where conditions change rapidly as $t \rightarrow t_s$ and an outer region which is not affected by the singularity being approached.

Nomenclature

$\mathbf{B}$	external magnetic field vector (Wb)
C	nanoparticle volume fraction
C_F	inertial resistance coefficient
Cp	specific heat capacity at constant pressure (J /kg /K)
$(\rho\ C_p)$	heat capacitance (J /K /m^3)
D_B	inertial resistance coefficient (m^{2}s^{-1})
D_T	thermophoretic diffusion coefficient(m^{2}s^{-1})
$\mathbf{E}$	electric field (V /m)
$\mathbf{g}$	gravitational acceleration vector (m/s^2)
$(\mathbf{i}, \mathbf{j})$	unit vector along $\mathbf{x}$ and $\mathbf{y}$ axes
j	microinertia density (m^2)
$\mathbf{j}$	electric current density
k	thermal conductivity (W /m /K)
k_{hnf}	thermal conductivity of the hybrid nanofluid (W /m /K)
k_{nf}	thermal conductivity of the nanofluid (W /m /K)
K	permeability of the porous medium (m^2)
$\mathbf{N}$	microrotation vector (s^{-1})
p	pressure (Pa)
q'''	volumetric rate of heat generation (W /m^3)

$\boldsymbol{t}$	dimensional time (s)
$\boldsymbol{T}$	dimensional fluid temperature (K)
$\boldsymbol{T_0}$	characteristic temperature (K)
$(\boldsymbol{u},\boldsymbol{v})$	velocity components along the $(\boldsymbol{x},\boldsymbol{y})$(m /s)
$\boldsymbol{v}$	velocity vector (m/s)
$<\boldsymbol{v}>$	Darcian velocity vector (m/s)
$(\boldsymbol{x},\ \boldsymbol{y})$	Cartesian coordinates (m)

Greek letters

α	thermal diffusivity (m^2/s)
$\alpha,\ \beta,\ \lambda$	new viscosity coefficients Eq. (1.27) for micropolar fluids
β	coefficient of thermal expansion (K^{-1})
ε	porosity
η	dimensionless similarity variable
γ	spin-gradient viscosity
κ	vortex viscosity (N s/m^2)
μ	dynamic viscosity (N s/m^2)
μ_0	magnetic permeability (H /m)
μ_{hnf}	dynamic viscosity of the hybrid nanofluid (N s/m^2)
$\widetilde{\mu}$	effective dynamic viscosity (N s/m^2)
ν	kinematic viscosity (m^2/s)
∇	vector differential operator
$\varphi,\ \varphi_1,\ \varphi_2$	nanoparticle volume fractions
Φ	viscous dissipation function
ρ	density of fluid (kg/m^3)
ρ_{hnf}	density of the hybrid nanofluid (kg/m^3)
σ	electrical conductivity
σ_{hnf}	electrical conductivity of the hybrid nanofluid ($LR^{-1}\ S^{-1}$)
ω	parameter Eq. (1.19), $=\frac{(\rho C_P)_{s_1}}{(\rho C_P)_f}$

Subscript

f	base fluid
hnf	hybrid nanofluid fluid
nf	nanofluid
$s_1,\ s_2$	solid phases

References

Ahmad, K., Nazar, R., Ishak, A., & Pop, I. (2012). Unsteady three-dimensional boundary layer flow due to a stretching surface in a micropolar fluid. *International Journal for Numerical Methods in Fluids, 68*(12), 1561–1573. https://doi.org/10.1002/fld.2543

Ariman, T., Turk, M. A., & Sylvester, N. D. (1973). Microcontinuum fluid mechanics—A review. *International Journal of Engineering Science, 11*(8), 905–930. https://doi.org/10.1016/0020-7225(73)90038-4

Ariman, T., Turk, M. A., & Sylvester, N. D. (1974). Applications of microcontinuum fluid mechanics. *International Journal of Engineering Science, 12*(4), 273–293. https://doi.org/10.1016/0020-7225(74)90059-7

Barletta, A., Rosi di Schio, E., & Celli, M. (2015). Convection and instability phenomena in nanofluid-saturated porous media. In V. Bianco, O. Manca, S. Nardini, & K. Vafai (Eds.), *Heat transfer enhancement with nanofluids* (pp. 341–364). CRC Press.

Bear, J. (2018). *Modeling phenomena of flow and transport in porous media.* Springer. https://doi.org/10.1007/978-3-319-72826-1

Bejan, A. (1987). Convective heat transfer in porous media. In S. Kakaç, R. K. Shah, & W. Aung (Eds.), *Handbook of Single-Phase Convective Heat Transfer.* New York: Wiley.

Bejan, A. (2013). Convection heat transfer. In *Convection heat transfer* (4th ed.). John Wiley & Sons, Inc. https://doi.org/10.1002/9781118671627

Bejan, A., Dincer, I., Lorente, S., Miguel, A., & Reis, H. (2004). *Porous and Complex Flow Structures in Modern Technologies.* New York: Springer-Verlag. https://doi.org/10.1007/978-1-4757-4221-3

Birkhoff, G. (1955). *Hydrodynamics.* Dover.

Blasius, H. (1908). Grenzschichten in Flussigkeiten mit kleiner Reibung. *Zeitschrift für Angewandte Mathematik und Physik*, 1–37.

Buongiorno, J. (2006). Convective transport in nanofluids. *Journal of Heat Transfer, 128.* https://doi.org/10.1115/1.2150834

Buongiorno, J., Venerus, D. C., Prabhat, N., McKrell, T., Townsend, J., Christianson, R., Tolmachev, Y. V., Keblinski, P., Hu, L. W., Alvarado, J. L., Bang, I. C., Bishnoi, S. W., Bonetti, M., Botz, F., Cecere, A., Chang, Y., Chen, G., Chen, H., Chung, S. J., … Zhou, S. Q. (2009). A benchmark study on the thermal conductivity of nanofluids. *Journal of Applied Physics, 106*, 094312. https://doi.org/10.1063/1.3245330

Butcher, J. C. (2008). Numerical methods for ordinary differential equations. In *Numerical methods for ordinary differential equations* (2nd ed.). John Wiley & Sons, Ltd. https://doi.org/10.1002/9780470753767

Cebeci, T. (2002). *Convective heat transfer* (Issue v. 2). Horizons Pub.

Cebeci, T., & Bradshaw, P. (2013). *Physical and computational aspects of convective heat transfer.* Springer Berlin Heidelberg.

Cheng, P. (1979). Heat transfer in geothermal systems. *Advances in Heat Transfer, 14*, 1–105. https://doi.org/10.1016/S0065-2717(08)70085-6

Cheng, P. (1985). Natural convection in a porous medium: External Flows. In S. Kakaç, W. Aung, & R. Viskanta (Eds.), *Natural Convection: Fundamentals and Applications.* Springer.

Das, S. K., Choi, S. U., Yu, W., & Pradeep, T. (2008). *Nanofluids: Science and technology.* Wiley.

Devi, S. S., & Devi, S. A. (2016a). Numerical investigation of hydromagnetic hybrid Cu-Al_2O_3/water nanofluid flow over a permeable stretching sheet with suction. *International Journal of Nonlinear Sciences and Numerical Stimulation, 17*, 249–257.

Devi, S. S., & Devi, S. A. (2016b). Numerical investigation of three-dimensional hybrid Cu-Al_2O_3/water nanofluid flow over a stretching sheet with effecting Lorentz force subject to Newtonian heating. *Canadian Journal of Physics, 94*, 490–496.

Devi, S. S. U., & Devi, S. P. A. (2017). Heat transfer enhancement of Cu-Al_2O_3 water hybrid nanofluid flow over a stretching sheet. *The Journal of the Nigerian Mathematical Society, 36*, 419–433.

Eringen, A. (1966). Theory of micropolar fluids. *Indiana University Mathematics Journal, 16*(1), 1–18. https://doi.org/10.1512/iumj.1967.16.16001

Eringen, A. (1972). Theory of thermomicropolar fluids. *Journal of Mathematical Analysis and Applications, 38*(2), 480–496. https://doi.org/10.1016/0022-247X(72)90106-0

Eringen, A. C. (2001). Microcontinuum field theories: II fluent media. In *Microcontinuum field theories*. New York: Springer. https://doi.org/10.1007/978-1-4612-0555-5

Gebhart, B., Jaluria, Y., Mahajan, R. L., & Sammakia, B. (1988). *Buoyancy- induced flows and transport*. Hemisphere.

Hansen, A. G. (1964). *Similarity analyses of boundary value problems in engineering*. Prentice-Hall.

Ingham, D. B., & Pop, I. (Eds.). (1998). *Transport Phenomena in Porous Media*. Pergamon. https://doi.org/10.1016/B978-0-08-042843-7.X5000-4

Ingham, D. B., & Pop, I. (Eds.). (2002). *Transport Phenomena in Porous Media II*. Pergamon. https://doi.org/10.1016/B978-0-08-043965-5.X5000-4

Ingham, D. B., & Pop, I. (Eds.). (2005). *Transport Phenomena in Porous Media III*. Elsevier Science. https://doi.org/10.1016/B978-0-08-044490-1.X5003-0

Keller, H. B. (1970). A new difference scheme for parabolic problems. In J. Bramble (Ed.), *Numerical solution of partial differential equations* (Vol. 2). Academic Press.

Keller, H. B. (1978). Numerical methods in boundary-layer theory. *Annual Review of Fluid Mechanics, 10*(1), 417–433. https://doi.org/10.1146/annurev.fl.10.010178.002221

Keller, H. B., & Cebeci, T. (1972). Accurate numerical methods for boundary-layer flows. II: Two dimensional turbulent flows. *AIAA Journal, 10*(9), 1193–1199. https://doi.org/10.2514/3.50349

Khanafer, K., & Vafai, K. (2019). Applications of nanofluids in porous medium. *Journal of Thermal Analysis and Calorimetry, 135*(2), 1479–1492. https://doi.org/10.1007/s10973-018-7565-4

Khanafer, K., Vafai, K., & Lightstone, M. (2003). Buoyancy-driven heat transfer enhancement in a two-dimensional enclosure utilizing nanofluids. *International Journal of Heat and Mass Transfer, 46*(19), 3639–3653. https://doi.org/10.1016/S0017-9310(03)00156-X

Kolpashchikov, V. L., Migun, N. P., & Prokhorenko, P. P. (1983). Experimental determination of material micropolar fluid constants. *International Journal of Engineering Science, 21*(4), 405–411. https://doi.org/10.1016/0020-7225(83)90123-4

Kuznetsov, A. V., & Nield, D. A. (2013). The Cheng–Minkowycz problem for natural convective boundary layer flow in a porous medium saturated by a nanofluid: A revised model. *International Journal of Heat and Mass Transfer, 65*, 682–685. https://doi.org/10.1016/j.ijheatmasstransfer.2013.06.054

Lie, S. (2015). Theory of transformation groups I. In , *A contemporary approach and translationTheory of transformation groups I: General properties of continuous transformation groups*. Berlin Heidelberg: Springer. https://doi.org/10.1007/978-3-662-46211-9

Ludlow, D. (2000). New similarity solutions of the unsteady incompressible boundary-layer equations. *Quarterly Journal of Mechanics & Applied Mathematics, 53*(2), 175–206. https://doi.org/10.1093/qjmam/53.2.175

Łukaszewicz, G. (1999). *Micropolar fluids: Theory and application*. Birkhäuser.

Migun, N. P. (1981). Experimental method of determining parameters characterizing the microstructure of micropolar liquids. *Journal of Engineering Physics, 41*(2), 832–835. https://doi.org/10.1007/BF00836545

Minkowycz, W. J., Sparrow, E. M., & Abraham, J. P. (2013). Nanoparticle heat transfer and fluid flow. In *Nanoparticle heat transfer and fluid flow*. CRC Press. https://doi.org/10.1201/b12983

Moler, C. (2014). *Ordinary differential equations solvers ODE23 and ODE45*. https://blogs.mathworks.com/cleve/2014/5/26/ordinary-differential-equation-solvers-ode23-and-ode45/.

Morgan, A. J. A. (1958). Possible similarity solutions for the laminar, incompressible, boundary layer equations. *Transactions of the American Society of Mechanical Engineers, 80*, 1559.

Na, T. Y. (1979). *Computational methods in engineering boundary value problems*. Academic Press.

Na, T. Y., & Hansen, A. G. (1967). Similarity solutions of a class of laminar three-dimensional boundary layer equations of power law fluids. *International Journal of Non-Linear Mechanics, 2*(4), 373–385. https://doi.org/10.1016/0020-7462(67)90005-4

Nield, D. A., & Bejan, A. (2017). Convection in porous media. In *Convection in porous media* (5th ed.). Springer International Publishing. https://doi.org/10.1007/978-3-319-49562-0

Oztop, H. F., & Abu-Nada, E. (2008). Numerical study of natural convection in partially heated rectangular enclosures filled with nanofluids. *International Journal of Heat and Fluid Flow, 29*, 1326–1336. https://doi.org/10.1016/j.ijheatfluidflow.2008.04.009

Pop, I., & Ingham, D. B. (2001). *Convective heat transfer: Mathematical and computational modelling of viscous fluids and porous media*. Pergamon.

Prandtl, L. (1904). Über Flüssigkeitsbewegung bei sehr kleiner Reibung ("Motion of Fluids with Very Little Viscosity"). *Internationalen Mathematiker-Kongresses Heidelberg. Gesammelte Abhandlungen, 2*, 484–491.

Rosenhead, L. (1963). *Laminar boundary layers*. Dover Publication Inc.

Saghir, M. Z., Ahadi, A., Mohamad, A., & Srinivasan, S. (2016). Water aluminum oxide nanofluid benchmark model. *International Journal of Thermal Sciences, 109*, 148–158. https://doi.org/10.1016/j.ijthermalsci.2016.06.002

Schlichting, H., & Gersten, K. (2000). *Boundary-layer theory*. Springer. https://doi.org/10.1007/978-3-662-52919-5

Shenoy, A., Sheremet, M., & Pop, I. (2016). *Convective flow and heat transfer from wavy surfaces*. CRC Press. https://doi.org/10.1201/9781315367637

Tiwari, R. K., & Das, M. K. (2007). Heat transfer augmentation in a two-sided lid-driven differentially heated square cavity utilizing nanofluids. *International Journal of Heat and Mass Transfer, 50*(9–10), 2002–2018. https://doi.org/10.1016/j.ijheatmasstransfer.2006.09.034

Vadasz, P. (2008). *Emerging Topics in Heat and Mass Transfer in Porous Media: From Bioengineering and Microelectronics to Nanotechnology*. Springer. https://doi.org/10.1007/978-1-4020-8178-1

Vafai, K. (Ed.). (2015). *Handbook of Porous Media* (3rd ed.). CRC Press. https://doi.org/10.1201/b18614

Vafai, K. (2010). *Porous Media: Applications in Biological Systems and Biotechnology*. CRC Press. https://doi.org/10.1201/9781420065428

CHAPTER 2

Viscous fluids

2.1 Unsteady mixed convection flow at a three-dimensional stagnation point

2.1.1 Introduction

Miklavčič and Wang (2006) have studied the steady viscous flow past a permeable shrinking sheet with suction and obtained a closed form exact solution. They found that the solutions may not be unique for certain suction rates. Physically there are two conditions for shrinking flow to exist physically, i.e., either imposed adequate suction on the boundary (Fang & Zhang, 2009) or added stagnation flow which contains the vorticity (Wang, 2008). This new type of shrinking sheet flow is essentially a backward flow as discussed by Goldstein (1965).

2.1.2 Governing equations

Let us consider the unsteady mixed convection flow near the stagnation point of a permeable three-dimensional body placed in a viscous fluid with a constant wall temperature T_w. It was assumed that the temperature of the ambient fluid is T_∞, where $T_w > T_\infty$ (heated wall) corresponds to assisting flow, while $T_w < T_\infty$ (cooled wall) corresponds to opposing flow, respectively. A locally Cartesian orthogonal system (x, y, z) is chosen with the origin N at the nodal stagnation point as shown in Fig. 2.1, where the $x-$ and $y-$ coordinates are measured along the body surface, while the $z-$ coordinate is measured normal to the body surface. Under these assumptions along with the boundary layer and Boussinesq approximations, the governing equations can be written as (Eswara & Nath, 1999; Noor et al., 2021)

$$\frac{\partial u}{\partial x} + \frac{\partial v}{\partial y} + \frac{\partial w}{\partial w} = 0 \tag{2.1}$$

Similarity Solutions for the Boundary Layer Flow and Heat Transfer of Viscous Fluids, Nanofluids, Porous Media, and Micropolar Fluids
ISBN 978-0-12-821188-5
https://doi.org/10.1016/B978-0-12-821188-5.00008-4

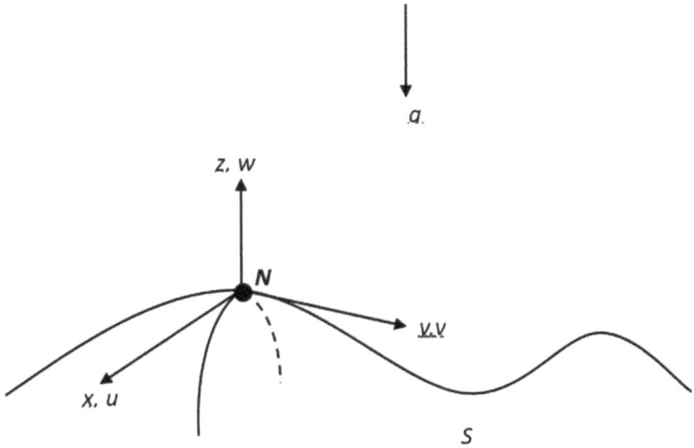

Figure 2.1 Physical model and coordinate system. *(From Noor, A., Nazar, R., Naganthran, K., & Pop, I. (c. 2021). Unsteady mixed convection flow at a three-dimensional stagnation point.* International Journal of Numerical Methods for Heat & Fluid Flow, *31(1), 236–250. https://doi.org/10.1108/HFF-03-2020-0138 247)*

$$\frac{\partial u}{\partial t}+u\frac{\partial u}{\partial x}+v\frac{\partial u}{\partial y}+w\frac{\partial u}{\partial z}=\frac{\partial u_e}{\partial t}+u_e\frac{\partial u_e}{\partial x}+v\frac{\partial^2 u}{\partial z^2}+g\beta(T-T_\infty) \tag{2.2}$$

$$\frac{\partial v}{\partial t}+u\frac{\partial v}{\partial x}+v\frac{\partial v}{\partial y}+w\frac{\partial v}{\partial z}=\frac{\partial v_e}{\partial t}+v_e\frac{\partial v_e}{\partial x}+v\frac{\partial^2 v}{\partial z^2}+g\beta(T-T_\infty) \tag{2.3}$$

$$\frac{\partial T}{\partial t}+u\frac{\partial T}{\partial x}+v\frac{\partial T}{\partial y}+w\frac{\partial T}{\partial z}=\alpha\frac{\partial^2 T}{\partial z^2} \tag{2.4}$$

subject to the initial and boundary conditions

$$\begin{aligned}
&t<0:\ u=0,\quad v=0,\quad w=0,\quad T=T_\infty \ \text{ for any } x,y,z=0\\
&t\geq 0:\ u=0,\quad v=0,\quad w=w_0,\quad T=T_w \ \text{ at } z=0\\
&u=u_e(x)\rightarrow ax,\quad v=v_e(y)\rightarrow by,\quad T\rightarrow T_\infty \ \text{ as } z\rightarrow\infty
\end{aligned} \tag{2.5}$$

Here $u_e(x,t)$ and $v_e(y,t)$ are the velocity components of the outer (inviscid) flow in the $x-$ and $y-$ directions and $c=b/a$ with a and b being the parameters of the principal curvatures at N, of the body. There is no loss of generality in requiring that $|a|\geq|b|$ with $a>0$. Clearly $b=0$ corresponds to the plane stagnation flow case, while $b=a$ is the axisymmetric case. The case when a and b are positive, the solutions of the resulting equations lead to stagnation points which are nodal points of attachment, i.e., $0\leq c\leqslant 1$. However, when a or b are negative, it leads to saddle points of attachment, i.e., $-1 \leq c \leqslant 0$. We assume that $T_w(x) = T_\infty + T_0(x/L)$, where T_0 is the characteristic temperature of the surface of the body with $T_0 > 0$ for assisting flow and $T_0 < 0$ for opposing flow, and L is the characteristic body's surface.

2.1.3 Solution

A little inspection shows that Eqs. (2.1)–(2.4) along with the boundary conditions Eq. (2.5) admit a semisimilar solution of the form

$$u = \frac{u_e(x)}{1-\delta_1 t} f'(\eta), \quad v = \frac{v_e(y)}{1-\delta_1 t} h'(\eta), \quad w = -\sqrt{\frac{a\nu}{1-\delta_1\, t}}[f(\eta) + c\, h(\eta)],$$
$$\theta(\eta) = (T - T_\infty)/(T_w - T_\infty), \quad \eta = \sqrt{\frac{a}{\nu(1-\delta_1 t)}} z \tag{2.6}$$

where it is assumed that $\delta t < 1$. Substituting variables Eq. (2.6) into Eqs. (2.2)–(2.4), we obtain the following ordinary (similarity) differential equations:

$$f''' + \left(f + ch - A\frac{\eta}{2}\right)f'' - (f' + A)f' + 1 + A + \lambda\theta = 0 \tag{2.7}$$

$$h''' + \left(f + ch - A\frac{\eta}{2}\right)h'' - (ch' + A)h' + c + A + \lambda\theta = 0 \tag{2.8}$$

$$\theta'' + Pr\left(f\theta' - f'\theta + ch\,\theta' - A\frac{\eta}{2}\theta'\right) = 0 \tag{2.9}$$

and the boundary conditions Eq. (2.5) become

$$\begin{aligned} &f(0) = s, \quad f'(0) = 0, \quad h(0) = 0, \quad h'(0) = 0, \quad \theta(0) = 1, \\ &f'(\eta) \to 1, \quad h'(\eta) \to 1, \quad \theta(\eta) \to 0 \text{ as } \eta \to \infty \end{aligned} \tag{2.10}$$

Here $Pr = \nu/\alpha$ is the Prandtl number, $s = -w_0/\sqrt{a\nu}$ is the constant parameter of suction ($s > 0$) or injection ($s < 0$), and $A = \delta_1/a$ is the unsteadiness parameter with $A > 0$ corresponding to an accelerating flow, $A < 0$ corresponds to a decelerating flow, and $A = 0$ corresponds to the steady-state flow, respectively. Furthermore, λ is the mixed convection parameter, which is defined as

$$\lambda = \frac{Gr_x}{Re_x^2} \tag{2.11}$$

where $Gr_x = g\beta(T_w - T_\infty)x^3/\nu^2$ is the local Grashof number and $Re_x = u_e(x)x/\nu$ is the local Reynolds number. It is worth mentioning that for $\lambda > 0$ the flow is assisting, for $\lambda < 0$ the flow is opposing, and for $\lambda = 0$, we have a forced convection stagnation-point flow. It is also worth mentioning that the two-dimensional case can been recovered when $c = 0$ so that $h(\eta) = 0$, while the axisymmetric case can be recovered when $c = 1$ so that $f(\eta) = h(\eta)$.

The physical quantities of interest are the skin friction coefficients C_{fx} and C_{fy}, and the local Nusselt number Nu_x, which are defined as

$$C_{fx} = \frac{\mu}{\rho u_e^2(x)}\left(\frac{\partial u}{\partial z}\right)_{z=0}, \quad C_{fy} = \frac{\mu}{\rho v_e^2(y)}\left(\frac{\partial v}{\partial z}\right)_{z=0}, \quad Nu_x = -\frac{x}{(T_w - T_\infty)}\left(\frac{\partial T}{\partial z}\right)_{z=0} \tag{2.12}$$

where ρ is the density, k is the thermal conductivity of the fluid.

Using Eqs. (2.6) and (2.12), we get

$$(1-\delta t)^{3/2} Re_x^{1/2} C_f = f''(0), \quad (1-t)^{3/2}[v_e(y)/u_e(x)] Re_x^{1/2} = g''(0)$$

$$Re_x^{-\frac{1}{2}} Nu_x = -\theta'(0) \tag{2.13}$$

where $Re_x = u_e(x)\,x/\nu$, the local Reynolds number.

2.1.4 Steady-state solution

In this case $A = 0$, so that Eqs. (2.7)–(2.9) reduce to

$$f''' + (f + ch)f'' + 1 - f'^2 + \lambda\theta = 0 \tag{2.14}$$

$$h''' + (f + ch)h'' + c\left(1 - h'^2\right) + \lambda\theta = 0 \tag{2.15}$$

$$\theta'' + Pr(f\theta' - f'\theta + ch\,\theta') = 0 \tag{2.16}$$

subjected to the boundary conditions Eq. (2.10).

2.1.5 Stability analysis

It has been shown in some papers (Rosca & Pop, 2013; Weidman et al., 2006) that dual (lower and upper branch) solutions exist. In order to determine which of these solutions are physically realizable in practice, a stability analysis of the problem Eqs. (2.7)–(2.10) is necessary. Following Weidman et al. (2006), a dimensionless time variable τ has to be introduced. The use of τ is associated with an initial value problem and this is consistent with the question of which solution will be obtained in practice (physically realizable). Thus, the new variables for the unsteady problem are

$$u = \frac{u_e(x)}{1-\delta_1 t}\frac{\partial f}{\partial \eta}(\eta,\tau),\quad v = \frac{v_e(y)}{1-\delta_1 t}\frac{\partial h}{\partial \eta}(\eta,\tau),$$

$$w = -\sqrt{\frac{a\nu}{1-\delta_1 t}}\,[f(\eta,\tau)+ch(\eta,\tau)],\quad \theta(\eta,\tau) = (T-T_\infty)/(T_w-T_\infty),$$

$$\eta = \sqrt{\frac{a}{\nu(1-\delta_1\, t)}}z,\quad \tau = at \tag{2.17}$$

Substituting Eq. (2.17) into Eqs. (2.2)–(2.4), we obtain

$$\frac{\partial^3 f}{\partial \eta^3}+\left(f+ch-A\frac{\eta}{2}\right)\frac{\partial^2 f}{\partial \eta^2}-\left(A+\frac{\partial f}{\partial \eta}\right)^2+1+A+\lambda\theta-\frac{\partial^2 f}{\partial \eta\partial \tau}=0 \tag{2.18}$$

$$\frac{\partial^3 h}{\partial \eta^3}+\left(f+ch-A\frac{\eta}{2}\right)\frac{\partial^2 h}{\partial \eta^2}-\left(A+c\frac{\partial h}{\partial \eta}\right)\frac{\partial h}{\partial \eta}+A+c+\lambda\theta-\frac{\partial^2 h}{\partial \eta\partial \tau}=0 \tag{2.19}$$

$$\frac{1}{Pr}\frac{\partial^2 \theta}{\partial \eta^2}+\left(f+ch-A\frac{\eta}{2}\right)\frac{\partial \theta}{\partial \eta}-\frac{\partial f}{\partial \eta}\theta-\frac{\partial \theta}{\partial \tau}=0 \tag{2.20}$$

subject to the boundary conditions

$$f\,(0,\tau)=s,\quad \frac{\partial f}{\partial \eta}(0,\tau)=0,\quad h(0,\,\tau)=0,\quad \frac{\partial h}{\partial \eta}(0,\tau)=0,\quad \theta(0,\tau)=1$$

$$\frac{\partial f}{\partial \eta}(\eta,\tau)\to 1,\quad \frac{\partial h}{\partial \eta}(\eta,\tau)\to 1,\quad \theta(\eta,\tau)\to 0 \text{ as } \eta\to\infty \tag{2.21}$$

To determine the stability of the solution $f=f_0(\eta)$, $h=h_0(\eta)$, and $\theta=\theta_0(\eta)$, satisfying the boundary value problem Eqs. (2.7)–(2.10), we write (see Weidman et al., 2006 or Rosca & Pop, 2013)

$$\begin{gathered} f(\eta,\tau)=f_0(\eta)+e^{-\gamma\tau}F(\eta,\tau)\ h(\eta,\tau)=h_0(\eta)+e^{-\gamma\tau}H(\eta,\tau) \\ \theta(\eta,\tau)=\theta_0(\eta)+e^{-\gamma\tau}G(\eta,\tau) \end{gathered} \tag{2.22}$$

where γ is an unknown eigenvalue parameter, and $F(\eta,\tau)$, $H(\eta,\tau)$, and $G(\eta,\tau)$ are small relative to $f_0(\eta)$, $h_0(\eta)$, and $\theta_0(\eta)$. Substituting Eq. (2.22) into Eqs. (2.18)–(2.20), we obtain the following linearized problem:

$$\frac{\partial^3 F}{\partial \eta^3}+\left(f_0+ch_0-A\frac{\eta}{2}\right)\frac{\partial^2 F}{\partial \eta^2}+\left(\gamma-A-2f_0'\right)\frac{\partial F}{\partial \eta}+(F+cH)f_0''+\lambda G=0 \tag{2.23}$$

$$\frac{\partial^3 H}{\partial \eta^3}+\left(f_0+ch_0-A\frac{\eta}{2}\right)\frac{\partial^2 H}{\partial \eta^2}+\left(\gamma-A-2ch_0'\right)\frac{\partial H}{\partial \eta}+(F+cH)h_0''+\lambda G=0 \tag{2.24}$$

$$\frac{1}{Pr}\frac{\partial^2 G}{\partial \eta^2}+\left(f_0+ch_0-A\frac{\eta}{2}\right)\frac{\partial G}{\partial \eta}+\gamma G+(F+cH)\theta_0'=0 \tag{2.25}$$

subject to the boundary conditions

$$F(0,\tau)=0,\quad \frac{\partial F}{\partial \eta}(0,\tau)=0,\quad H(0,\ \tau)=0,\quad \frac{\partial H}{\partial \eta}(0,\tau)=0,$$
$$G(0,\tau)=0\quad \frac{\partial F}{\partial \eta}(\eta,\tau)\to 0,\quad \frac{\partial H}{\partial \eta}(\eta,\tau)\to 0,\quad G(\eta,\tau)\to 0\ \text{as}\ \eta\to\infty \tag{2.26}$$

As suggested by Weidman et al. (2006), we investigate the stability of the steady flow and heat transfer solution $f_0(\eta)$, $h_0(\eta)$, and $\theta_0(\eta)$ by setting $\tau=0$, and hence $F=F_0(\eta)$, $H=H_0(\eta)$, and $G=G_0(\eta)$ in Eqs. (2.23)–(2.25) to identify initial growth or decay of the solution Eq. (2.22). To test our numerical procedure, we have to solve the following linear eigenvalue problem:

$$F_0'''+\left(f_0+ch_0-A\frac{\eta}{2}\right)F_0''+\left(\gamma-A-2f_0'\right)F_0'+(F_0+cH_0)f_0''+\lambda G_0=0 \tag{2.27}$$

$$H_0'''+\left(f_0+ch_0-A\frac{\eta}{2}\right)H_0''+\left(\gamma-A-2c\,h_0'\right)H_0'+(F_0+cH_0)h_0''+\lambda G_0=0 \tag{2.28}$$

$$\frac{1}{Pr}G_0''+\left(f_0+ch_0-A\frac{\eta}{2}\right)G_0'+\left(\gamma-f_0'\right)G_0+(F_0+cH_0)\theta_0'=0 \tag{2.29}$$

with the boundary conditions

$$\begin{aligned} &F_0(0)=0,\ \ F_0'(0)=0,\quad H_0(0)=0,\quad H_0'(0)=0,\quad G_0(0)=0 \\ &F_0'(\eta)\to 0,\ \ H_0'(\eta)\to 0,\quad G_0(\eta)\to 0 \text{ as } \eta\to\infty \end{aligned} \tag{2.30}$$

It should be mentioned that for particular values of λ, s, c, and A the corresponding steady flow solution $f_0(\eta)$, $h_0(\eta)$, and $\theta_0(\eta)$, the stability of the steady flow solution is determined by the smallest eigenvalue γ_1. Solving the eigenvalue problem Eqs. (2.27)−(2.30), an infinite number of eigenvalues $\gamma_1 < \gamma_2 < \gamma_3 < \ldots$ have been obtained. If the smallest eigenvalue is positive, the flow is stable, and if the smallest eigenvalue is negative, the flow is unstable. According to Harris et al. (2009), the range of possible eigenvalues can be determined by relaxing a boundary condition on $F_0'(\eta)$, $H_0'(\eta)$ or $G_0'(\eta)$. For the present problem, we relax the condition that $H_0'(\eta)\to 0$ as $\eta\to\infty$, and for a fixed value of γ we solve the system Eqs. (2.27)−(2.30) along with the new boundary condition $H_0''(0)=1$.

In order to determine which solution is more stable, a stability analysis is performed using bvp4c in Matlab. Before we can find the smallest eigenvalue γ_1 we have to find the solutions for ordinary differential Eqs. (2.7)−(2.9) subject to the boundary conditions Eq. 2.10 using bvp4c. Since the present problem has more than one (dual) solution, the bvp4c function requires an initial guess which should satisfy the boundary conditions and reveals the behavior of the desired solution. While it is easy to find the first (upper branch) solution even using a poor initial guess, it is very difficult, however, to find the second (lower branch) solution. To overcome this difficulty, we begin with a set of parameter values for which the problem is easy to be solved. The obtained results are then used as initial guess for the solution of the problem with small variation of the parameters. This step is repeated until the desired values of parameters are reached. This technique is called continuation (Shampine et al., 2003).

Using the numerical results obtained, we then compute the smallest eigenvalue γ_1 as shown in Tables 2.1 and 2.2. The results show that all the upper branch solutions have positive eigenvalue γ_1, while all the lower branch solutions have negative eigenvalue γ_1. Therefore, we can conclude that the upper branch solution is stable, while the lower branch is unstable over time. It should be noted that as λ and A approach critical values λ_c and A_c, respectively, the smallest eigenvalue becomes closer to zero. In fact the eigenvalue signs changed from positive to negative at the turning point of the upper and lower branches.

Table 2.1 Smallest eigenvalue γ_1 for several values of λ and s when $A = 0$ (steady case).

		Upper branch solution	Lower branch solution
s	λ	γ_1	γ_1
−0.5	2	2.5485	−2.8015
	1	2.1326	−2.3898
	0	2.3104	−1.7889
	−1	0.6279	−0.5157
	−1.2	0.1569	−0.145
	−1.21	0.0967	−0.092
0	2	2.8429	−2.8555
	1	2.4589	−2.4551
	0	2.7697	−1.919
	−1	1.2854	−1.0496
	−1.7	0.3627	−0.3171
	−1.78	0.0188	−0.0186
0.5	2	3.1829	−3.0135
	1	2.8539	−2.6471
	0	3.3435	−2.1928
	−1	1.9383	−1.5682
	−2	1.1346	−0.8403
	−2.6	0.367	−0.3313
	−2.68	0.0839	−0.082

Table 2.2 Smallest eigenvalue γ_1 for several values of $A \neq 0$ (unsteady case) when $\lambda = -0.5$ and $s = 0.5$

	Upper branch solution	Lower branch solution
A	γ_1	γ_1
1	1.5522	-
0.5	1.9563	−1.7018
0	2.3078	−1.9103
−1	2.6158	−2.0646
−2	2.4769	−2.0142
−3	2.2029	−1.8577
−4	1.8451	−1.6103
−5	1.3725	−1.2472
−6	0.5838	−0.5623
−6.22	0.0866	−0.0861

2.1.6 Comments

The nonlinear ordinary differential Eqs. (2.7)–(2.9) subject to boundary conditions Eq. (2.10) have been solved numerically using the Runge–Kutta–Fehlberg method with shooting technique for different values of suction/injection parameter s, mixed convection parameter λ, and unsteadiness parameter A. The Prandtl number Pr is set equal to 0.7 throughout the paper, while c is set equal to 0.5 (nodal point of attachment flow) unless stated otherwise. The numerical results for the reduced skin friction coefficients $f''(0)$ and $h''(0)$ and the reduced local Nusselt number $\theta'(0)$ are illustrated in Figs. 2.2–2.7. Figs. 2.2–2.4 show the existence of dual (first and second) solutions obtained by setting different initial guesses for the missing values of $f''(0), h''(0)$, and $\theta'(0)$, where all profiles satisfy the far field boundary conditions Eq. (2.10) asymptotically. Both steady ($A = 0$) and unsteady ($A \neq 0$) cases were considered. The solutions exist for both opposing ($\lambda < 0$) and assisting flow ($\lambda > 0$) up to some critical value λ_c beyond which the boundary layer separate from the surface and the solutions based upon the boundary layer approximations are not possible.

In order to validate our numerical results, a comparison of reduced skin friction coefficient has been made for steady-state forced convection case $A = 0$, $\lambda = 0$ with zero injection ($f_0 = 0$) as shown in Table 2.3. The present results show an excellent agreement with those reported by Eswara and Nath (1999), Davey (1961) and Jeng and Williams (1973), thus give us confidence in our numerical results presented in this chapter.

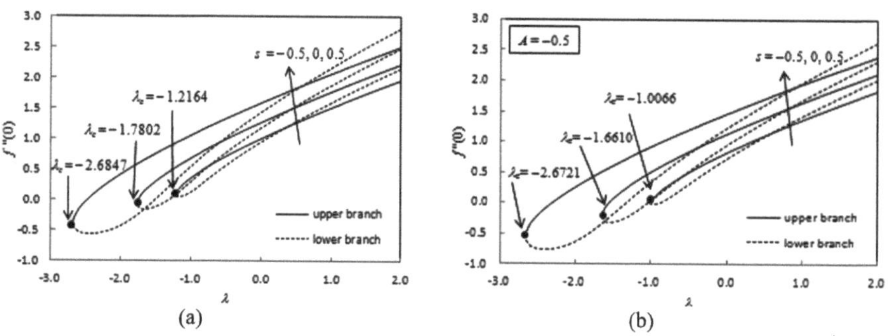

Figure 2.2 Variation of reduced skin-friction coefficient $f''(0)$ with λ for different values of s when (A) the flow is steady ($A = 0$) and (B) the flow is unsteady ($A = -0.5$). *(From Noor, A., Nazar, R., Naganthran, K., & Pop, I. (c. 2021). Unsteady mixed convection flow at a three-dimensional stagnation point.* International Journal of Numerical Methods for Heat & Fluid Flow, 31*(1), 236–250. https://doi.org/10.1108/HFF-03-2020-0138 247.)*

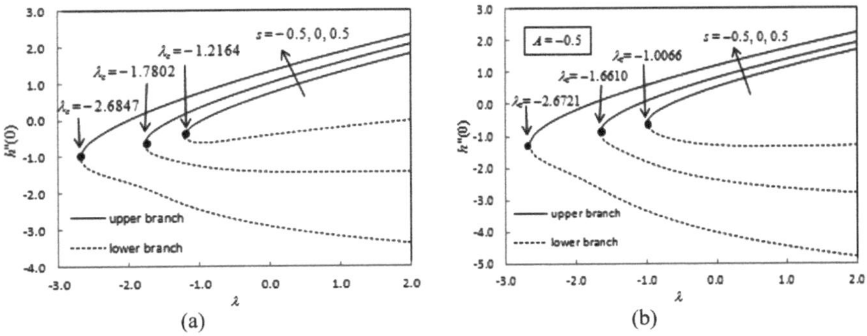

Figure 2.3 Variation of reduced skin-friction coefficient $h''(0)$ with λ for different values of s when (A) the flow is steady ($A=0$) and (B) the flow is unsteady ($A=-0.5$). *(From Noor, A., Nazar, R., Naganthran, K., & Pop, I. (c. 2021). Unsteady mixed convection flow at a three-dimensional stagnation point.* International Journal of Numerical Methods for Heat & Fluid Flow, 31*(1), 236–250. https://doi.org/10.1108/HFF-03-2020-0138 247)*

Variations of reduced skin-friction coefficient $f''(0)$ and $h''(0)$ with λ for different value of suction/injection parameter s are shown in Figs. 2.2 and 2.3, respectively. It can be seen that $f''(0)$ and $h''(0)$ increases with suction/injection parameter s. Fig. 2.4 shows the variation of $-\theta'(0)$ with λ for different values of the suction/injection parameter s. It can be seen that $-\theta'(0)$ increases as s increases. The variation of $f''(0)$ and $h''(0)$, and $-\theta'(0)$ with λ exhibits similar patterns for both steady ($A=0$) and unsteady cases

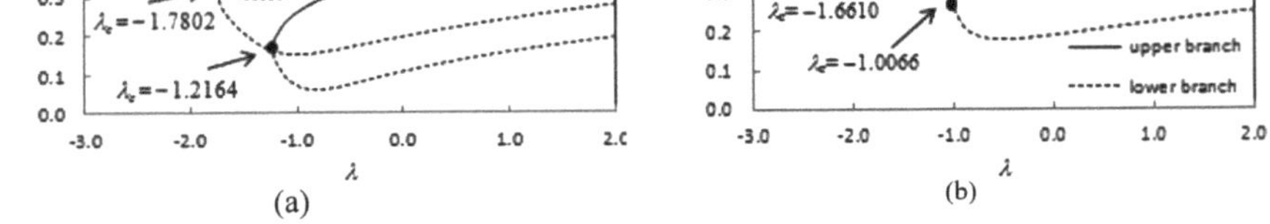

Figure 2.4 Variation of reduced local Nusselt $-\theta'(0)$ number with λ for different values of s when (A) the flow is steady ($A=0$) and (B) the flow is unsteady ($A=-0.5$). *(From Noor, A., Nazar, R., Naganthran, K., & Pop, I. (c. 2021). Unsteady mixed convection flow at a three-dimensional stagnation point.* International Journal of Numerical Methods for Heat & Fluid Flow, 31*(1), 236–250. https://doi.org/10.1108/HFF-03-2020-0138 247)*

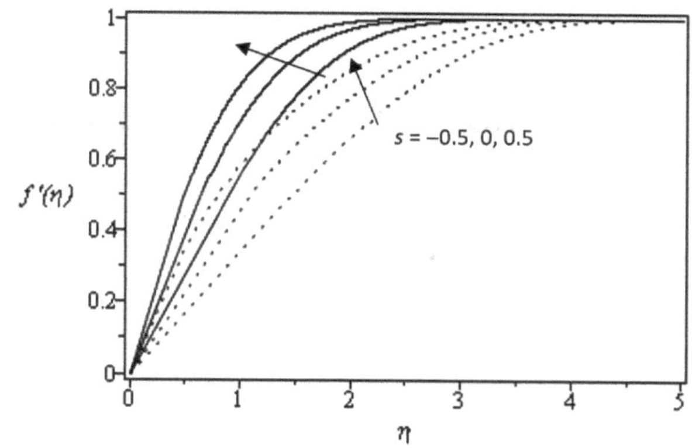

Figure 2.5 Velocity profiles $f'(\eta)$ for different values of s. *(From Noor, A., Nazar, R., Naganthran, K., & Pop, I. (c. 2021). Unsteady mixed convection flow at a three-dimensional stagnation point.* International Journal of Numerical Methods for Heat & Fluid Flow, 31*(1), 236–250. https://doi.org/10.1108/HFF-03-2020-0138 247)*

$(A \neq 0)$. However, it is observed that the solution domain for the unsteady case is noticeably smaller compared to those of the steady case.

Finally, Figs. 2.5–2.7 present the velocity $f'(\eta)$ and $h(\eta)$ temperature profiles $\theta(\eta)$ for different values of s when $A = 0$ (steady case). In these figures, the solid lines indicate the upper branch solution, while the dot lines refer to the lower branch solution, respectively. It is evident from these figures that the far field boundary conditions Eq. (2.10) are approached asymptotically. Thus, it supports the numerical results obtained for the boundary value problem Eqs. (2.8)–(2.10). It can be also noticed from

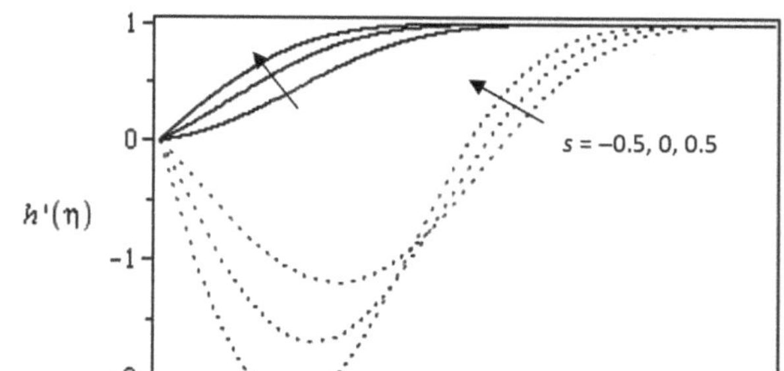

Figure 2.6 Velocity profiles $h'(\eta)$ for different values of s. *(From Noor, A., Nazar, R., Naganthran, K., & Pop, I. (c. 2021). Unsteady mixed convection flow at a three-dimensional stagnation point.* International Journal of Numerical Methods for Heat & Fluid Flow, 31*(1), 236–250. https://doi.org/10.1108/HFF-03-2020-0138 247)*

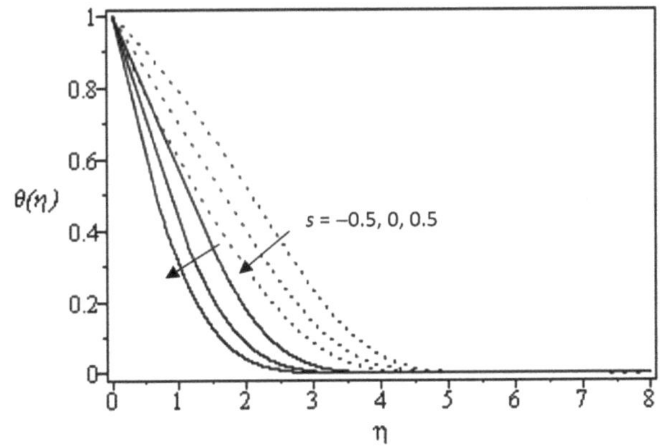

Figure 2.7 Temperature profiles $\theta(\eta)$ for different values of s. *(From Noor, A., Nazar, R., Naganthran, K., & Pop, I. (c. 2021). Unsteady mixed convection flow at a three-dimensional stagnation point.* International Journal of Numerical Methods for Heat & Fluid Flow, 31*(1), 236–250. https://doi.org/10.1108/HFF-03-2020-0138 247)*

Figs. 2.5–2.7 that for both velocity and temperature profiles, the upper branch solution displays a thinner boundary layer thickness compared to the lower branch solution.

2.2 Mixed convection boundary layer flow near the stagnation point on a vertical surface with slip

2.2.1 Introduction

In this paper, Aman et al. (2011) studied the similarity solution for the problem of a mixed convection boundary layer flow near the stagnation point on a vertical surface with slip at the boundary. The effects of the buoyancy and slip parameters on the reduced skin friction coefficient and the heat transfer rate at the surface are analyzed and discussed.

2.2.2 Problem formulation

Consider a steady, two-dimensional laminar boundary layer flow of a viscous and incompressible fluid near the stagnation point on an impermeable vertical surface as it is shown in Fig. 2.8, where the Cartesian coordinates x and y are measured along the plate and normal to it, respectively. It is assumed that the velocity of the free stream is $u_e(x)$, the temperature of the plate is $T_w(x)$, and the temperature of the ambient fluid is T_∞.

Table 2–3 Comparison of reduced skin friction coefficient for steady-state forced convection case ($\lambda = 0$) with zero injection ($f_0 = 0$) when $Pr = 0.7$

	Present results			Eswara and Nath (1999)			Jeng and Williams (1973) for $0 \le c \le 1$ and Davey (1961) for $-1 \le c \le 0$		
c	$f''(0)$	$h''(0)$	$-\theta'(0)$	$f''(0)$	$h''(0)$	$-\theta'(0)$	$f''(0)$	$h''(0)$	$-\theta'(0)$
1.00	1.31194	1.31194	0.66538	1.3128	1.3128	—	1.3119	1.3119	—
0.75	1.28863	1.16432	0.62308	1.2885	1.1642	—	1.2887	1.1643	—
0.5	1.26687	0.99811	0.57967	1.2677	0.998	—	1.2669	0.9981	—
0.25	1.24761	0.80514	0.53621	1.2475	0.805	—	1.2476	0.8051	—
0	1.23259	0.57047	0.49587	1.2324	0.5706	—	1.2326	0.5705	—
−0.25	1.22513	0.26795	0.46778	1.2249	0.2671	—	1.2251	0.268	—
−0.5	1.2302	−0.1115	0.47059	1.2302	−0.111	—	1.2302	−0.1115	—
−0.75	1.24732	−0.48219	0.50755	1.2489	−0.4975	—	1.2473	−0.4821	—
−1	1.27277	−0.8095	0.56595	1.2762	−0.8226	—	1.2729	−0.8112	—

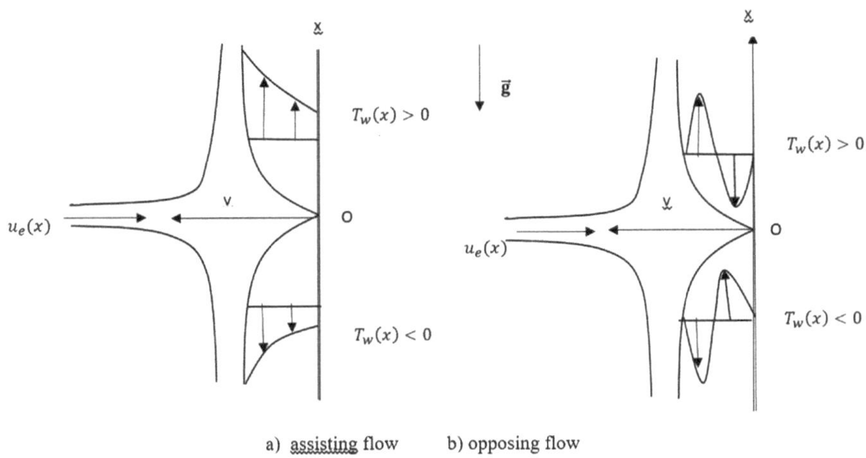

Figure 2.8 Physical model and coordinate system.

The boundary layer equations are

$$\frac{\partial u}{\partial x} + \frac{\partial v}{\partial y} = 0 \tag{2.31}$$

$$u\frac{\partial u}{\partial x} + v\frac{\partial u}{\partial y} = u_e\frac{du_e}{dx} + \nu\frac{\partial^2 u}{\partial y^2} + g\beta(T - T_\infty), \tag{2.32}$$

$$u\frac{\partial T}{\partial x} + v\frac{\partial T}{\partial y} = \alpha\frac{\partial^2 T}{\partial y^2}, \tag{2.33}$$

subject to the boundary conditions given by Mukhopadhyay and Andersson (2009).

$$\begin{gathered} u = \nu\, N_1\frac{\partial u}{\partial y}, \quad v = 0, \quad T = T_w(x) + N_2\frac{\partial T}{\partial y} \ \text{ at } \ y = 0 \\ u \to u_e(x), \quad T \to T_\infty \quad \text{as } \ y \to \infty \end{gathered} \tag{2.34}$$

where N_1 is the slip length and N_2 is a proportionality constant. Further, we assume that

$$u_e(x) = ax, \quad T_w(x) = T_\infty + T_0(x/L) \tag{2.35}$$

where a is a positive constant, T_0 is a characteristic temperature with $T_0 > 0$ for assisting flow and $T_0 < 0$ for opposing flow, respectively, and L is a characteristic length of the plate.

To obtain similarity solutions, we introduce the following transformations (Aman et al., 2011):

$$\eta = (a/\nu)^{1/2} y, \quad \psi = (a\nu)^{1/2} x f(\eta), \quad \theta(\eta) = (T - T_\infty)/(T_w - T_\infty) \tag{2.36}$$

where ψ is the stream function defined as $u = \partial\psi/\partial y$ and $v = -\partial\psi/\partial x$ which identically satisfy the continuity Eq. (2.31). Substituting Eq. (2.36) into Eqs. (2.32) and (2.33), we get the following nonlinear ordinary (similarity) differential equations:

$$f''' + ff'' - f'^2 + 1 + \lambda\theta = 0 \tag{2.37}$$

$$\frac{1}{Pr}\theta'' + f\theta' - f'\theta = 0, \tag{2.38}$$

subject to the boundary conditions

$$\begin{aligned} &f(0) = 0, \ f'(0) = \delta f''(0), \quad \theta(0) = 1 + \gamma\theta'(0), \\ &f'(\eta) \to 1, \quad \theta(\eta) \to 0 \text{ as } \eta \to \infty \end{aligned} \tag{2.39}$$

where $Pr = \nu/\alpha$ is the Prandtl number, $\delta = N_1\sqrt{a\nu}$ is the velocity slip parameter, $\gamma = N_2\sqrt{a/\nu}$ is the thermal slip parameter, and $\lambda = g\beta T_0/a^2$ is the mixed convection parameter. It is worth mentioning that $\lambda > 0$ corresponds to the assisting flow, $\lambda < 0$ corresponds to the opposing flow, and $\lambda = 0$ corresponds to the forced convection flow.

The physical quantities of interest are the reduced skin friction coefficient C_f and the reduced local Nusselt number Nu_x, which are proportional to $f''(0)$ and $-\theta'(0)$, respectively.

2.2.3 Comments

The boundary value problem Eqs. (2.37)–(2.39) has been solved numerically by Aman et al. (2011) using the shooting method for some values of the velocity slip parameter δ, thermal slip parameter γ, the buoyancy or mixed convection parameter λ, and the Prandtl number Pr. Comparisons of $f''(0)$ and $-\theta'(0)$ with those obtained by Ramachandran et al. (1988), Devi et al. (1991), Hassanien and Gorla (1990), Lok et al. (2006) for several values of Pr when the slip effect is absent $\delta = 0$ and $\gamma = 0$ are listed in Tables 2.4 and 2.5, respectively. It is observed that the results show a very good agreement.

Table 2.4 Values of $f''(0)$ for different values of Pr when $\lambda = 1$, $\delta = 0$, and $\gamma = 0$.

Pr	Ramachandran et al. (1988)	Devi et al. (1991)	Lok et al. (2006)	Hassanien and Gorla (1991)	Aman et al. (2011)
0.7	1.7063	1.7064	1.7064	1.70632	1.7063
1	—	—	—	—	1.6754
7	1.5179	1.518	1.518	—	1.5179
10	—	—	—	1.49284	1.4928
20	1.4485	1.4485	1.4486	—	1.4485
40	1.4101	—	1.4102	—	1.4101
50	—	—	—	1.40686	1.3989
60	1.3903	1.3903	1.3903	—	1.3903
80	1.3774	—	1.3773	—	1.3774
100	1.368	1.368	1.3677	1.38471	1.368

The effect of the governing parameters on $f''(0)$ and $-\theta'(0)$ are shown in Figs. 2.9–2.12, for several values of the governing parameters. The parameter δ on $f''(0)$ and $-\theta'(0)$ are shown in Figs. 2.9 and 2.10, respectively, while the effect of γ on $f''(0)$ and $-\theta'(0)$ are shown, respectively, in Figs. 2.11 and 2.12. It is evident from these four figures that dual (upper and lower) solutions exist for buoyancy opposing flow case ($\lambda < 0$). For the assisting flow ($\lambda > 0$), the solution is unique. The upper branch solution has a higher value of $f''(0)$ and $-\theta'(0)$ for a given λ, than the lower branch solution. It can be seen that for the upper branch solution, as the buoyancy parameter λ increases, both $f''(0)$ and $-\theta'(0)$ increase, due to the increase of velocity caused by the buoyancy forces. The opposite

Table 2.5 Values of $-\theta'(0)$ for different values of Pr when $\lambda = 1$, $\delta = 0$, and $\gamma = 0$.

Pr	Ramachandran et al. (1988)	Devi et al. (1991)	Lok et al. (2006)	Hassanien and Gorla (1991)	Aman et al. (2011)
0.7	0.7641	0.7641	0.7641	0.76406	0.7641
1	—	—	—	—	0.8708
7	1.7224	1.7223	1.7226	—	1.7224
10	—	—	—	1.94461	1.9446
20	2.4576	2.4574	2.4577	—	2.4576
40	3.1011	—	3.1023	—	3.1011
50	—	—	—	3.34882	3.3415
60	3.5514	3.5517	3.556	—	3.5514
80	3.9095	—	3.9195	—	3.9095
100	4.2116	4.2113	4.2289	4.23372	4.2116

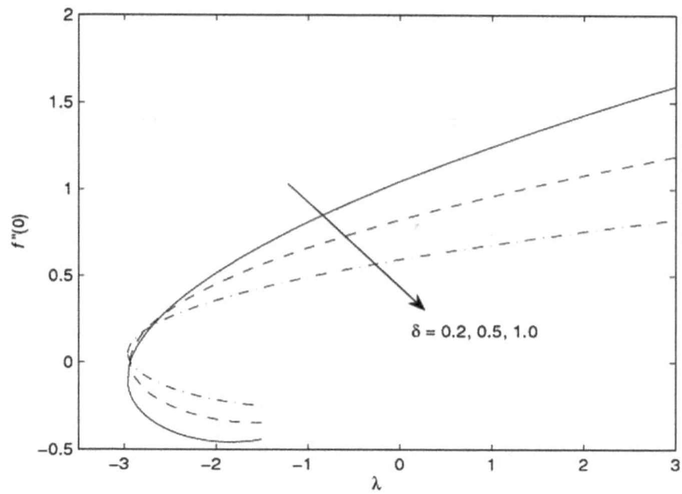

Figure 2.9 Variation of $f''(0)$ with λ for different values of δ when $Pr = 0.7, \gamma = 1$. *(From Aman, F., Ishak, A., & Pop, I. (c. 2011). Mixed convection boundary layer flow near stagnation-point on vertical surface with slip.* Applied Mathematics and Mechanics, 32*(12), 1599—1606. https://doi.org/10.1007/s10483-011-1526-x 1603)*

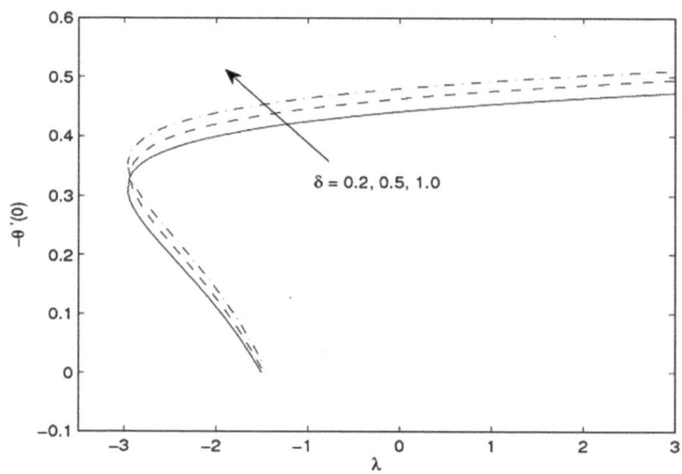

Figure 2.10 Variation of $-\theta'(0)$ with λ for different values of δ when $Pr = 0.7$, $\gamma = 1$. *(From Aman, F., Ishak, A., & Pop, I. (c. 2011). Mixed convection boundary layer flow near stagnation-point on vertical surface with slip.* Applied Mathematics and Mechanics, 32*(12), 1599—1606. https://doi.org/10.1007/s10483-011-1526-x 1603)*

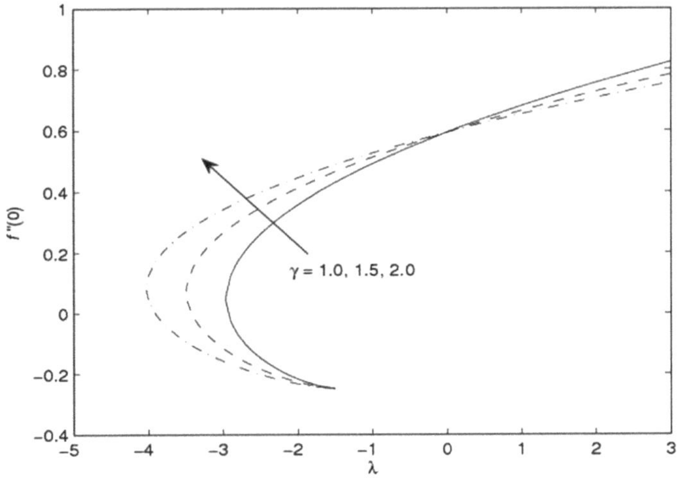

Figure 2.11 Variation of $f''(0)$ with λ for different values of γ when $Pr = 0.7$, $\delta = 1$. *(From Aman, F., Ishak, A., & Pop, I. (c. 2011). Mixed convection boundary layer flow near stagnation-point on vertical surface with slip.* Applied Mathematics and Mechanics, *32(12), 1599–1606. https://doi.org/10.1007/s10483-011-1526-x 1603)*

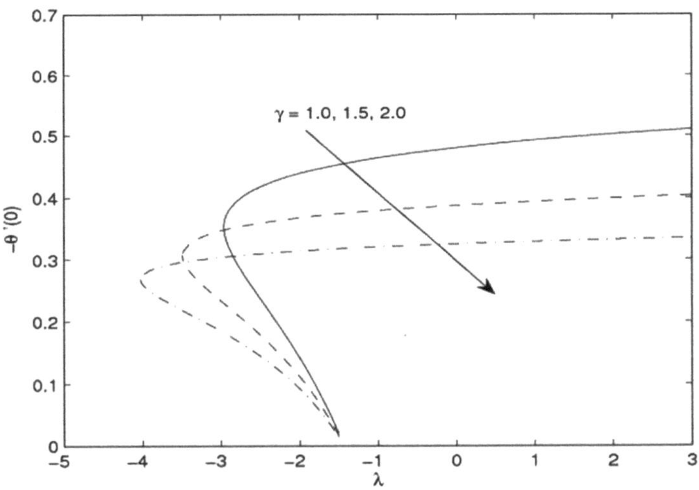

Figure 2.12 Variation of $-\theta'(0)$ with λ for different values of γ when $Pr = 0.7$, $\delta = 1$. *(From Aman, F., Ishak, A., & Pop, I. (c. 2011). Mixed convection boundary layer flow near stagnation-point on vertical surface with slip.* Applied Mathematics and Mechanics, *32(12), 1599–1606. https://doi.org/10.1007/s10483-011-1526-x 1603)*

trend can be observed for the lower branch solution. In general, for the upper branch solution, an increase in the velocity slip parameter δ is to decrease $f''(0)$, but increase $-\theta'(0)$. In contrast, Fig. 2.12. shows that $-\theta'(0)$ decreases with increasing values of δ. It is seen from Figs. 2.10 and 2.12 that $-\theta'(0) > 0$ for all values of parameters considered in this study, which means the heat is transferred from the hot surface to the cool fluid. Furthermore, it is seen that for each selected values of δ and γ (see Figs. 2.9–2.12), there is indeed a critical value λ_c of λ for which the solution exists. Based on the present computations, it is found that $\lambda_c = -2.9600$, -2.9361 and -2.9613, respectively, for $\delta = 0.2$, 0.5, and 1.0 (Figs. 2.9 and 2.10), and $\lambda_c = -2.9613, -3.4890$ and -4.0266, respectively, for $\gamma = 1.0$, 1.5, and 2.0 (Figs. 2.11 and 2.12). It is worth mentioning that the computations have been performed until the point where the solution does not converge, and the calculations were terminated at that point. It may also be pointed out here that the effect of slip is to reduce the range of λ for which the solution exists. More results on this problem can be found in Aman et al. (2011).

2.3 Mixed convection nonaxisymmetric Homann stagnation-point flow

2.3.1 Introduction

Stagnation-point flows, describing the fluid motion near the stagnation region at the front of a blunt-nosed body, exists on all solid bodies moving in a fluid. The stagnation region encounters the highest pressure, the highest heat transfer, and the highest rates of mass deposition. Hiemenz (1911) was the first to study two-dimensional stagnation flows using a similarity transform to reduce the Navier–Stokes equations to nonlinear ordinary differential equations. The axisymmetric case was later solved by Homann (1936). Mixed convection over flat surfaces represents another important problem, which is often encountered in the environment and in engineering devices. Therefore, the knowledge of heat transfer phenomena over this kind of geometry is also very important in developing compact and highly efficient heat exchanger components.

2.3.2 Basic equations

Lok et al. (2017) have considered recently the steady mixed convection nonaxisymmetric Homann stagnation-point flow (Homann, 1936) (over a vertical flat wall placed in a viscous and incompressible fluid. It was assumed that the velocity components of the external flow, or potential flow, are $u_e(x) = (a+b)x$, $v_e(y) = (a-b)y$, and $w_e(z) = -2az$, and the surface temperature is taken as $T_w(x) = T_\infty + T_0 x$. Here x, y, and z are Cartesian coordinates, with the z-axis measured in the normal direction to the wall and the axis measured in the vertical direction, respectively; a is the strain rate of the stagnation-point flow, b is the shear rate, and T_0 is a characteristic temperature scale, with for $T_0 > 0$ a heated surface (aiding flow) and $T_0 < 0$ for a cooled surface (opposing flow). Under these assumptions, the governing equations are, in Cartesian form (see Lok et al., 2017).,

$$\frac{\partial u}{\partial x} + \frac{\partial v}{\partial y} + \frac{\partial w}{\partial z} = 0 \tag{2.40}$$

$$u\frac{\partial u}{\partial x} + v\frac{\partial u}{\partial y} + w\frac{\partial u}{\partial z} = -\frac{1}{\rho}\frac{\partial p}{\partial x} + \nu\left(\frac{\partial^2 u}{\partial x^2} + \frac{\partial^2 u}{\partial y^2} + \frac{\partial^2 u}{\partial z^2}\right) + g\beta(T - T_\infty) \tag{2.41}$$

$$u\frac{\partial v}{\partial x} + v\frac{\partial v}{\partial y} + w\frac{\partial v}{\partial z} = -\frac{1}{\rho}\frac{\partial p}{\partial y} + \nu\left(\frac{\partial^2 v}{\partial x^2} + \frac{\partial^2 v}{\partial y^2} + \frac{\partial^2 v}{\partial z^2}\right) \tag{2.42}$$

$$u\frac{\partial w}{\partial x} + v\frac{\partial w}{\partial y} + w\frac{\partial w}{\partial z} = -\frac{1}{\rho}\frac{\partial p}{\partial z} + \nu\left(\frac{\partial^2 w}{\partial x^2} + \frac{\partial^2 w}{\partial y^2} + \frac{\partial^2 w}{\partial z^2}\right) \tag{2.43}$$

$$u\frac{\partial T}{\partial x} + v\frac{\partial T}{\partial y} + w\frac{\partial T}{\partial z} = \alpha\left(\frac{\partial^2 T}{\partial x^2} + \frac{\partial^2 T}{\partial y^2} + \frac{\partial^2 T}{\partial z^2}\right) \tag{2.44}$$

subject to the boundary conditions

$$\begin{gathered} u = v = w = 0, \quad T = T_w \text{ on } z = 0 \\ u \to u_e(x), \quad v \to v_e(y), \quad w \to w_e(z), \quad T \to T_\infty, \quad p \to p_0 \text{ as } z \to \infty \end{gathered} \tag{2.45}$$

where u, v, and w are, respectively, the velocity components along x, y, and z directions, p_0 is the constant far-field pressure, and the physical meaning of the other quantities is mentioned in the Nomenclature.

2.3.3 Solution

Following Weidman (2012), the solutions to Eqs. (2.40)–(2.44) assumed of the form

$$u = (a+b)xf'(\eta), v = (a-b)yg'(\eta), w = -\sqrt{\frac{\nu}{a}}[(a+b)f(\eta) + (a-b)g(\eta)]$$
$$T = T_\infty + T_0(x/L), \quad \eta = z\sqrt{a/\nu} \tag{2.46}$$

Substituting Eq. (2.46) into Eqs. (2.40)−(2.44), the following ordinary (similarity) equations are obtained

$$f''' + (1+\gamma)\left(ff'' + 1 - f'^2\right) + (1-\gamma)gf'' + (1+\gamma)\,\theta\,\lambda = 0 \tag{2.47}$$

$$g''' + (1-\gamma)\left(gg'' + 1 - g'^2\right) + (1+\gamma)fg'' = 0 \tag{2.48}$$

$$\frac{1}{Pr}\,\theta'' + (1+\gamma)(f\theta' - f'\theta) + (1-\gamma)g\theta' = 0 \tag{2.49}$$

along with the boundary conditions

$$\left.\begin{array}{l} f(0) = g(0) = 0,\, f'(0) = g'(0) = 0, \quad \theta(0) = 1 \\ f'(\eta) \to 1, \quad g'(\eta) \to 1, \quad \theta(\eta) \to 0 \ \text{ as } \eta \to \infty \end{array}\right\} \tag{2.50}$$

where Pr is the Prandtl number, defined in the Nomenclature, $\gamma = b/a$, and λ is the mixed convection parameter, which is defined as

$$\lambda = \frac{g\beta T_0}{(a+b)^2} \tag{2.51}$$

The main physical quantities of interest are the values of the reduced skin friction coefficients $f''(0)$ and $g''(0)$ as well as the surface temperature gradient $\theta'(0)$. The Prandtl number Pr has been fixed at $Pr = 7$.

2.3.4 Comments

The boundary value problem given by Eqs. (2.47)−(2.50) was solved numerically by Lok et al. (2017) using bvp4c in Matlab (Shampine et al., 2003) and D02AGF in the NAG (Numerical Algorithm Group) library (www.nag.co.uk). When λ= (forced convection flow), the present problem reduces to that considered by Weidman (2012). Weidman also pointed out that, when $\lambda = 0$ and $\gamma = 1$, the problem reduces to (Hiemenz, 1911) flow for $f(\eta)$ and when $\gamma = -1$ it reduces to $g(\eta)$. Also, when $\lambda = \gamma = 0$, Homann's axisymmetric stagnation-point flow (1936) flow is recovered.

Variation of $f''(0)$, $g''(0)$, and $\theta'(0)$ is shown in Fig. 2.13 when $\lambda = 0$ (forced convection flow). The results given by Weidman (2012) are also

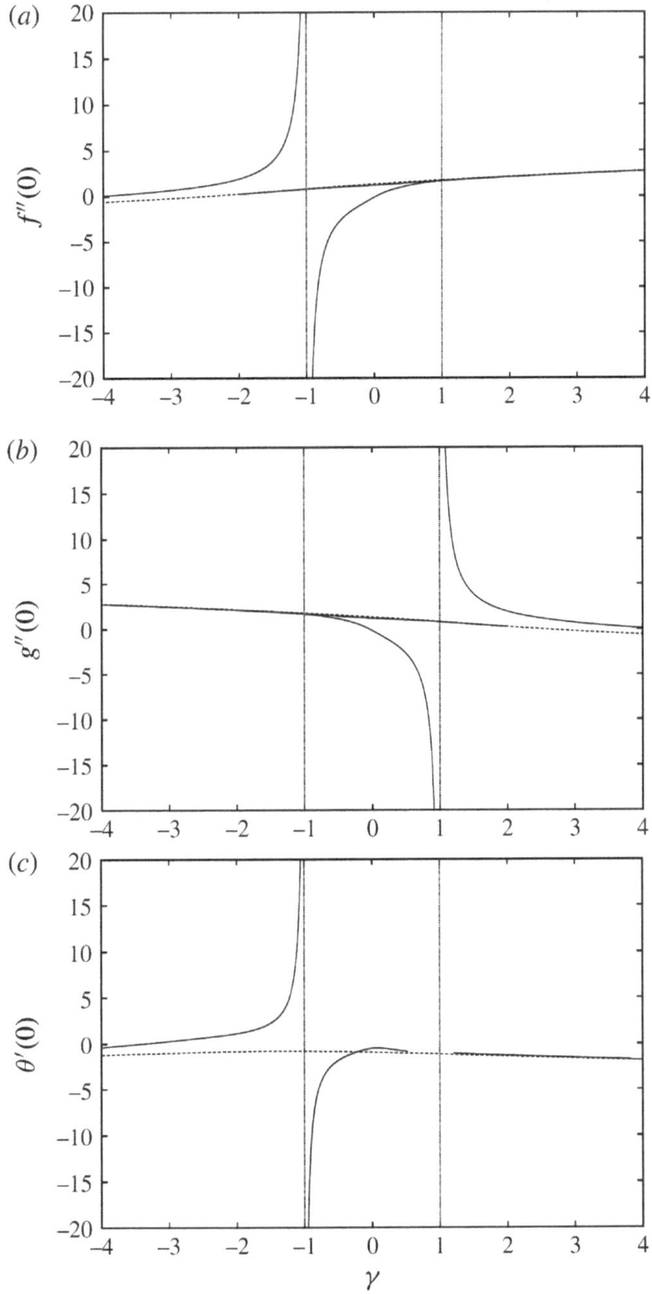

Figure 2.13 Variation of $f''(0)$, $g''(0)$, and $\theta'(0)$ with γ when $\lambda = 0$ (forced convection flow). The results given by Weidman (2012) when $\lambda = 0$ are shown by the broken lines. *(From Lok, Y. Y., Merkin, J. H., & Pop, I. (c. 2017). Mixed convection non-axisymmetric Homann stagnation-point flow.* Journal of Fluid Mechanics, 812, *418–434. https://doi.org/10.1017/jfm.2016.809 423)*

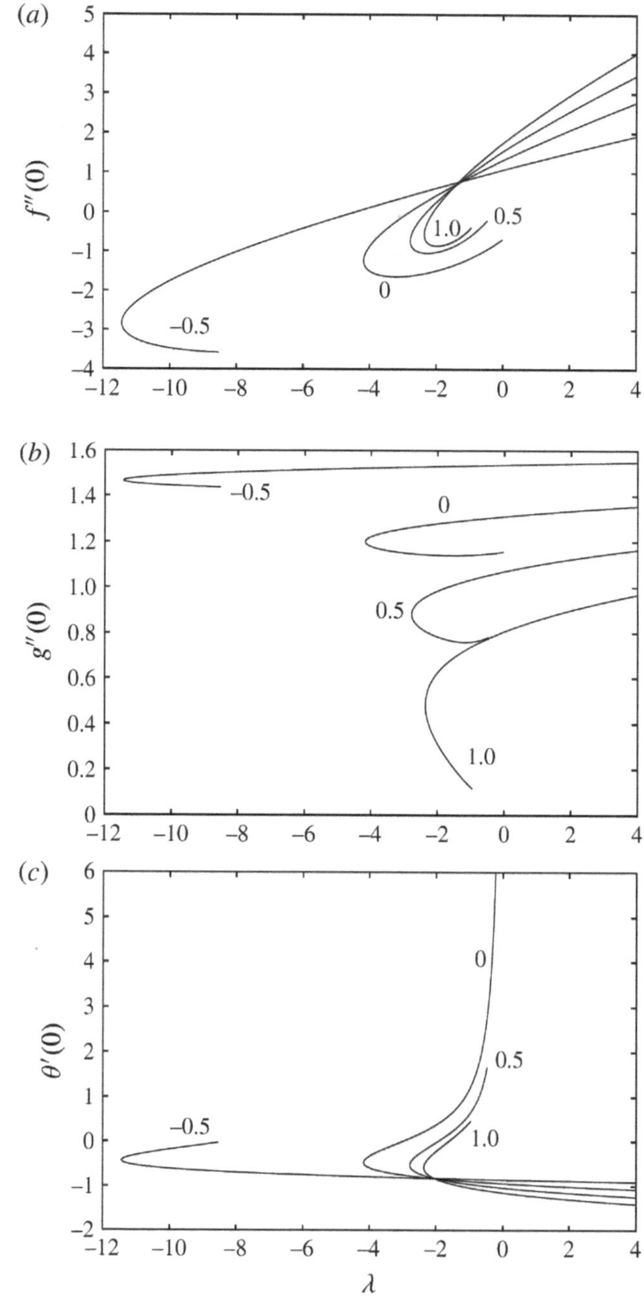

Figure 2.14 Variation of $f''(0)$, $g''(0)$, and $\theta'(0)$ with λ when $\gamma = -0.5$, 0.0, 0.5, and 1.0 obtained from the numerical solution of Eqs. (2.47)–(2.50). *(From Lok, Y. Y., Merkin, J. H., & Pop, I. (c. 2017). Mixed convection non-axisymmetric Homann stagnation-point flow.* Journal of Fluid Mechanics, 812, *418–434. https://doi.org/10.1017/jfm.2016.809 423)*

shown by the broken lines. The singular nature of $f''(0)$ as $\gamma \to -1$ and of $g''(0)$ as $\gamma \to 1$ is clearly seen in Fig. 2.13. For large $|\gamma|$ the solutions follow closely the solution given by (Weidman, 2012). The results for the corresponding heat transfer problem $\theta'(0)$ are shown in Fig. 2.13C where the broken lines are the results presented by Weidman (2012). One can see that $\theta'(0)$ becomes singular as $\gamma \to -1$ arising from the singular nature of the solution for $f(\eta)$. However, the singularity in $g(\eta)$ as $\gamma \to 1$ does not seem to affect the solution for $\theta(\eta)$, with the values of $\theta'(0)$ appearing to progress smoothly through this value of γ.

Plots of $f''0)$, $g''(0)$, and $\theta'(0)$ against λ for representative values of $\gamma > -1$, noting for $\gamma = -1$ that the solution is independent of λ are illustrated in Fig. 2.14. A significant feature of these results is the existence of a critical value λ_c of λ with the saddle-node bifurcation at $\lambda = \lambda_c$ producing two solution branches in $\lambda \geq \lambda_c$. These critical values depend on γ with $|\lambda_c|$ increasing noticeably as γ approaches -1. The upper solution branch, as seen in Fig. 2.14, continues to large positive λ with the values of $f''(0)$ and $g''(0)$ increasing and $\theta'(0)$ decreasing. The lower branch solution terminates at a finite value of λ; the present numerical results suggest that this might be as $\lambda \to 0$ from below with $\theta'(0)$ becoming increasingly larger. More very important results on this problem can be found in Lok et al. (2017).

Nomenclature

a, b, c	constants
A	unsteadiness parameter
C_{fx}, C_{fy}	skin friction coefficients
$f(\eta), h(\eta)$	dimensionless functions related to velocity
$F(\eta, \tau), H(\eta, \tau), G(\eta, \tau)$	functions
g	gravitational acceleration vector (m/s^2)
Gr_x	local Grashof number
k	thermal conductivity $(W/m/K)$
L	characteristic length (m)
N_1	slip length
N_2	constant
Nu_x	local Nusselt number
Pr	Prandtl number
Re_x	local Reynolds number
s	suction parameter
t	dimensional time (s)
T	dimensional fluid temperature (K)
T_0	characteristic temperature (K)

T_w	wall temperature (K)
T_∞	ambient temperature (K)
(u, v, w)	velocity vector components (m/s)
u_e, v_e	velocity components of the outer flow (m/s)
w_0	mass flux velocity
(x, y, z)	Cartesian coordinates (m)

Greek letters

α	thermal diffusivity (m^2s^{-1})
β	coefficient of thermal expansion (K^{-1})
$\gamma, \gamma_1, \gamma_2, \ldots$	eigenvalues (Section 2.1)
γ	thermal slip parameter (Section 2.2)
γ	constant ($= b/a$) (Section 2.3)
δ_1	constant
δ	velocity slip parameter
λ	mixed convection parameter
η	dimensionless similarity variable
$\theta(\eta)$	dimensionless temperature function
ν	kinematic viscosity (m^2s^{-1})
ρ	density of fluid (kg/m^3)
τ	dimensionless time variable

Subscript

e	outer free stream flow
∞	far field boundary condition
w	condition at the wall

Superscript

$'$	differentiation with respect to η

References

Aman, F., Ishak, A., & Pop, I. (2011). Mixed convection boundary layer flow near stagnation-point on vertical surface with slip. *Applied Mathematics and Mechanics, 32*(12), 1599–1606. https://doi.org/10.1007/s10483-011-1526-x

Davey, A. (1961). Boundary-layer flow at a saddle point of attachment. *Journal of Fluid Mechanics, 10*(04), 593. https://doi.org/10.1017/S0022112061000391

Devi, C. D. S., Takhar, H. S., & Nath, G. (1991). Unsteady mixed convection flow in stagnation region adjacent to a vertical surface. *Wärme- Und Stoffübertragung, 26*(2), 71–79. https://doi.org/10.1007/BF01590239

Eswara, A. T., & Nath, G. (1999). Effect of large injection rates on unsteady mixed convection flow at a three-dimensional stagnation point. *International Journal of Non-Linear Mechanics, 34*(1), 85–103. https://doi.org/10.1016/S0020-7462(98)00004-3

Fang, T., & Zhang, J. (2009). Closed-form exact solutions of MHD viscous flow over a shrinking sheet. *Communications in Nonlinear Science and Numerical Simulation, 14*(7), 2853–2857. https://doi.org/10.1016/j.cnsns.2008.10.005

Goldstein, S. (1965). On backward boundary layers and flow in converging passages. *Journal of Fluid Mechanics, 21*(01), 33. https://doi.org/10.1017/S0022112065000034

Harris, S. D., Ingham, D. B., & Pop, I. (2009). Mixed convection boundary-layer flow near the stagnation point on a vertical surface in a porous medium: Brinkman model with slip. *Transport in Porous Media,* 77(2), 267–285. https://doi.org/10.1007/s11242-008-9309-6

Hassanien, I. A., & Gorla, R. S. R. (1990). Heat transfer to a micropolar fluid from a non-isothermal stretching sheet with suction and blowing. *Acta Mechanica, 84*(1–4), 191–199. https://doi.org/10.1007/BF01176097

Hiemenz, K. (1911). Die Grenzschicht an einem in der gleichförmigen Flüssigkeitsstrom eingetauchten geraden Kreiszylinder. *Dinglers Polytechn. J., 326*, 321–324.

Homann, F. (1936). Der Einfluß großer Zähigkeit bei der Strömung um den Zylinder und um die Kugel. *ZAMM - Zeitschrift Für Angewandte Mathematik Und Mechanik, 16*(3), 153–164. https://doi.org/10.1002/zamm.19360160304

Jeng, D. R., & Williams, D. W. (1973). Transpiration cooling in three-dimensional laminar boundary-layer flow near a stagnation point. *AIAA Journal, 11*(11), 1560–1562. https://doi.org/10.2514/3.50628

Lok, Y. Y., Amin, N., & Pop, I. (2006). Non-orthogonal stagnation point flow towards a stretching sheet. *International Journal of Non-Linear Mechanics, 41*(4), 622–627. https://doi.org/10.1016/j.ijnonlinmec.2006.03.002

Lok, Y. Y., Merkin, J. H., & Pop, I. (2017). Mixed convection non-axisymmetric Homann stagnation-point flow. *Journal of Fluid Mechanics, 812*, 418–434. https://doi.org/10.1017/jfm.2016.809

Miklavčič, M., & Wang, C. Y. (2006). Viscous flow due to a shrinking sheet. *Quarterly of Applied Mathematics, 64*(2), 283–290. https://doi.org/10.1090/S0033-569X-06-01002-5

Noor, A., Nazar, R., Naganthran, K., & Pop, I. (2021). Unsteady mixed convection flow at a three-dimensional stagnation point. *International Journal of Numerical Methods for Heat & Fluid Flow, 31*(1), 236–250. https://doi.org/10.1108/HFF-03-2020-0138

Ramachandran, N., Chen, T. S., & Armaly, B. F. (1988). Mixed convection in stagnation flows adjacent to vertical surfaces. *Journal of Heat Transfer, 110*(2), 373–377. https://doi.org/10.1115/1.3250494

Rosca, A. V., & Pop, I. (2013). Flow and heat transfer over a vertical permeable stretching/shrinking sheet with a second order slip. *International Journal of Heat and Mass Transfer, 60*, 355–364. https://doi.org/10.1016/j.ijheatmasstransfer.2012.12.028

Shampine, L. F., Gladwell, I., & Thompson, S. (2003). Solving ODEs with MATLAB. In *Solving ODEs with MATLAB.* Cambridge University Press. https://doi.org/10.1017/CBO9780511615542

Wang, C. Y. (2008). Stagnation flow towards a shrinking sheet. *International Journal of Non-Linear Mechanics, 43*(5), 377–382. https://doi.org/10.1016/j.ijnonlinmec.2007.12.021

Weidman, P. D. (2012). Non-axisymmetric Homann stagnation-point flows. *Journal of Fluid Mechanics, 702*, 460–469. https://doi.org/10.1017/jfm.2012.197

Weidman, P. D., Kubitschek, D. G., & Davis, A. M. J. (2006). The effect of transpiration on self-similar boundary layer flow over moving surfaces. *International Journal of Engineering Science, 44*(11–12), 730–737. https://doi.org/10.1016/j.ijengsci.2006.04.005

CHAPTER 3

Stretching/shrinking sheets near a stagnation-point flow in viscous fluids

3.1 Introduction

The study of the dynamics of fluid flow against a surface is known as stagnation-point flow. The stagnation point marks the location where the oncoming flow divides to pass on either side of the surface. Hiemenz (1911) was the first to discover that the stagnation-point flow can be analyzed exactly by the Navier–Stokes equations using similarity transformation. The axisymmetric three-dimensional stagnation-point flow toward an infinite flat plate was studied by Homann (1936). Generalized three-dimensional stagnation-point flow which is composed of the nonlinear interaction between two orthogonal Hiemenz flows of arbitrary strain rate was reported by Howarth (1951). After these pioneering work, various studies on stagnation-point flows have been investigated. One of the fields of studies is the stagnation-point flow over a stretching or shrinking surface.

3.2 Unsteady separated stagnation-point flow toward stretching/shrinking sheet

3.2.1 Introduction

Unsteady separated stagnation-point (USSP) flow is one of the group-invariant similarity solution classes derived by Ma and Hui (1990) while solving unsteady two-dimensional laminar boundary layer equations by using the Lie group transformation. Dholey and Gupta (2013) have obtained similarity solutions of USSP flow over a moving permeable plate which is subject to continuous suction or blowing. Ma and Hui's USSP problem was then extended to the case of stretching or shrinking sheet by Lok and Pop (2014). Here, the stretching or shrinking axis and the

Similarity Solutions for the Boundary Layer Flow and Heat Transfer of Viscous Fluids, Nanofluids, Porous Media, and Micropolar Fluids
ISBN 978-0-12-821188-5
https://doi.org/10.1016/B978-0-12-821188-5.00009-6

stagnation flow can be either aligned or nonaligned. Other publications on USSP flow can be referred to Dholey (2015, 2016, 2018a, 2018b, 2019) and Nandy et al. (2015).

3.2.2 Problem formulation

Consider the unsteady two-dimensional stagnation-point flow of a viscous fluid over a continuously stretching or shrinking sheet as shown in Fig. 3.1. Following Ma and Hui (1990), it is assumed that the velocity of the sheet is $u_w(x,t) = \left(\frac{b}{t}\right)(x + c)$, where $b > 0$ represents the stretching rate, while $b < 0$ represents the shrinking rate, respectively. The sheet origin is located at $-c$ and t is time. The time-dependent free stream velocity is assumed to be $u_e(x,t) = (a/t)x$ and $v_e(y,t) = -(a/t)y$, where a is a positive constant. The x-axis is measured along the sheet and the y-axis is perpendicular to it.

The basic equations of this problem are given by

$$\frac{\partial u}{\partial x} + \frac{\partial v}{\partial y} = 0 \tag{3.1}$$

$$\frac{\partial u}{\partial t} + u\frac{\partial u}{\partial x} + v\frac{\partial u}{\partial y} = -\frac{1}{\rho}\frac{\partial p}{\partial x} + \nu\left(\frac{\partial^2 u}{\partial x^2} + \frac{\partial^2 u}{\partial y^2}\right) \tag{3.2}$$

$$\frac{\partial v}{\partial t} + u\frac{\partial v}{\partial x} + v\frac{\partial v}{\partial y} = -\frac{1}{\rho}\frac{\partial p}{\partial y} + \nu\left(\frac{\partial^2 v}{\partial x^2} + \frac{\partial^2 v}{\partial y^2}\right) \tag{3.3}$$

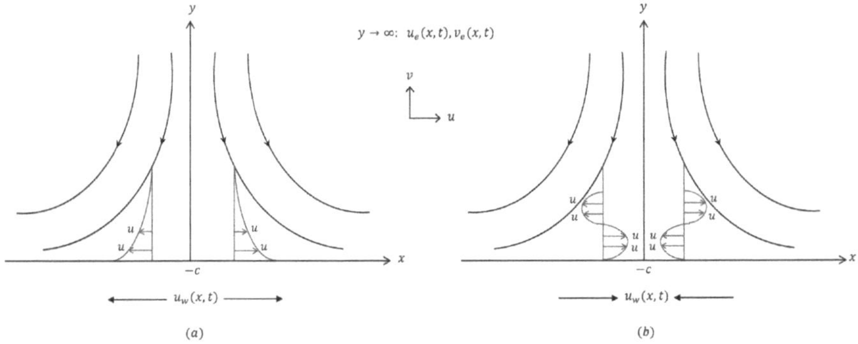

Figure 3.1 Physical model and coordinate system for (A) stretching sheet; (B) shrinking sheet.

where ν is the kinematic viscosity of the fluid and the physical meaning of other variables is mentioned in the Nomenclature. The initial and boundary conditions are

$$\begin{gathered} t=0: u=v=0 \text{ for any } x, y \\ t>0: u=u_w(x,t), \quad v=0 \text{ at } y=0 \\ u \rightarrow u_e(x,t), \quad v \rightarrow v_e(x,t) \text{ as } y \rightarrow \infty \end{gathered} \tag{3.4}$$

3.2.3 Similarity transformation

The similarity solutions are in the following form (Ma & Hui, 1990; Wang, 2008):

$$u=\frac{a}{t} x f'(\eta)+\frac{b}{t} c g(\eta), \quad v=-a \sqrt{\frac{\nu}{t}} f(\eta), \quad \eta=y \sqrt{\frac{1}{\nu t}} \tag{3.5}$$

where prime denotes differentiation with respect to η. From Eqs. (3.2)–(3.5), the following expression for the pressure p is obtained:

$$p=p_0+\frac{\rho a}{2 t^2}\left[(1-a) x^2-(1+a) y^2\right] \tag{3.6}$$

where p_0 is the stagnation pressure.

Substituting Eqs. (3.5) and (3.6) into Eqs. (3.1–3.3), the continuity equation Eq. (3.1) is satisfied automatically, while the momentum equations Eqs. (3.2) and (3.3) become

$$f'''+a\left(f f''-f'^2+1\right)+\underbrace{\left(\frac{\eta}{2} f''+f'-1\right)}_{\text{unsteady effect}}=0 \tag{3.7}$$

$$g''+a\left(f g'-f' g\right)+\underbrace{\left(\frac{\eta}{2} g'+g\right)}_{\text{unsteady effect}}=0 \tag{3.8}$$

and
the boundary conditions Eq. 3.4 now become

$$\begin{gathered} f(0)=0, \quad f'(0)=\alpha, \quad g(0)=1 \\ f'(\infty)=1, \quad g(\infty)=0 \end{gathered} \tag{3.9}$$

where $\alpha=b/a$ with $\alpha>0$ is for the stretching parameter while $\alpha<0$ is for the shrinking parameter.

By comparing Eqs. (3.7) and (3.8) with Wang (2008), it is found that the unsteady effect arises from the terms in the second bracket.

The skin friction coefficient C_f is defined as

$$C_f = \frac{\tau_w}{\rho u_e^2} \tag{3.10}$$

where τ_w is the shear stress that is given by

$$\tau_w = \mu \left(\frac{\partial u}{\partial y} \right)_{y=0} \tag{3.11}$$

where μ is the dynamic viscosity. Apply the similarity variables Eq. (3.5) into Eq. (3.11), the local skin friction coefficient Eq. (3.10) becomes

$$(aRe_x)^{1/2} C_f = f''(0) + \frac{bc}{ax} g'(0) \tag{3.12}$$

where $Re_x = u_e x/\nu$ is the local Reynolds number. The point of reattachment x_s can be obtained by considering zero skin friction, hence Eq. (3.12) becomes

$$x_s = -\frac{bcg'(0)}{af''(0)} \tag{3.13}$$

The streamlines are given by

$$\frac{\psi}{\sqrt{\nu}} = \frac{1}{t^{1/2}} \left(axf(\eta) + bc \int_0^\eta g(s)\, ds \right) \tag{3.14}$$

where ψ is streamfunction that is defined as $u = \partial\psi/\partial y$ and $v = -\partial\psi/\partial x$.

3.2.4 Comments

The ordinary differential equations Eqs. (3.7) and (3.8), subject to boundary conditions Eq. (3.9), have been solved numerically by Lok and Pop (2014) using the program *bvp4c*, a BVP solver in Matlab. Figs. 3.2 and 3.3 show the variation of the initial values $f''(0)$ and $g'(0)$ with $\alpha > 0$ (stretching sheet) and $\alpha < 0$ (shrinking sheet) for $a = 0.5, 1$ and 1.5, respectively. It is noticed that dual solutions exist for both stretching and shrinking sheets of USSP flow. The solid line represents the first solution (upper branch solution), whereas the dotted line represents the second solution (lower branch solution). No solution exists for $\alpha < \alpha_c$. Fig. 3.2 shows that for the upper branch solution when $\alpha > 1$, the values of $f''(0)$ decrease as a

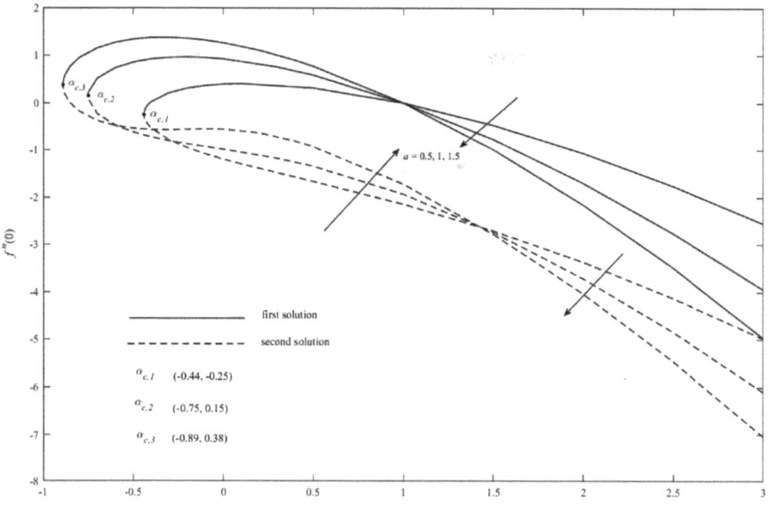

Figure 3.2 Initial values of $f''(0)$ versus α for $a = 0.5$, 1, and 1.5. *(From Lok, Y. Y., & Pop, I. (2014). Stretching or shrinking sheet problem for unsteady separated stagnation-point flow.* Meccanica, *49(6), 1479–1492. https://doi.org/10.1007/s11012-014-9932-y).*

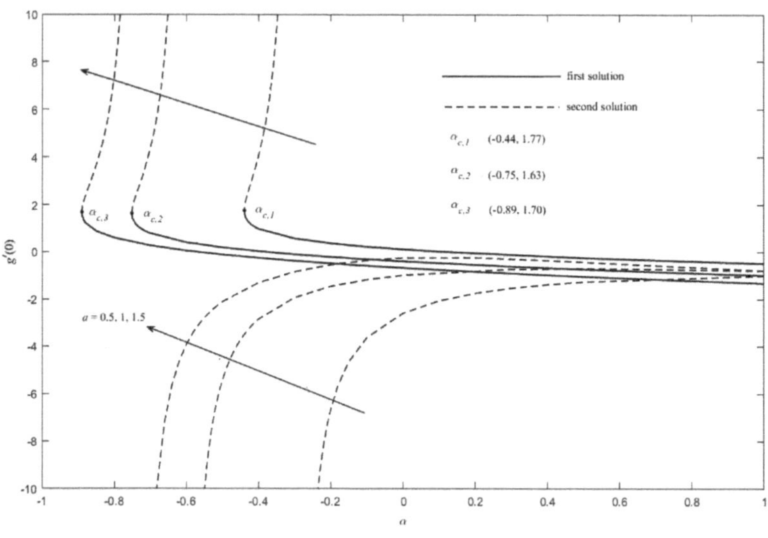

Figure 3.3 Initial values of $g'(0)$ versus α for $a = 0.5$, 1, and 1.5. *(From Lok, Y. Y., & Pop, I. (2014). Stretching or shrinking sheet problem for unsteady separated stagnation-point flow.* Meccanica, *49(6), 1479–1492. https://doi.org/10.1007/s11012-014-9932-y).*

increases; the opposite trend is observed for $\alpha < 1$. In Fig. 3.3, the range of solutions increases as a increases, and there is a vertical asymptote (discontinuity) within the range of the second solution.

Fig. 3.4 shows the functions $f'(\eta)$ and $g(\eta)$ for $a = 0.5$ when $\alpha = 0.5$ (stretching sheet) and $\alpha = -0.2$ (shrinking sheet), respectively. As expected, boundary layer thickness for the second solution is larger than those of the first solution. It can be seen that the second solution reveals regions of reverse flow.

The solutions for the case $a > 1$ is more complicated than the case of $a < 1$, more than two solutions are obtained for each value of α considered. For example, for the case $a = 2$, six possible solutions are found when considering $\alpha = 0$ (fixed plate), while ten possible solutions are found when taking $\alpha = -0.5$, see Fig. 3.5. It has to me mentioned here that some values are too close to each other, therefore only four solution curves are shown in Fig. 3.5. In particular, the velocity profiles and their corresponding streamlines for both aligned and nonaligned case that presented in Fig. 3.6 show different profile development and boundary layer thickness. More important results can be found in the paper by Lok and Pop (2014).

Clearly, multiple solutions exist for the problem of USSP flow, either toward a stretching or a shrinking sheet, with the stretching or shrinking axis can be aligned or nonaligned with the stagnation flow. However, nonalignment will destroy the symmetry line of the stagnation-point flow and sheet.

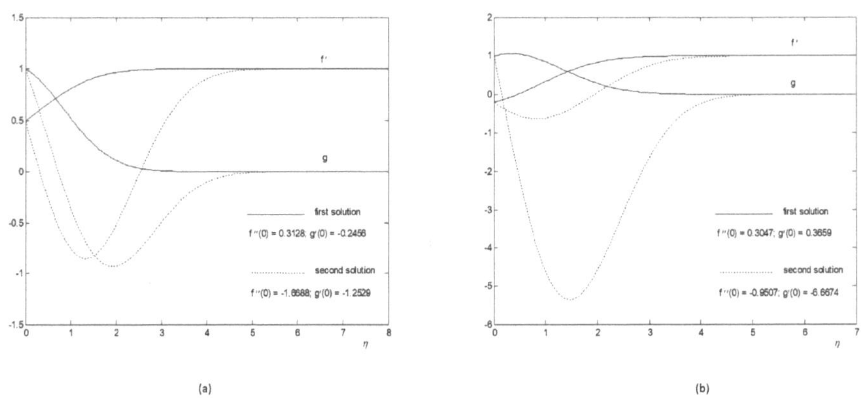

Figure 3.4 The functions $f'(\eta)$ and $g(\eta)$ for $a = 0.5$ when (A) $\alpha = 0.5$; (B) $\alpha = -0.2$. *(From Lok, Y. Y., & Pop, I. (2014). Stretching or shrinking sheet problem for unsteady separated stagnation-point flow.* Meccanica, *49(6), 1479–1492. https://doi.org/10.1007/s11012-014-9932-y).*

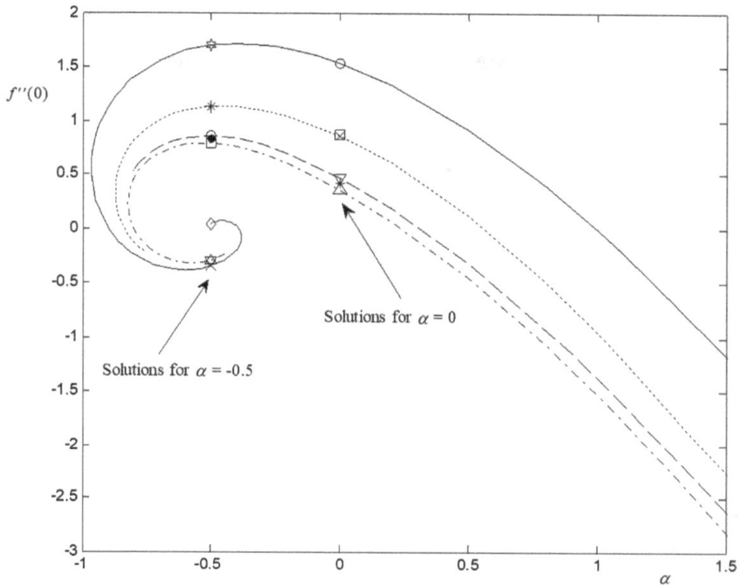

Figure 3.5 Initial values of $f''(0)$ against α for the case $a = 2$. *(From Lok, Y. Y., & Pop, I. (2014). Stretching or shrinking sheet problem for unsteady separated stagnation-point flow.* Meccanica, *49(6), 1479–1492. https://doi.org/10.1007/s11012-014-9932-y).*

3.3 Axisymmetric rotational stagnation-point flow over a permeable stretching/shrinking rotating disk

3.3.1 Introduction

The study of motion driven by rotating disk has wide applications in many areas, such as rotating machinery, aerodynamic system, and power generating system. Lately, many researches are now focusing on new materials, with the aim either to increase the performance of device (higher conductivity, elasticity, and strength) or to reduce the size of device (micro or nano sizes). Hence, many studies of fluid mechanics and heat transfer over solid material have now extended to studies over flexible and stretchable material. One of the areas is the study of stagnation-point flow over a stretching or shrinking rotating disk.

Hannah (1947) studied the forced flow against a rotating disk, while Agrawal (1957) discovered the axisymmetric rotational stagnation-point flow which impinging normally to an infinite plane wall. Turkyilmazoglu (2012) investigated the three-dimensional rotational stagnation-point flow of an electrically conducting fluid, in the presence of a magnetic

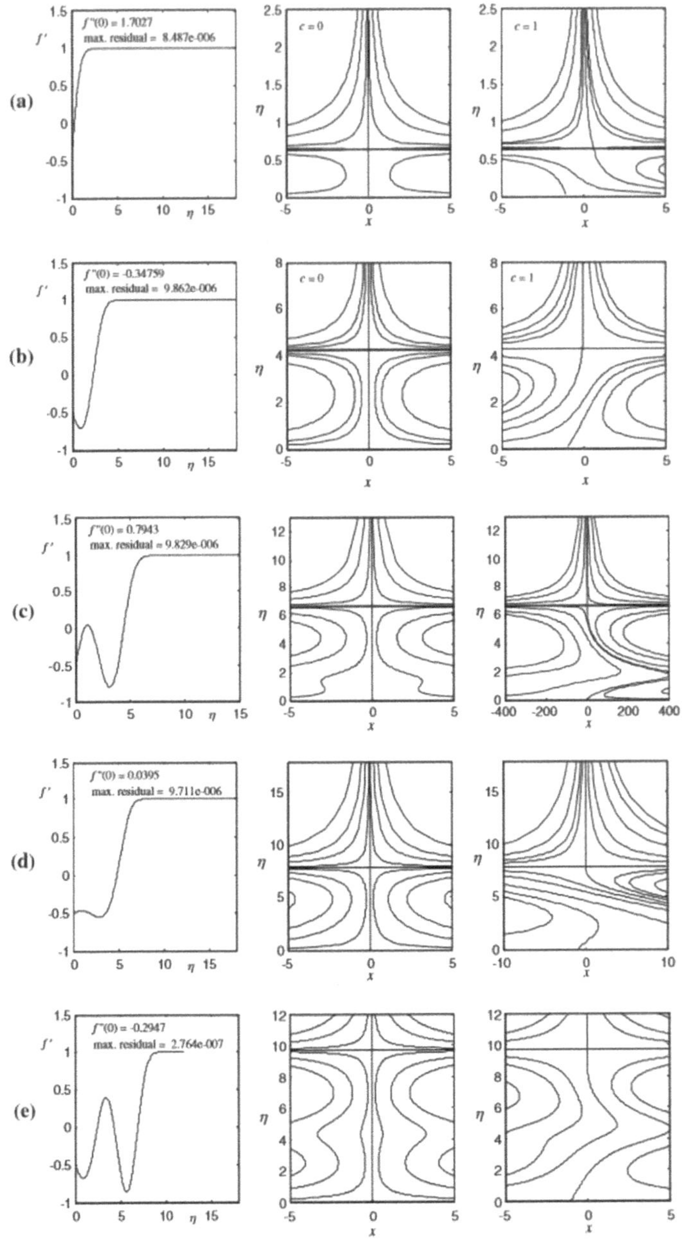

Figure 3.6 Plots of ten different velocity profiles and the corresponding streamlines (both aligned and nonaligned) for $a = 2$ and $\alpha = -0.5$ (shrinking sheet). *(From Lok, Y. Y., & Pop, I. (2014). Stretching or shrinking sheet problem for unsteady separated stagnation-point flow.* Meccanica, *49(6), 1479–1492. https://doi.org/10.1007/s11012-014-9932-y).*

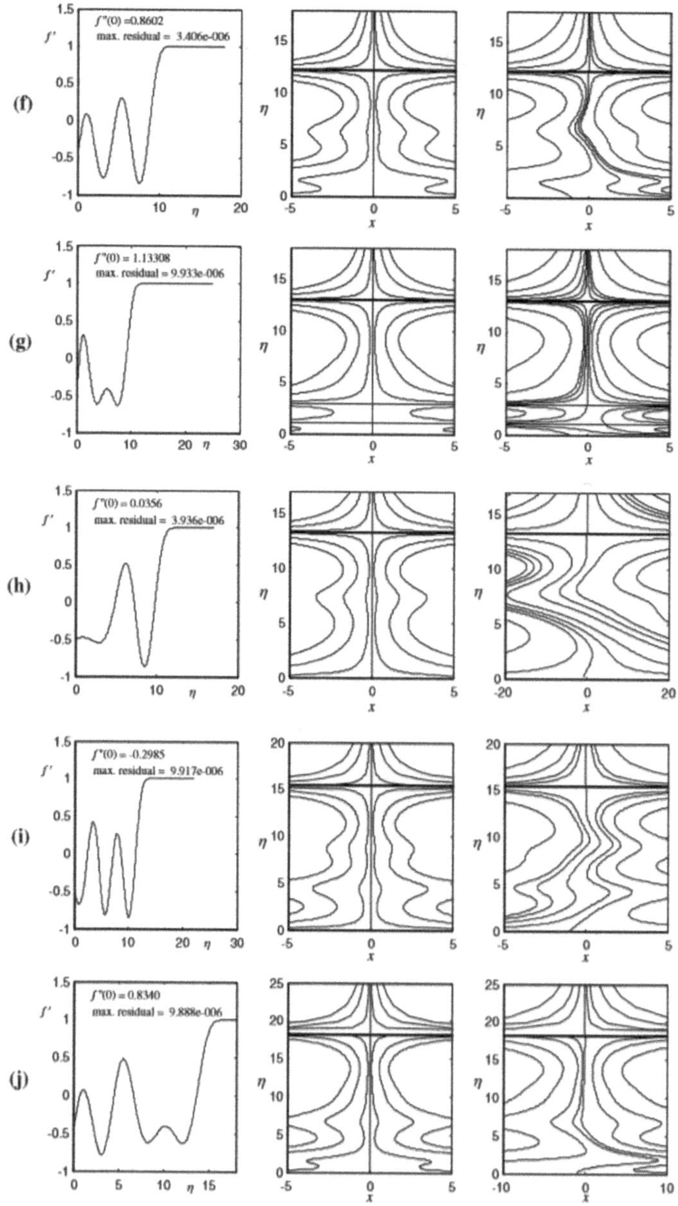

Figure 3.6 Cont'd.

field, on a radially stretchable rotating disk. He found that the stagnation velocities and shear stresses are strongly dependent upon the rotation parameter, stretching parameter, and magnetic field parameter. Weidman (2015) extended Agrawal's stagnation-point flow to the case of rotating disk. Later, Weidman (2016a) considered Agrawal's rotational stagnation-point flow but impinging on quiescent liquid layer. Results are given for both upper and lower layers interfacial shear stress as well as velocity induced in the lower liquid layer. Also, Weidman (2016b) has analyzed the axisymmetric rotational stagnation-point flow impinging on a radially stretching sheet. Dual solutions are found, and the critical point is identified. Rosca and Pop (2017) extended Weidman's problem to consider the flow on radially permeable stretching or shrinking surface in nanofluids, while Lok et al. (2018) consider such flow in viscous fluid.

3.3.2 Problem formulation

Consider the steady axisymmetric rotational stagnation-point flow impinging on a permeable stretching/shrinking rotating disk in a viscous fluid. The problem is formulated using the cylindrical coordinates (r, θ, z) measured in the radial, azimuthal, and axial directions, respectively. The flow is symmetric to the $z = 0$ plane and is axisymmetric about the z-axis, hence, independent of θ. The stagnation line is at $z = 0$, and the domain of the flow is in the upper half plane.

The governing equations are given by (Lok et al., 2018)

$$\frac{1}{r}\frac{\partial(ru)}{\partial r} + \frac{\partial w}{\partial z} = 0 \tag{3.15}$$

$$u\frac{\partial u}{\partial r} + w\frac{\partial u}{\partial z} - \frac{v^2}{r} = -\frac{1}{\rho}\frac{\partial p}{\partial r} + \nu\left(\frac{\partial^2 u}{\partial r^2} + \frac{1}{r}\frac{\partial u}{\partial r} + \frac{\partial^2 u}{\partial z^2} - \frac{u}{r^2}\right) \tag{3.16}$$

$$u\frac{\partial v}{\partial r} + w\frac{\partial v}{\partial z} + \frac{uv}{r} = \nu\left(\frac{\partial^2 v}{\partial r^2} + \frac{1}{r}\frac{\partial v}{\partial r} + \frac{\partial^2 v}{\partial z^2} - \frac{v}{r^2}\right) \tag{3.17}$$

$$u\frac{\partial w}{\partial r} + w\frac{\partial w}{\partial z} = -\frac{1}{\rho}\frac{\partial p}{\partial z} + \nu\left(\frac{\partial^2 w}{\partial r^2} + \frac{1}{r}\frac{\partial w}{\partial r} + \frac{\partial^2 w}{\partial z^2}\right) \tag{3.18}$$

The physical meaning of the variables is given in the Nomenclature.

It is assumed that the velocity components at the wall are $u_w = a^{2/3}\nu^{1/3}r$ and $v_w = \omega r$, while the velocity components of the ambient fluid (inviscid flow) are $u_e = 2arz$ and $v_e = 0$ (since the flow is axisymmetric), where a is a parameter measuring the strength of the stagnation-point flow and ω is the constant angular velocity. Therefore, the boundary conditions are given by

$$u = \lambda u_w, \quad v = v_w, \quad w = w_0 \text{ on } z = 0$$
$$\frac{\partial u}{\partial z} \to \frac{\partial u_e}{\partial z}, \quad v \to v_e \text{ as } z \to \infty \tag{3.19}$$

where λ is the dimensionless parameter measuring the radial stretching/shrinking rate, with $\lambda > 0$ for stretching surface and $\lambda < 0$ for shrinking surface; w_0 is the velocity of fluid through the wall, with $w_0 < 0$ for suction and $w_0 > 0$ for injection of the fluid.

Introducing the similarity variables (Weidman, 2015)

$$u = a^{2/3}\nu^{1/3} r f'(\eta), \quad v = \omega r g(\eta), \quad w = -2a^{1/3}\nu^{2/3} f(\eta), \quad \eta = (a/\nu)^{1/3} z \tag{3.20}$$

Eqs. (3.15)–(3.18) can be reduced to similarity form

$$f''' + 2ff'' - f'^2 + \alpha g^2 = 0 \tag{3.21}$$

$$g'' + 2(fg' - f'g) = 0 \tag{3.22}$$

and the boundary conditions Eq. (3.19) become

$$f(0) = S, \quad f'(0) = \lambda, \quad g(0) = 1$$
$$f'' \to 2, \quad g \to 0 \text{ as } \eta \to \infty \tag{3.23}$$

where $S = -\dfrac{w_0}{2a^{1/3}\nu^{2/3}}$ is the dimensionless mass flux parameter, with $S > 0$ for suction and $S < 0$ for injection, $\alpha = \dfrac{\omega^2}{a^{4/3}\nu^{2/3}} \geq 0$ is the dimensionless parameter of rotation rate of disk. The wall shear stress in radial and azimuthal directions, denoted by τ_r and τ_θ, are given by

$$\tau_r = \mu\left(\frac{\partial u}{\partial z}\right)_{z=0} = a\rho\nu r\ f''(0), \quad \tau_\theta = \mu\left(\frac{\partial v}{\partial z}\right)_{z=0} = \rho\omega a^{1/3}\nu^{2/3} r g'(0) \tag{3.24}$$

3.3.3 Comments (impermeable surface, $S = 0$)

The problem for the case when the rotating disk is impermeable ($S = 0$) and fixed ($\lambda = 0$) has been solved by Weidman (2015). Weidman (2016b) investigated the case when $S = 0$ and $\lambda > 0$ (stretching disk), while Lok et al. (2018) further extended the problem to the case of $\lambda < 0$ (shrinking disk) where dual solutions are found and linear temporal stability analysis is performed.

3.3.3.1 $S = 0$, $\alpha = 0$, $\lambda = 0$

When $S = \alpha = \lambda = 0$, Eq. (3.21) has the trivial solution

$$f(\eta) = \eta^2 \tag{3.25}$$

while Eq. (3.22) has the solution g that can be expressed in terms of confluent hypergeometric functions as

$$g(\eta) = \frac{\Gamma\left(\frac{5}{3}\right)}{\Gamma\left(\frac{1}{3}\right)} \exp(-2\eta^3/3) U\left(\frac{4}{3}; \frac{2}{3}; \frac{2\eta^3}{3}\right) \tag{3.26}$$

giving $g'(0) = -6\left(\frac{2}{3}\right)^{\frac{1}{3}} \left(\frac{\Gamma\left(\frac{2}{3}\right)}{\Gamma\left(\frac{1}{3}\right)}\right)^2 \simeq -1.3392$ on using the notation given in Slater (1960).

3.3.3.2 $S = 0$, $\alpha = 0$, $\lambda \in R$

Fig. 3.7 shows the plots of $f''(0)$ and $g'(0)$ against λ when $S = \alpha = 0$. It is found that there exists a critical value for λ of $\lambda_c \simeq -0.6571$ with dual solution for $\lambda > \lambda_c$. There exists a singularity in g at $\lambda \simeq -0.0786$.

3.3.3.3 $S = 0$, $\alpha > 0$, $\lambda \in R$

Figs. 3.8 and 3.9 show the plots of $f''(0)$ and $g'(0)$ against λ for representative values of α when $S = 0$, respectively. The solutions start from large negative λ and then break into three branches of solution as λ increases. For the upper branch of solutions, two critical points are found for smaller values of α, i.e., $\alpha = 0.1$, 0.3 in the figure, giving a range of values of λ that contains three possible solutions. For larger values of α, i.e., $\alpha = 0.5$, 1.0 in the figure, the solution is monotone with only one solution for a given value of λ. Further investigation found that the hysteresis bifurcation is at $\alpha \simeq 0.4622$, as can be seen in Fig. 3.10 for the plot λ_c against α, where the two critical values are arising from this bifurcation point. It is evident from Fig. 3.10 that only one critical point, $\lambda_c \simeq -0.6571$ for $\alpha = 0$, consistent with the plot shown in Fig. 3.7.

From Fig. 3.8, the upper branch of solutions appearing to approach the same limit that is independent of α as $\lambda \to +\infty$. For the lower branch of solutions, it is found that two solutions are branching from the first branch

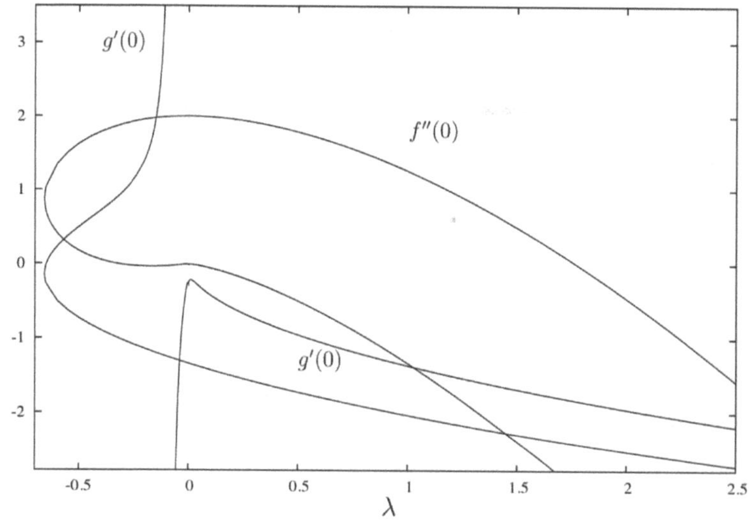

Figure 3.7 Plots of $f''(0)$ and $g'(0)$ against λ for $\alpha = 0$ when $S = 0$. *(From Lok, Y. Y., Merkin, J. H., & Pop, I. (2018). Axisymmetric rotational stagnation-point flow impinging on a permeable stretching/shrinking rotating disk.* European Journal of Mechanics - B/Fluids, *72, 275–292. https://doi.org/10.1016/j.euromechflu.2018.05.013).*

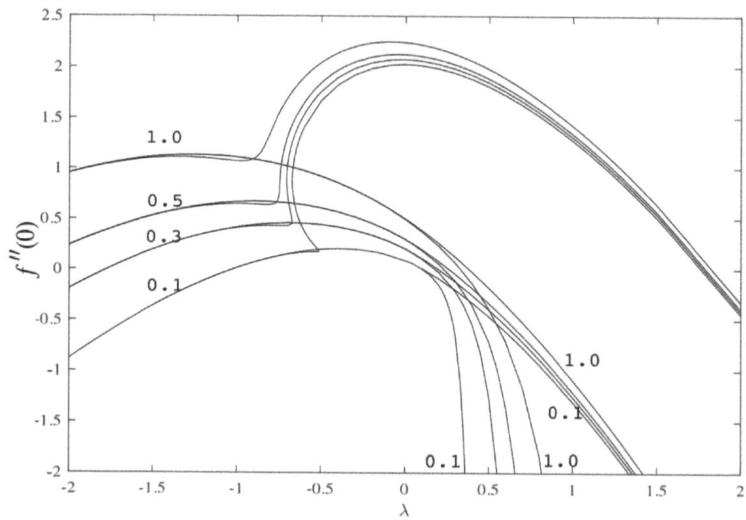

Figure 3.8 Plots of $f''(0)$ against λ for $\alpha = 0.1, 0.3, 0.5, 1.0$ when $S = 0$. *(From Lok, Y. Y., Merkin, J. H., & Pop, I. (2018). Axisymmetric rotational stagnation-point flow impinging on a permeable stretching/shrinking rotating disk.* European Journal of Mechanics - B/Fluids, *72, 275–292. https://doi.org/10.1016/j.euromechflu.2018.05.013).*

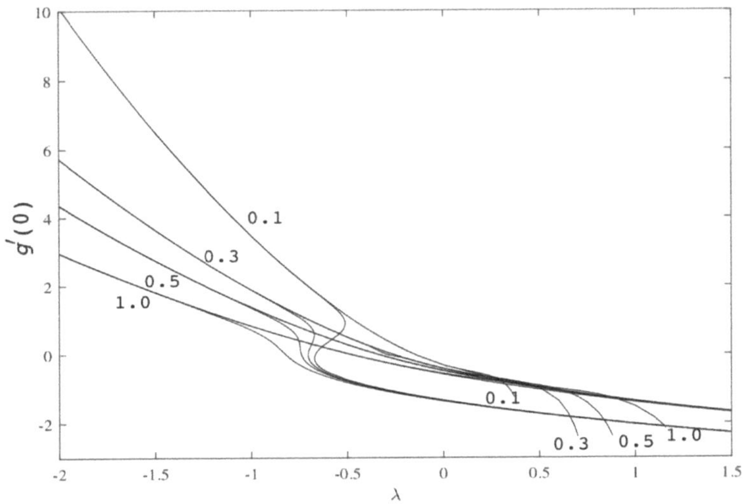

Figure 3.9 Plots of $g'(0)$ against λ for $\alpha = 0.1, 0.3, 0.5, 1.0$ when $S = 0$. *(From Lok, Y. Y., Merkin, J. H., & Pop, I. (2018). Axisymmetric rotational stagnation-point flow impinging on a permeable stretching/shrinking rotating disk.* European Journal of Mechanics - B/Fluids, *72, 275–292. https://doi.org/10.1016/j.euromechflu.2018.05.013).*

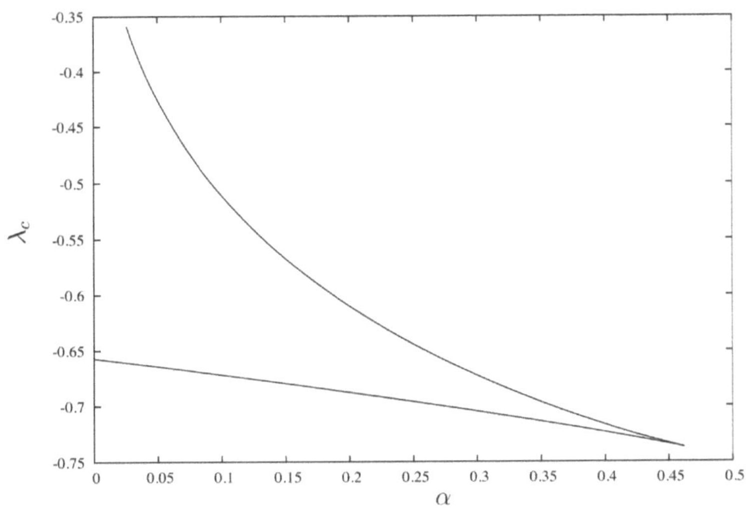

Figure 3.10 Plots of critical value λ_c against α for $S = 0$. *(From Lok, Y. Y., Merkin, J. H., & Pop, I. (2018). Axisymmetric rotational stagnation-point flow impinging on a permeable stretching/shrinking rotating disk.* European Journal of Mechanics - B/Fluids, *72, 275–292. https://doi.org/10.1016/j.euromechflu.2018.05.013).*

of solution, one of which proceeds to a limit value independent of α for large positive λ; however, this limit value is lower than that for the upper branch of solutions. The third branch of solutions becomes singular at a positive value of λ, more easily seen in the plots of $g'(0)$ in Fig. 3.9.

3.3.3.4 Solution for λ large

To investigate the behavior of the solution for λ large, put $f = \lambda^{1/2}F$, $\zeta = \lambda^{1/2}\eta$ in Eqs. (3.21)–(3.23), leaving g unscaled and assuming that α is of O (1), the leading-order problem is

$$F_0''' + 2F_0F_0'' - F_0'^2 = 0, \quad g_0'' + 2\left(F_0g_0' - F_0'g_0\right) = 0$$
$$F_0(0) = 0, \quad F_0'(0) = 1, \quad g_0(0) = 1, \quad F_0' \to 0, \quad g_0 \to 0 \text{ as } \zeta \to \infty \tag{3.27}$$

where primes now denote differentiation with respect to ζ. A numerical integration of Eq. (3.27) gives the asymptotic solutions for λ large

$$f''(0) \sim -1.17372\lambda^{3/2} + \cdots, \quad g'(0) \sim -1.45407\lambda^{1/2} + \cdots \text{ as } \lambda \to \infty \tag{3.28}$$

It is noticed that the leading-order problem Eq. (3.27) is independent of α, consistent with the results seen in Figs. 3.8 and 3.9 for the solution branches that continue to large positive λ. For α to play a role it must be large, of O (λ^2).

3.3.3.5 Solution for α large

For a fixed λ, $f''(0)$ increases as α is increased, while $g'(0)$ decrease slowly as α increases. To see how the solution behaves for α large, put $f = \alpha^{1/4}F$, $\xi = \alpha^{1/4}\eta$ in Eqs. (3.21)–(3.23), leaving g unscaled, the asymptotic solutions for α large are

$$f''(0) \sim 0.51023\alpha^{3/4} + \cdots, \quad g'(0) \sim -0.61592\alpha^{1/4} + \cdots \text{ as } \alpha \to \infty \tag{3.29}$$

For the effect of the moving wall to influence the leading-order problem requires λ to be large, of O $(\alpha^{1/2})$, consistent with the solution for large λ. For wall transpiration to have a role, S must also be large, of O $(\alpha^{1/4})$.

3.3.4 Comments (permeable surface, $S \neq 0$)

3.3.4.1 $S > 0$, $\alpha > 0$, $\lambda \in R$

The initial values of $f''(0)$ and $g'(0)$ against λ for a range of positive values of α when $S = 1.0$ are plotted in Fig. 3.11 and 3.12, respectively. For each value of α that is considered, there are two sets of solutions. The upper set of solutions, as seen in Fig. 3.11, the profiles are similar to those seen in Fig. 3.7, though the critical values of λ are somewhat higher. Both upper and lower sets of solutions continue to large positive λ and large negative λ, appearing to approach the same asymptotic limit for λ large and positive. For the lower set of solutions, there are critical points for the larger values of α, giving a range of λ where there are three solutions with this range increasing as α is increased. The critical values λ_c plotted against α is also similar to that in Fig. 3.10 for $S = 0$, but now the two branches starting from the hysteresis bifurcation at $\alpha = \alpha_0 \simeq 2.5$, $\lambda_c \simeq -2.6$ giving the possibility of having three solutions in $0 < \alpha < \alpha_0$ with only monotone solutions in $\alpha > \alpha_0$. This value of α_0 is somewhat greater than the corresponding value of $\alpha_0 \simeq 0.4622$ for $S = 0$, see Fig. 3.10.

In Fig. 3.13, the critical values λ_c for upper set of solutions that shown in Fig. 3.11 and 3.12, are plotted against positive S for $\alpha = 1.0$. For $S = 1.0$

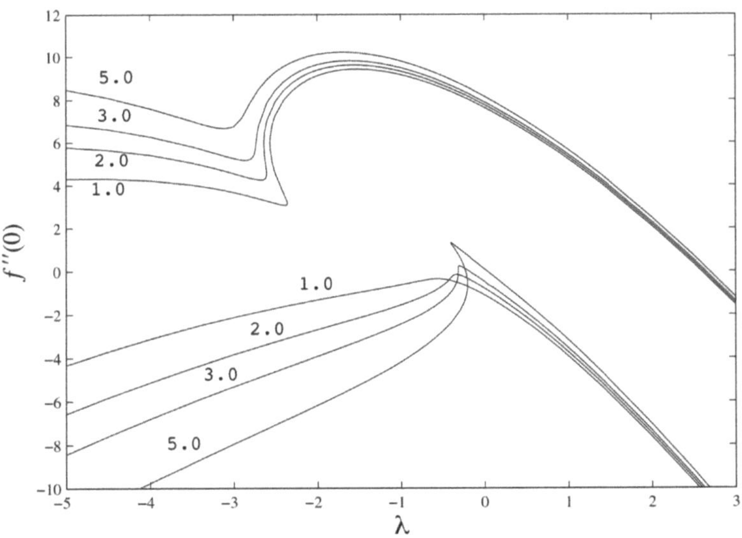

Figure 3.11 Plots of $f''(0)$ against λ for $\alpha = 1.0, 2.0, 3.0, 5.0$ when $S = 1.0$. *(From Lok, Y. Y., Merkin, J. H., & Pop, I. (2018). Axisymmetric rotational stagnation-point flow impinging on a permeable stretching/shrinking rotating disk.* European Journal of Mechanics - B/Fluids, *72, 275–292. https://doi.org/10.1016/j.euromechflu.2018.05.013).*

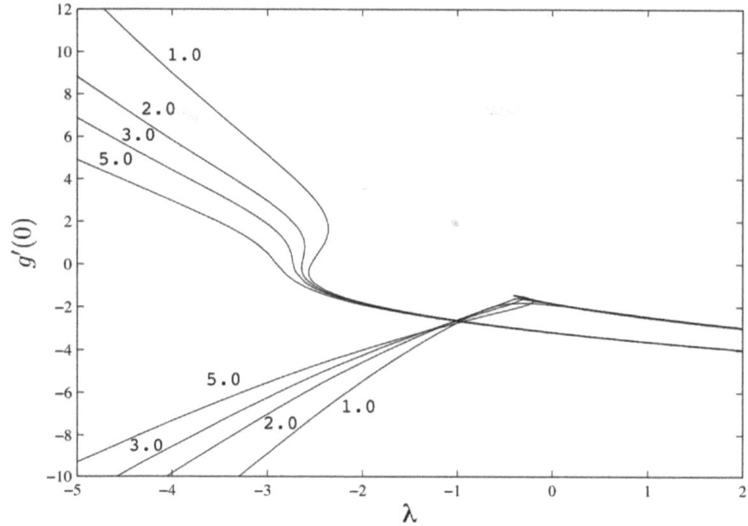

Figure 3.12 Plots of $g'(0)$ against λ for $\alpha = 1.0, 2.0, 3.0, 5.0$ when $S = 1.0$. *(From Lok, Y. Y., Merkin, J. H., & Pop, I. (2018). Axisymmetric rotational stagnation-point flow impinging on a permeable stretching/shrinking rotating disk.* European Journal of Mechanics - B/Fluids, *72, 275–292. https://doi.org/10.1016/j.euromechflu.2018.05.013).*

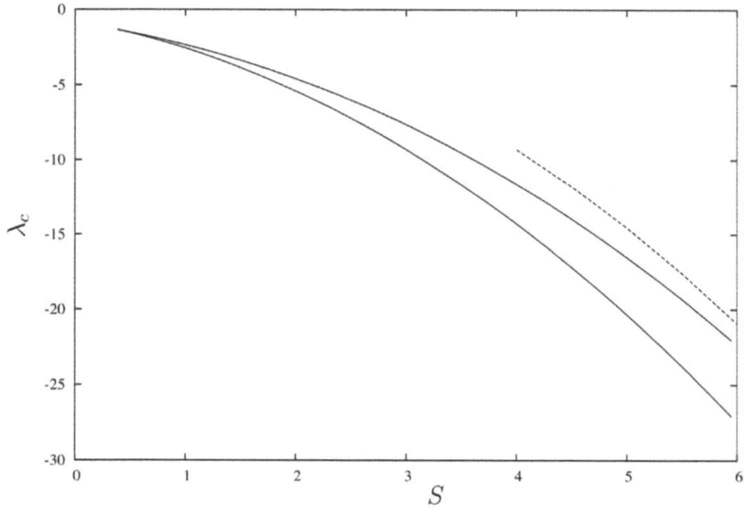

Figure 3.13 Plots of critical value λ_c against S for $\alpha = 1.0$. *(From Lok, Y. Y., Merkin, J. H., & Pop, I. (2018). Axisymmetric rotational stagnation-point flow impinging on a permeable stretching/shrinking rotating disk.* European Journal of Mechanics - B/Fluids, 72, *275–292. https://doi.org/10.1016/j.euromechflu.2018.05.013).*

there are two values of λ_c, namely -2.3574 and -2.5631, giving a range of λ where there are three possible solutions, as seen in Fig. 3.11 and 3.12. This range of λ where there are three solutions increases as S is increased from the hysteresis bifurcation at $\lambda_c \simeq -1.3527$, $S \simeq 0.38$.

3.3.4.2 Solution for large S (strong fluid suction)

To obtain a solution valid for S large, we apply $f = S + S^{-1/2}F$, $Y = S\eta$ in Eqs. (3.21)–(3.23), leaving g unscaled, where F and g have an expansion of the form

$$F = F_0 + S^{-1/2}F_1 + S^{-1}F_2 + S^{-3/2}F_3 + \cdots,$$

$$g = g_0 + S^{-1/2}g_1 + S^{-1}g_2 + S^{-3/2}g_3 + \cdots \tag{3.30}$$

The asymptotic solution for large S is given by

$$f''(0) \sim S^{3/2}\left(\pm 4 - 2\lambda S^{-1/2} + \cdots\right), \quad g'(0) \sim -2S + \cdots \quad \text{as } S \to \infty \tag{3.31}$$

giving two possibilities for $f''(0)$ as $S \to \infty$. It is noted that, to the order given in Eq. (3.31), the asymptotic forms for large S are independent of the parameter λ and α.

To see how the critical values behave for large S, λ is required to be of $O\,(S^2)$. In this case, put $\lambda = S^2\mu$, assuming that μ and α are of $O\,(1)$, and introduce the scaling $f = S\overline{F}$, leaving Y and g unchanged. On calculating the critical value μ_c of μ arising from the leading-order problem for $\overline{F}$, as in Merkin and Mahmood (1989) and Merkin and Pop (2010), it is found that $\mu_c = -0.58116$ with corresponding values $\overline{F}_c''(0) = 0.657548$, $\overline{g}_c' =$ $(0) = -0.70641$. Hence,

$$\lambda_c \sim -0.58116S^2 + \cdots \quad \text{as } S \to \infty \tag{3.32}$$

consistent with the upper curve seen in Fig. 3.13. The expression Eq. (3.32) is shown in Fig. 3.13 by a broken line.

3.3.4.3 $S < 0$, $\alpha > 0$, $\lambda \in R$

Figs. 3.14 and 3.15 show that the plots of $f''(0)$ and $g'(0)$ are monotone and do not appear to have any critical points for the case of $S = -1.0$ (fluid injection) and values of α and λ that are considered. There are two sets of almost identical solutions, one shown by a full line, the other by a dotted line, which separate when λ becomes large and positive for a given value of α.

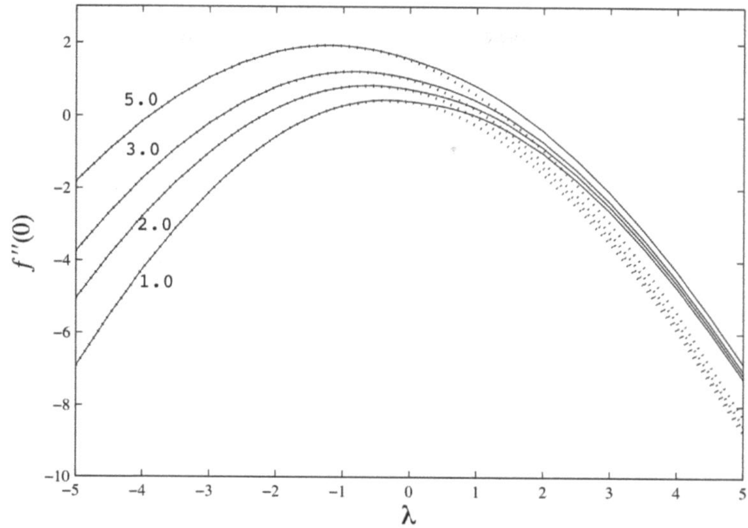

Figure 3.14 Plots of $f''(0)$ against λ for $\alpha = 1.0, 2.0, 3.0, 5.0$ when $S = -1.0$. *(From Lok, Y. Y., Merkin, J. H., & Pop, I. (2018). Axisymmetric rotational stagnation-point flow impinging on a permeable stretching/shrinking rotating disk.* European Journal of Mechanics - B/Fluids, *72, 275–292. https://doi.org/10.1016/j.euromechflu.2018.05.013).*

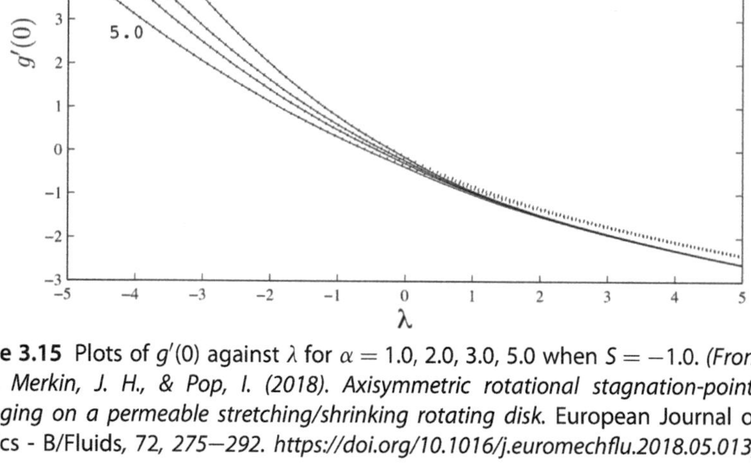

Figure 3.15 Plots of $g'(0)$ against λ for $\alpha = 1.0, 2.0, 3.0, 5.0$ when $S = -1.0$. *(From Lok, Y. Y., Merkin, J. H., & Pop, I. (2018). Axisymmetric rotational stagnation-point flow impinging on a permeable stretching/shrinking rotating disk.* European Journal of Mechanics - B/Fluids, *72, 275–292. https://doi.org/10.1016/j.euromechflu.2018.05.013).*

The solution also proceeds to large negative λ but now the behavior does appear to depend on the value of α.

3.3.4.4 $S \in R$, $\alpha > 0$, $\lambda > 0$

The initial values $f''(0)$ and $g'(0)$ are plotted against S in Figs. 3.16 and 3.17, respectively, for $\alpha = 1.0$ and for a range of positive values of λ. From Fig. 3.16, there are two sets of solutions, one with $f''(0) > 0$ (shown by a full line) and the other with $f''(0) < 0$ (shown by a dotted line), both approaching the asymptotic limit given by Eq. (3.31) for S large. The corresponding values of $g'(0)$ are shown in Fig. 3.17. It is clearly shown in Fig. 3.16 and 3.17 that for $S < 0$, the two sets of solutions approach the same limit as $S \rightarrow -\infty$.

3.3.4.5 $S \in R$, $\alpha > 0$, $\lambda < 0$

The initial values of $f''(0)$ and $g'(0)$ against S for a range of negative values of λ when $\alpha = 1.0$ are plotted in Fig. 3.18 and 3.19, respectively. As can be seen from Fig. 3.18, there are two sets of solutions where the upper set continues to large negative S with the values of $f''(0)$ and $g'(0)$ appearing to approach the asymptotic limit for $|S|$ large. There are two critical values for the larger values of $|\lambda|$, at $S = 0.7399$, 0.7958 for $\lambda = -2.0$ and at $S = 1.3293$, 1.1610 for $\lambda = -3.0$. These results are consistent with that

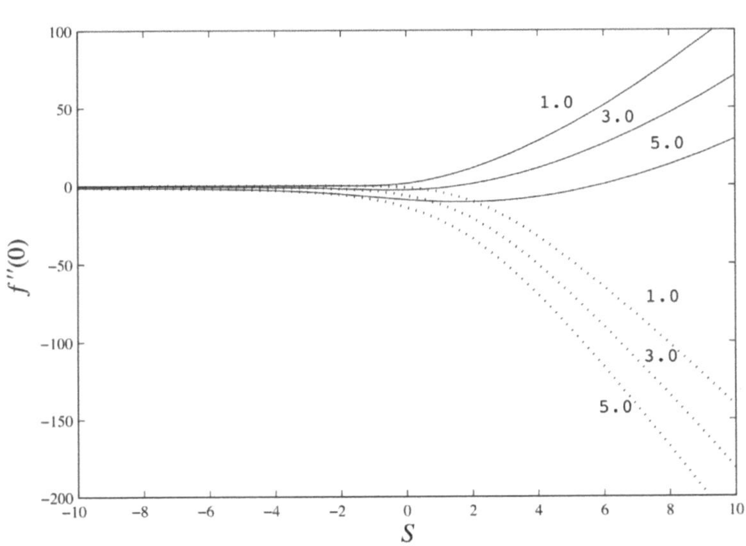

Figure 3.16 Plots of $f''(0)$ against S for $\lambda = 1.0, 3.0, 5.0$ when $\alpha = 1.0$. *(From Lok, Y. Y., Merkin, J. H., & Pop, I. (2018). Axisymmetric rotational stagnation-point flow impinging on a permeable stretching/shrinking rotating disk.* European Journal of Mechanics - B/Fluids, *72, 275–292. https://doi.org/10.1016/j.euromechflu.2018.05.013).*

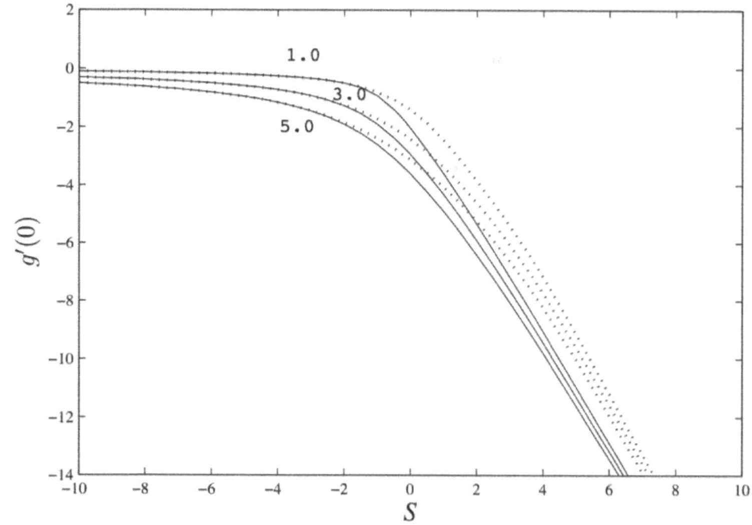

Figure 3.17 Plots of $g'(0)$ against S for $\lambda = 1.0, 3.0, 5.0$ when $\alpha = 1.0$. *(From Lok, Y. Y., Merkin, J. H., & Pop, I. (2018). Axisymmetric rotational stagnation-point flow impinging on a permeable stretching/shrinking rotating disk.* European Journal of Mechanics - B/Fluids, *72, 275–292. https://doi.org/10.1016/j.euromechflu.2018.05.013).*

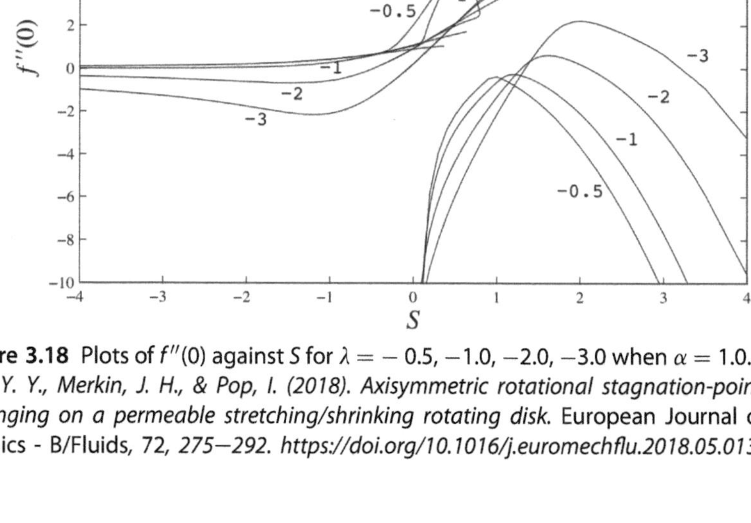

Figure 3.18 Plots of $f''(0)$ against S for $\lambda = -0.5, -1.0, -2.0, -3.0$ when $\alpha = 1.0$. *(From Lok, Y. Y., Merkin, J. H., & Pop, I. (2018). Axisymmetric rotational stagnation-point flow impinging on a permeable stretching/shrinking rotating disk.* European Journal of Mechanics - B/Fluids, *72, 275–292. https://doi.org/10.1016/j.euromechflu.2018.05.013).*

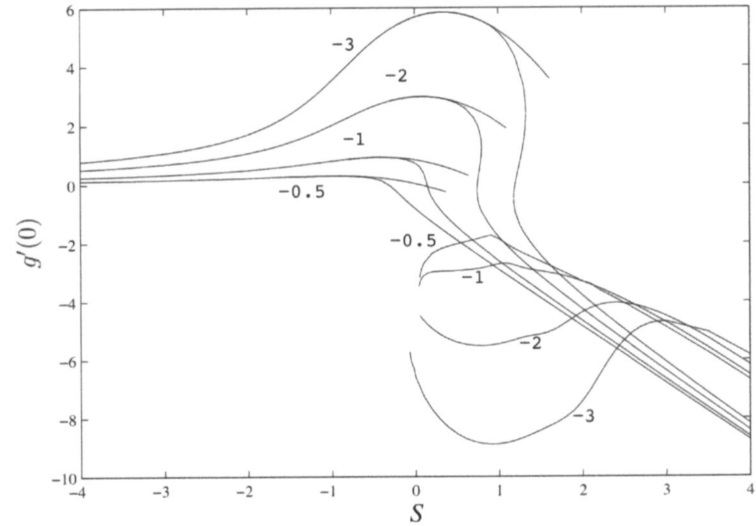

Figure 3.19 Plots of $g'(0)$ against S for $\lambda = -0.5, -1.0, -2.0, -3.0$ when $\alpha = 1.0$. *(From Lok, Y. Y., Merkin, J. H., & Pop, I. (2018). Axisymmetric rotational stagnation-point flow impinging on a permeable stretching/shrinking rotating disk.* European Journal of Mechanics - B/Fluids, 72, *275–292. https://doi.org/10.1016/j.euromechflu.2018.05.013).*

plotted in Fig. 3.13, noting the hysteresis bifurcation at $\lambda_c \simeq -1.3527$ proved the existence of two critical values for $\lambda > \lambda_c$. The upper solution branch continues to large positive S approaching the asymptotic limit Eq. (3.31) as $S \to \infty$, on taking the positive sign for $f''(0)$. From Fig. 3.19, the lower set of solutions of $g'(0)$ are negative, have a local minimum, increasing as λ increases, and a local maximum, more easily seen for the larger values of λ. This lower set of solutions also continues to its large S limit Eq. (3.31) and becomes singular as S decreases toward zero.

3.3.4.6 Solution for large |S| (strong fluid injection)

To obtain a valid solution when $S < 0$ and $|S|$ is large, for inner region put $f = |S|H, \ \varsigma = \eta|S|^{-1}$ in Eqs. (3.21)–(3.23), leaving g unscaled and the outer boundary conditions being relaxed. This leads to the asymptotic solutions for inner region

$$f''(0) \sim -\frac{\lambda^2 - \alpha}{2}|S|^{-1} + \cdots, \quad g'(0) \sim -\lambda|S|^{-1} + \cdots \quad \text{as } |S| \to \infty \tag{3.33}$$

The solution is incomplete without considering outer region. Using the scaling $\eta = \frac{2|S|}{\lambda} + \overline{\eta}$ gives us

$$f \sim -\frac{\lambda^2 \overline{\eta}^2}{4|S|} + \cdots, \quad g \sim \frac{\lambda^2 \overline{\eta}^2}{4|S|} + \cdots \tag{3.34}$$

where $\varsigma = 2/\lambda$. We now expand

$$f\left(\overline{\eta}; |S|\right) = f_0\left(\overline{\eta}\right) + |S|^{-1/2} f_1\left(\overline{\eta}\right) + |S|^{-1} f_2\left(\overline{\eta}\right) + \cdots \tag{3.35}$$

The resulting solution gives, after matching with inner region Eq. (3.34),

$$\lambda^2 = \frac{C_1^2}{C_0} \tag{3.36}$$

where $C_0(>0)$ and C_1 are constant. Based on Eq. 3.36, an important consequence of the above analysis is the requirement that we must have $\lambda > 0$ (stretching surface) to obtain a solution in the limit as $S \to -\infty$, as is also the case when $\alpha > 0$.

3.3.5 Comments (stability analysis)

Following Merkin (1980) and Weidman et al. (2006), an unsteady case of Eqs. (3.21)–(3.22), in terms of nondimensional time variable t, is considered

$$\frac{\partial^3 f}{\partial \eta^3} + 2f\frac{\partial^2 f}{\partial \eta^2} - \left(\frac{\partial f}{\partial \eta}\right)^2 + \alpha g^2 = \frac{\partial^2 f}{\partial t\, \partial \eta} \tag{3.37}$$

$$\frac{\partial^2 g}{\partial \eta^2} + 2\left(f\frac{\partial g}{\partial \eta} - g\frac{\partial f}{\partial \eta}\right) = \frac{\partial g}{\partial t} \tag{3.38}$$

subject to boundary conditions Eq. (3.23) and some suitable initial conditions. To investigate the temporal stability of the multiple solutions, let

$$f(\eta, t) = f_0(\eta) + \exp(\gamma t) f_1(\eta), \quad g(\eta, t) = g_0(\eta) + \exp(\gamma t) g_1(\eta) \tag{3.39}$$

where γ is the eigenvalue, f_0 and g_0 are the solutions to Eqs. (3.21)–(3.23). Substitute Eq. (3.39) into Eqs. (3.37) and (3.38), assume that f_1 and g_1 are small to obtain the linear problem

$$f_1''' + 2f_0 f_1'' - \left(2f_0' + \gamma\right) f_1' + 2f_0'' f_1 + 2\alpha g_0 g_1 = 0 \tag{3.40}$$

$$g_1'' + 2\left(f_0 g_1' + f_1 g_0' - f_0' g_1 - f_1' g_0\right) - \gamma g_1 = 0 \tag{3.41}$$

subject to

$$f_1(0) = f_1'(0) = 0,\ f_1''(0) = 1,\ g_1(0) = 0,$$
$$f_1'' \to 0,\ g_1 \to 0 \text{ as } \eta \to \infty \tag{3.42}$$

As t approaching infinity and γ is negative, there is an initial decay of disturbance hence the solution is physically stable. The solution is unstable if γ is positive due to the initial growth of disturbance.

A specific case from Fig. 3.11 for $S = \alpha = 1.0$ is illustrated to discuss the stability of the multiple solutions. Fig. 3.20 shows the plot of γ against λ for the upper set of solutions. For upper curve, it can be seen that γ passes through zero at the critical point at $\lambda_c \simeq -2.5631$ (shown by the broken line). The lower curve in Fig. 3.20 has $\gamma < 0$, with the values decrease to approach the other critical point at $\lambda_c \simeq -2.3574$ (shown by the broken line) and there appears to be a jump in the value of γ as the solution proceeds through this point. This might be accounted for by the cusp-like nature of this critical point. Thus, it can be concluded that those solution branches seen in Fig. 3.7, 3.8, 3.9, 3.11, 3.12, 3.14 and 3.15 that proceed to large positive values of λ are temporally stable. This stability continues to the critical point where there is a change in stability at the saddle—node bifurcation (when $\gamma = 0$), with the second branch emerging from this bifurcation then being temporally unstable.

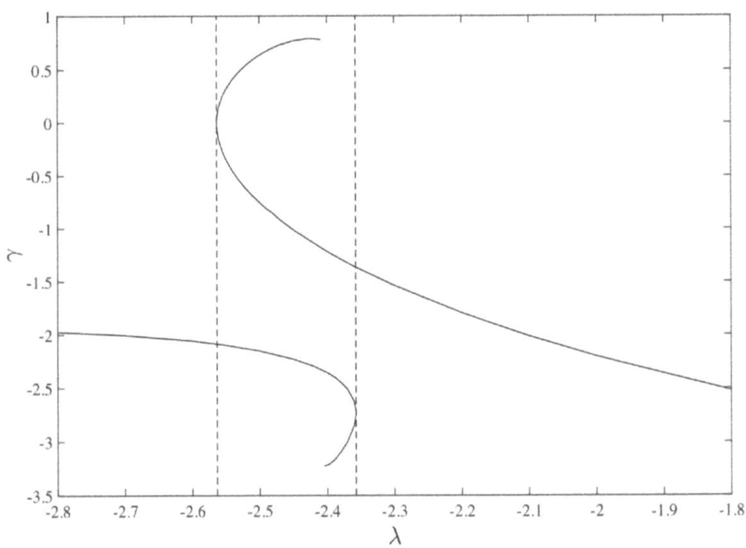

Figure 3.20 Plot of eigenvalue γ against λ for the case $S = \alpha = 1.0$. *(From Lok, Y. Y., Merkin, J. H., & Pop, I. (2018). Axisymmetric rotational stagnation-point flow impinging on a permeable stretching/shrinking rotating disk.* European Journal of Mechanics - B/Fluids, *72, 275—292. https://doi.org/10.1016/j.euromechflu.2018.05.013).*

3.4 Magnetohydrodynamic oblique stagnation-point flow toward a stretching/shrinking surface

3.4.1 Introduction

Oblique stagnation-point flow appears when a jet of viscous fluid impinges on a wall obliquely. Such flow is also common when a separated flow reattaches a wall. The mechanism of flow separation and reattachment is not yet fully understood. Consequently, the study of the separated stagnation-point flow is essential to understand how a boundary layer begins to develop, and therefore it is of great importance to study the flow evolution from the stagnation-point location.

Stuart (1959), Tamada (1979), Takemitsu and Matunobu (1979), Dorrepaal (1986), (Liu, 1992), Labropulu et al. (1996), Dorrepaal (2000), Blyth and Pozrikidis (2005) and Tooke and Blyth (2008) have considered the case of oblique stagnation-point flow in viscous fluid over a fixed plane, while Mahapatra and Gupta (2002), Reza and Gupta (2005), and Mahapatra et al. (2007) have considered the two-dimensional oblique stagnation-point flow toward a surface which is stretched with a velocity proportional to the distance from a fixed point. It is also worth mentioning that Weidman and Putkaradze (2003) have considered the case of oblique axisymmetric stagnation-point flow on a circular cylinder. A very good description of the oblique stagnation-point flow can be found in the books by Pozrikidis (1997) and Drazin and Riley (2006). They proposed the oblique stagnation-point flow model which consists of an orthogonal stagnation-point flow and a shear shear whose vorticity is fixed at infinity. A free parameter associated with the superimposed shear flow component is also introduced.

Grosan et al. (2009) and Borrelli et al. (2012) have studied the effect of an external electromagnetic field on the oblique stagnation-point flow over a rigid fixed wall in viscous fluid. They analyzed three cases which are significant from a physical point of view. In the first two cases, they showed that an oblique stagnation-point flow exists when either the electric or magnetic field is imposed parallel to the wall. In the third case, they show that solutions exist only if the electric field is neglected and with the magnetic field parallel to the dividing streamline. Lok et al. (2015) applied the third case of Borrelli et al. (2012) problem to stretching or shrinking surface. They found that dual solutions exist, and the critical value is dependent on magnetic parameter. The stagnation line is displaced due to the reversed flow region being observed near to the shrinking surface.

Other studies on shrinking sheets with magnetohydrodynamic (MHD) flows in viscous fluid that have been published are Sajid et al. (2008) and Fang and Zhang (2009). Sajid et al. (2008) discussed MHD rotating fluid flows over shrinking surfaces. They showed that the fluid flows are more stable in the MHD flows with shrinking surfaces compared to normal hydrodynamic flows. An exact analytical solution for MHD flow over a permeable shrinking sheet with multiple solution branches for certain mass suction parameter has been reported by Fang and Zhang (2009).

3.4.2 Problem formulation

Consider the steady two-dimensional MHD oblique flow of an electrically conducting fluid near a stagnation point on a stretching/shrinking surface in the presence of a uniform external magnetic field. Assume that the body forces are absent, the strength of the constant magnetic field is **H**, and the magnetic Reynolds number is small hence the induced magnetic field is negligible. The governing equations in vector form are (Borrelli et al., 2012)

$$\nabla \cdot \mathbf{V} = 0 \tag{3.43}$$

$$\rho(\mathbf{V} \cdot \nabla)\mathbf{V} = -\nabla p + \mu \nabla^2 \mathbf{V} + \mu_e (\nabla \times \mathbf{H}) \times \mathbf{H} \tag{3.44}$$

$$\nabla \cdot \mathbf{H} = 0 \tag{3.45}$$

$$\nabla \times \mathbf{H} = \mu_e \sigma_e \mathbf{V} \times \mathbf{H} \tag{3.46}$$

where **V** is the velocity vector, μ_e is the magnetic permeability, σ_e is the electrical conductivity, $\nabla^2 = \frac{\partial^2}{\partial x^2} + \frac{\partial^2}{\partial y^2}$, and the physical meaning of other variables are given in the Nomenclature.

Following Borrelli et al. (2012), the external magnetic field $\mathbf{H} = H_0(\cos\theta\, \mathbf{i} + \sin\theta\, \mathbf{j})$ is taken to be parallel to the dividing streamlines where H_0 is the constant characteristic magnetic field strength, θ is the angle between the dividing line and x-axis with $0 < \theta < \pi$, (i, j) are the unit vector along x and y axes.

The steady oblique stagnation-point flow has an outer flow, given in terms of the stream function as (Drazin & Riley, 2006; Tooke & Blyth, 2008)

$$\psi = \kappa_0 x y + \frac{1}{2}\xi_0 y^2 \tag{3.47}$$

with κ_0, ξ_0 being positive constants. Setting the stream function Eq. (3.47) to zero, it is found that such fluid is possible only if $\tan\,\theta = -2\kappa_0/\xi_0$, while the velocity of the external (inviscid) flow and the external magnetic field are given by

$$\mathbf{V} = (\kappa_0 x + \xi_0 y)\mathbf{i} - \kappa_0 y\mathbf{j},\ \ \mathbf{H} = \frac{H_0}{\sqrt{\left(4\kappa_0^2 + \xi_0^2\right)}}(-\xi_0\mathbf{i} + 2\kappa_0\mathbf{j}) \tag{3.48}$$

The boundary conditions are

$$\psi = 0,\ \ \frac{\partial\psi}{\partial y} = u_0 x,\ \text{ on } y = 0;\ \ \frac{\partial\psi}{\partial y} \rightarrow \kappa_0 x + \xi_0 y,\ \ \frac{\partial\psi}{\partial x} \rightarrow \kappa_0 y \ \text{ as } y \rightarrow \infty \tag{3.49}$$

where u_0 corresponds to the velocity of moving surface.

3.4.3 Similarity transformation

The form of the outer flow Eq. (3.47) and bounday conditions Eq. (3.49) suggests that the streamfunction is in the form Drazin and Riley (2006), Tooke and Blyth (2008)

$$\psi = (\nu\kappa_0)^{1/2} x f(\eta) + \xi_0\left(\frac{\nu}{\kappa_0}\right)\int_0^{\eta} g(s)\,ds \tag{3.50}$$

where $\eta = y\left(\frac{\kappa_0}{\nu}\right)^{1/2}$ and ν is the kinematic viscosity.

By substituting Eqs. (3.46) and (3.48) into Eq. (3.44) and eliminating the pressure p from Eq. (3.44), we can use the solution assumption in Eq. (3.50) to obtain

$$f^{(4)} + ff''' - f'f'' - Mf'' = 0 \tag{3.51}$$

$$g''' + fg'' - f''g + M(f' - g') = 0 \tag{3.52}$$

by simply equating the terms in x and those independent of x. The derivatives in Eqs. (3.51) and (3.52) are differentiation with respect to η where $M = \frac{4\kappa_0\sigma_e\mu_e^2 H_0^2}{\rho\left(4\kappa_0^2 + \xi_0^2\right)}$ is the magnetic parameter or Hartmann number with $M \geq 0$.

The boundary conditions Eq. (3.49) become

$$f(0) = 0,\ \ f'(0) = \lambda,\ \ g(0) = 0;\ f' \rightarrow 1, g' \rightarrow 1 \ \text{ as } \eta \rightarrow \infty \tag{3.53}$$

where $\lambda = \mu_0/\kappa_0$ is the surface velocity parameter, with $\lambda = 0$ for fixed surface, $\lambda > 0$ for stretching surface, $\lambda < 0$ for shrinking surface.

Eq. (3.51) can be integrated once to obtain, on using boundary conditions Eq. (3.53),

$$f''' + ff'' + 1 - f'^2 + M(1 - f') = 0; \; f(0) = 0,$$
$$f'(0) = \lambda; \; f' \to 1 \text{ as } \eta \to \infty \tag{3.54}$$

From Eq. (3.54), it is found that $f \sim \eta - \alpha$ where $\alpha = \alpha(\lambda, M)$ is a constant. On using this and noting that $g \sim \eta - \beta$, integrate Eq. (3.52) will get

$$g'' + fg' - f'g + M(f - g) + (1 + M)(\alpha - \beta) = 0;$$
$$g(0) = 0; \; g' \to 1 \text{ as } \eta \to \infty \tag{3.55}$$

As pointed out in Drazin and Riley (2006) and Tooke and Blyth (2008), the term β in Eq. (3.55) is an arbitrary constant and leads to an additional constant pressure gradient in the x-direction proportional to $(\beta - \alpha)$. In the above equations, Eq. (3.54) corresponds to orthogonal flow, while Eq. (3.55) corresponds to oblique flow.

3.4.4 Comments (numerical solutions)

For the case $\lambda = 1$, Eq. (3.54) has the simple solution $f = \eta$, so that $\alpha = 0$. Eq. (3.55) then can be solved in terms of confluent hypergeometric functions as $g = \eta - \beta + \beta \dfrac{\Gamma\left(\frac{3+M}{2}\right)}{\sqrt{\pi}} \exp(-\eta^2/2) U\left(\frac{2+M}{2}; \frac{1}{2}; \frac{\eta^2}{2}\right)$ giving $g'(0) = 1 - \beta \dfrac{\sqrt{2}\,\Gamma\left(\frac{3+M}{2}\right)}{\Gamma\left(\frac{2+M}{2}\right)}$ on using the notation defined in Slater (1960).

Fig. 3.21 shows the plot of $f''(0)$ against λ for $M = 0$, 1, and 5 on the range $\lambda \le 1$, by considering $\alpha = \beta$. It is noticed that the plot for $M = 0$ is the same as given by Merchant and Davis (1989). There is a critical value λ_c of λ, dependent on M, with $|\lambda_c|$ increase as M is increased thus extending the range of similarity solutions in the opposing flow regime. The upper solution branch can be continued into $\lambda > 1$ to large values of λ. There is a range of λ where dual solutions exist with the lower solution branch

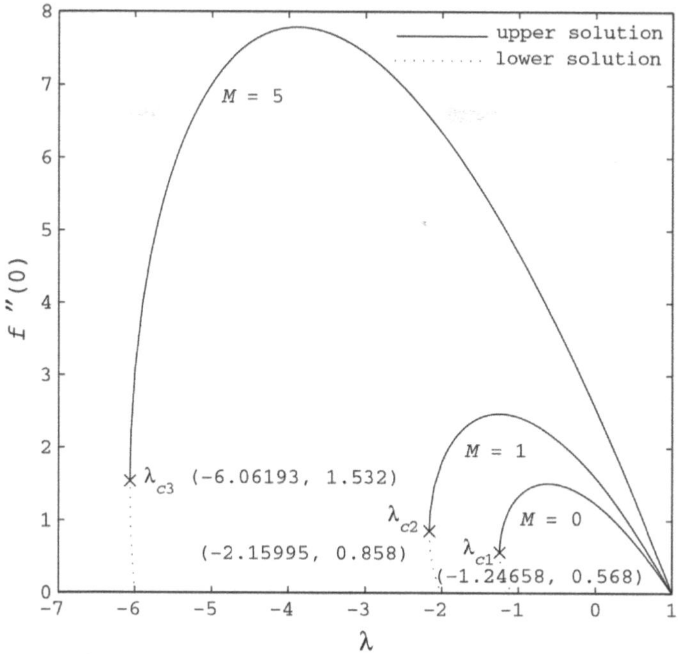

Figure 3.21 Plots of $f''(0)$ against λ for $M = 0, 1, 5$, taking $\alpha = \beta$. *(From Lok, Y. Y., Merkin, J. H., & Pop, I. (2015). MHD oblique stagnation-point flow toward a stretching/shrinking surface.* Meccanica, *50(12), 2949–2961. https://doi.org/10.1007/s11012-015-0188-y).*

appearing to terminate at $f''(0) = 0$ as $\lambda \to -(1 + M)$. This can be explained as follows: from Eq. (3.54), $f'''(0) = (\lambda - 1)(\lambda + M + 1)$, giving $f'''(0) = 0$ when $\lambda + M + 1 = 0$. As a consequence, with f'' and all the higher derivatives (obtained by successive differentiation of Eq. 3.54) are zero at $\eta = 0$, then there is no solution which satisfies all the boundary conditions, thus limiting the range of the lower solution branch. It is further found that there is a marked difference in the g' profiles, depending on whether $M = 0$ or $M > 0$. For $M = 0$, $g'(0) \to 0$ as $\lambda \to -(1 + M)$. However, for $M > 0$, the value of $g'(0) \to -(1 + M)$ as lambda $\lambda \to -(1 + M)$.

The effect of changing the value of β is examined by plotting the g profiles, see Fig. 3.22. For the case $M = 1$ and $\lambda = -0.5$, solve for Eq. (3.54) yields $\alpha = 0.8899$. Use this value of α and take arbitrary values of β while solving Eq. (3.55), it is found that, for all $\beta \leq \alpha$, the profiles change monotonically from wall value to their outer value, $g \sim \eta - \beta$. However, for $\beta > \alpha$, there is an initial range of g where $g < 0$.

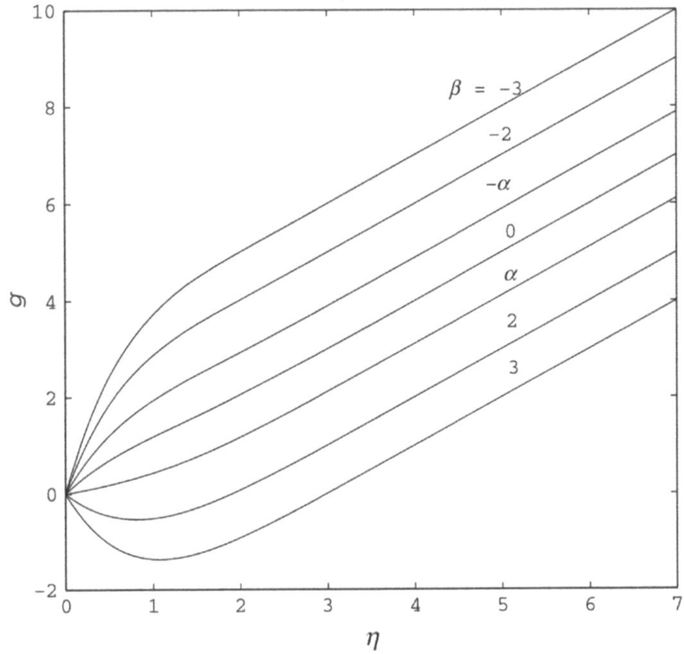

Figure 3.22 Plots of g against η for $M = 1$, $\lambda = -0.5$ (shrinking surface) for different values of β. *(From Lok, Y. Y., Merkin, J. H., & Pop, I. (2015). MHD oblique stagnation-point flow toward a stretching/shrinking surface.* Meccanica, *50(12), 2949–2961. https://doi.org/10.1007/s11012-015-0188-y).*

The dimensionless stream function can be obtained from Eq. (3.50) as $\overline{\psi} = \frac{\psi}{\nu} = \xi f(\eta) + \left(\frac{\xi_0}{\kappa_0}\right) \int_0^{\eta} g(s)ds$, where $\xi = \left(\frac{\kappa_0}{\nu}\right)^{1/2} x$. Figs. 3.23 and 3.24 show the streamline plots $\overline{\psi}$ when $\alpha = \beta$, $M = 1$ for both shrinking surface ($\lambda = -0.5$) and stretching case $(\lambda = 1.5)$, respectively.

Following Drazin and Riley (2006) and Tooke and Blyth (2008), $\frac{\xi_0}{\kappa_0}$ is taken as constant 2 in the streamline plots. It is seen from Fig. 3.23 that for shrinking surface, the stagnation line is displaced from the body surface leading to regions of reversed flow next to the wall. However, the flow is confined to a region next to the wall for the stretching surface, see Fig. 3.24.

3.4.5 Comments (asymptotic solutions)

3.4.5.1 Lower branch solutions as λ → −(1+ M), M = 0

To investigate the behavior of lower branch solutions as $\lambda \to -(1 + M)$, put $\lambda \to -(1 + M) - \delta$, where $0 < \delta \ll 1$, so that $f'(0) = -(1 + M) - \delta$,

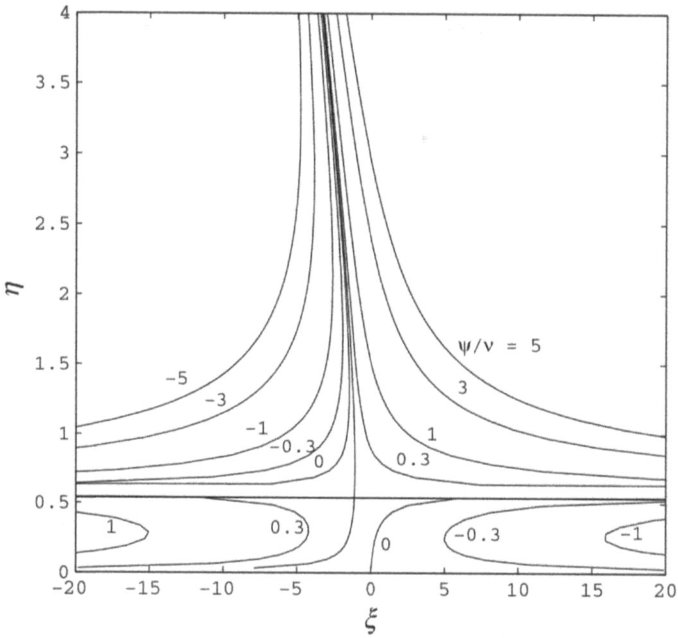

Figure 3.23 Plots of dimensionless streamlines when $M = 1$, $\alpha = \beta$ for $\lambda = -0.5$ (shrinking surface). *(From Lok, Y. Y., Merkin, J. H., & Pop, I. (2015). MHD oblique stagnation-point flow toward a stretching/shrinking surface.* Meccanica, *50(12), 2949–2961. https://doi.org/10.1007/s11012-015-0188-y).*

and look for a solution valid for δ small. When $M = 0$, Eq. (3.54) has the solution expanded in the form

$$f = -\eta + \delta f_1(\eta) + \cdots \tag{3.56}$$

where η is of O(1) with the solution $f_1 = \frac{1}{3}(\eta^3 - 3\eta)$ for inner boundary conditions.

Expansion Eq. (3.56) breaks down when $\eta \sim \delta^{-1/2}$. This leads to a further region where we put

$$f = \delta^{-1/2} F, \quad Y = \delta^{1/2}\eta \tag{3.57}$$

into Eq. (3.54), on matching with expansion Eq. (3.56) and looking for a solution by expanding in powers of δ, the leading-order term F_0 has the solution

$$F_0 = -\frac{1}{\sqrt{2}}\sin(\sqrt{2}\,Y), \quad F_0' = -\cos(\sqrt{2}\,Y) \tag{3.58}$$

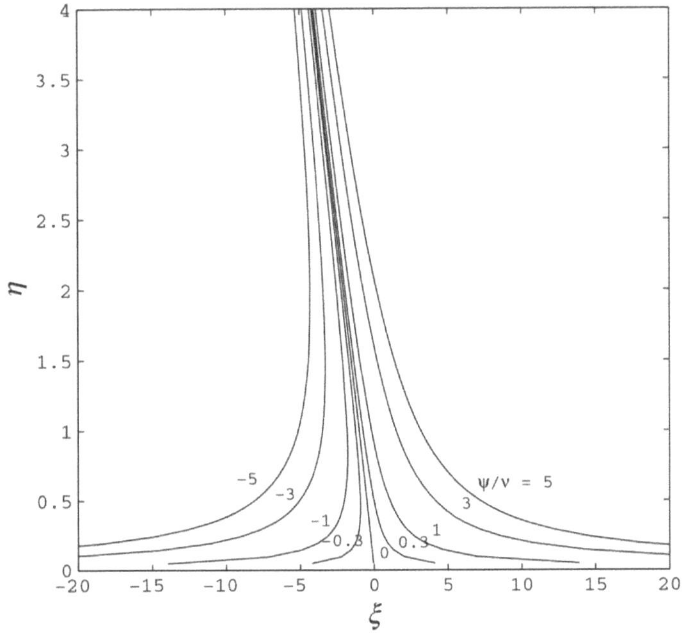

Figure 3.24 Plots of dimensionless streamlines when $M = 1$, $\alpha = \beta$ for $\lambda = 1.5$ (stretching surface). *(From Lok, Y. Y., Merkin, J. H., & Pop, I. (2015). MHD oblique stagnation-point flow toward a stretching/shrinking surface.* Meccanica, 50*(12), 2949–2961. https://doi.org/10.1007/s11012-015-0188-y).*

From Eq. (3.58), $F_0 = 0$ at $Y = Y_0 = \pi/\sqrt{2}$ or at $\eta = \eta_0 = \pi\delta^{-1/2}/\sqrt{2}$, where $F_0'(Y_0) = 1$, thus satisfying the outer boundary condition. To obtain further outer region through which the outer condition is satisfied smoothly we put

$$f = \overline{\eta} + \delta\overline{f}, \quad \eta = \eta_0 + \overline{\eta} = \frac{\pi}{\sqrt{2}\delta^{1/2}} + \overline{\eta} \tag{3.59}$$

into Eq. (3.54), lead to the solution, to leading order,

$$\overline{f}' = A_0\left(\overline{\eta}^2 + 1\right)\int_{\overline{\eta}}^{\infty} \frac{e^{-s^2/2}}{\left(s^2+1\right)^2}\,ds \tag{3.60}$$

where A_0 is constant and $A_0 \int_{-\infty}^{\infty} \frac{e^{-s^2/2}}{(s^2+1)^2}\,ds = -1$. From Eq. (3.59), it is obvious that the boundary layer thickness increases as δ is decreased.

To have a consistent asymptotic expansion for g, it is found that $g = 0$, at least to leading order, in the central region defined by Eq. (3.57), will change in the outer region, as given by Eq. (3.59), where the solution satisfied the leading order term

$$g = \frac{1}{\sqrt{2\pi}}\left[\exp\left(-\frac{\overline{\eta}^2}{2}\right) + \overline{\eta}\int_{-\infty}^{\overline{\eta}} \exp\left(-\frac{s^2}{2}\right) ds\right] \tag{3.61}$$

for the case when $\alpha = \beta$.

3.4.5.2 Solution for large M, $\lambda \neq 1$

Assume that λ is of $O(1)$, put

$$f = M^{-1/2}F, \quad \zeta = M^{1/2}\eta \tag{3.62}$$

into Eq. (3.54) suggest an expansion in powers of M^{-1}, lead to the initial value

$$f''(0) \sim (1-\lambda)M^{1/2} + \cdots \quad \text{as } M \to \infty \tag{3.63}$$

The sign of $f''(0)$ in Eq. (3.63) depends on whether $\lambda \gtrless 1$. The solution is $f \sim \eta - (1-\lambda)M^{-1/2} + \cdots$ as $\eta \to \infty$, giving $\alpha \sim (1-\lambda)M^{-1/2}$. Hence, assume that β is of $O(M^{-1/2})$, put $\beta = \overline{\beta}M^{-1/2}$ and $g = M^{-1/2}G$ into Eq. (3.55) lead to the initial value

$$g'(0) \sim -\overline{\beta} + \frac{(3-\lambda)}{2} + \cdots \quad \text{as } M \to \infty \tag{3.64}$$

This result also holds when $\beta = 0$, which is the case studied by Dorrepaal (1986).

When β is of $O(1)$ and leave g unscaled, on using transformation Eq. (3.62) into Eq. (3.55) lead to the solution for the leading-order, $g_0'(0) \sim -\beta + \cdots$ for M large. This tells us that, $g'(0)$ in general is independent of λ, unless β is related to α which can depend on λ.

When λ is of $O(M)$, putting $\lambda = \mu M$ and taking μ to be of $O(1)$. Modify transformation Eq. (3.62) to $f = M^{1/2}\phi$ where ϕ has an expansion in powers of M^{-1}. Substitute these expressions into Eq. (3.54) to give the solution, at leading order, $\phi_0 = a(1 - \exp(-c\zeta))$ for constants a and $c > 0$, while its boundary conditions give

$$c^2 = 1 + \mu \quad \text{giving } c = \sqrt{1+\mu}, \quad a = \frac{\mu}{\sqrt{1+\mu}} \tag{3.65}$$

Hence, the asymptotic limit is

$$f''(0) - \mu\sqrt{1+\mu}M^{3/2} + \cdots \text{ as } M \to \infty \tag{3.66}$$

Expression Eq. (3.65) holds only when $\mu > -1$ giving a critical value of $\mu_c = -1$, suggesting that

$$\lambda_c \sim -M + \cdots \text{ as } M \to \infty \tag{3.67}$$

Using Eq. (3.65), we have that $f \sim \frac{\mu}{\sqrt{1+\mu}} M^{1/2}$ as $\eta \to \infty$, giving $\alpha = -\frac{\mu}{\sqrt{1+\mu}} M^{1/2}$ at leading order. This suggest that, if considering $\alpha = \beta$, g is of $O(M^{1/2})$, hence putting $g = M^{1/2}H$ into Eq. (3.55) gives

$$g'(0) \sim \frac{\mu}{3+2\mu} M + \cdots \text{ as } M \to \infty \tag{3.68}$$

3.4.5.3 Solution for large λ

Assume that M is of $O(1)$ and put the transformation

$$f = \lambda^{1/2}\Phi, \quad \xi = \lambda^{1/2}\eta \tag{3.69}$$

in Eq. (3.54), with Φ has an expansion in powers of λ^{-1}, give the solution

$$f''(0) \sim -\lambda^{3/2} + \cdots \text{ as } \lambda \to \infty \tag{3.70}$$

The solution is $f \sim \lambda^{1/2} + \cdots$ as $\eta \to \infty$, giving $\alpha \sim -\lambda^{1/2}$. This solution, together with transformation Eq. (3.69), suggests that $g = \lambda^{-1/2}h(\xi) = \lambda^{-1/2}\left(h_0(\xi) + \cdots\right)$, lead to the asymptotic solution

$$g'(0) \sim (2 - M) + \cdots \text{ as } \lambda \to \infty \tag{3.71}$$

on assuming that β is of $O(1)$.

As a conclusion, the asymptotic limits for strong magnetic field (large M) reveals that its solution depended on whether λ is of $O(1)$ or $O(M)$. With λ of $O(1)$, the skin friction $f''(0)$ for orthogonal flow becomes larger, of $O(M^{1/2})$, see Eq. (3.63), than the skin friction $g'(0)$ for oblique flow being of $O(1)$, see Eq. (3.64). This suggests that the orthogonal flow dominates when there is a strong magnetic field. Large λ is needed to recover the critical point, see Eq. (3.67). In this case, the orthogonal flow still dominates with $f''(0)$ now being of $O(M^{3/2})$, see Eq. (3.66), though the oblique flow does have a stronger effect, with $g'(0)$ being of $O(M)$, see Eq. (3.68). On the other hand, with dominant stretching velocity of surface (λ large) and M of $O(1)$, the skin friction for orthogonal flow dominates, of $O(\lambda^{3/2})$, see Eq. (3.70), whereas the skin friction for oblique flow is smaller, of $O(1)$, see Eq. (3.71).

Nomenclature

Roman letters

a, b	positive constant (no unit)—Section 3.2
a	strength of stagnation-point flow ($m^{-1}s^{-1}$) —Section 3.3
c	location of sheet origin (m)
C_f	skin friction coefficient
$\mathbf{H}$	magnetic field
H_0	constant characteristic magnetic field strength
(i,j)	unit vector along x and y axes
M	magnetic parameter
p	pressure (Nm^{-2})
p_0	stagnation pressure (Nm^{-2})
Re_x	local Reynolds number
(r, θ, z)	cylindrical coordinates (m, rad, m)
S	dimensionless mass flux parameter
t	dimensional time (s)
(u, v), (u, v, w)	velocity components along the (x, y) or (r, θ, z), respectively, (ms^{-1})
u_0	velocity of moving surface (ms^{-1})
$\mathbf{V}$	velocity vector
w_0	velocity of fluid through the wall
(x, y)	Cartesian coordinates (m)
x_s	point of reattachment (m)

Greek symbols

α	stretching or shrinking parameter—Section 3.2
α	dimensionless rotation rate of disk—Section 3.3
α, β	constant—Section 3.4
λ	stretching or shrinking parameter
η	dimensionless similarity variable
γ	eigenvalue
κ_0, ξ_0	positive constant (s^{-1})
μ	dynamic viscosity (Nsm^{-2})
μ_e	magnetic permeability
ν	kinematic viscosity (m^2s^{-1})
ω	rotational velocity (s^{-1})
ρ	density of fluid (kgm^{-3})
ψ	stream function (m^2s^{-1})
σ_e	electrical conductivity
τ	shear stress (Nm^{-2})

Subscripts

e, ∞ condition outside of the boundary layer
c critical point
r in radial direction
θ in azimuthal direction
w condition at the wall

References

Agrawal, H. L. (1957). A new exact solution of the equations of viscous motion with axial symmetry. *Quarterly Journal of Mechanics & Applied Mathematics, 10*(1), 42–44. https://doi.org/10.1093/qjmam/10.1.42

Blyth, M. G., & Pozrikidis, C. (2005). Stagnation-point flow against a liquid film on a plane wall. *Acta Mechanica, 180*(1), 203–219. https://doi.org/10.1007/s00707-005-0240-4

Borrelli, A., Giantesio, G., & Patria, M. C. (2012). MHD oblique stagnation-point flow of a Newtonian fluid. *Zeitschrift für Angewandte Mathematik und Physik, 63*(2), 271–294. https://doi.org/10.1007/s00033-011-0174-8

Dholey, S. (2015). The boundary layers of an unsteady separated stagnation-point flow of a viscous incompressible fluid over a moving plate. *Fluid Dynamics Research, 47*, 035504. https://doi.org/10.1088/0169-5983/47/3/035504

Dholey, S. (2016). Magnetohydrodynamic unsteady separated stagnation-point flow of a viscous fluid over a moving plate. *ZAMM - Journal of Applied Mathematics and Mechanics/ Zeitschrift für Angewandte Mathematik und Mechanik, 96*(6), 707–720. https://doi.org/10.1002/zamm.201400218

Dholey, S. (2018a). Unsteady separated stagnation-point flow and heat transfer of a viscous fluid over a moving flat surface. *Physics of Fluids, 30*(4), 042105. https://doi.org/10.1063/1.5022545

Dholey, S. (2018b). Unsteady separated stagnation-point flow over a permeable surface. *Zeitschrift für Angewandte Mathematik und Physik, 70*(1), 10. https://doi.org/10.1007/s00033-018-1051-5

Dholey, S. (2019). An unsteady separated stagnation-point flow towards a rigid flat plate. *Journal of Fluids Engineering, 141*(2), 021202. https://doi.org/10.1115/1.4040572 (11 pages).

Dholey, S., & Gupta, A. S. (2013). Unsteady separated stagnation-point flow of an incompressible viscous fluid on the surface of a moving porous plate. *Physics of Fluids, 25*(2), 023601. https://doi.org/10.1063/1.4788713

Dorrepaal, J. M. (1986). An exact solution of the Navier-Stokes equation which describes non-orthogonal stagnation-point flow in two dimensions. *Journal of Fluid Mechanics, 163*, 141–147. https://doi.org/10.1017/S0022112086002240

Dorrepaal, J. M. (2000). Is two-dimensional oblique stagnation point flow unique? *Canadian Applied Mathematics Quarterly, 8*, 61–66.

Drazin, P. G., & Riley, N. (2006). The Navier-Stokes equations: A classification of flows and exact solutions. In *London mathematical society lecture note series* (Vol. 334 (Issue v. 13)) Cambridge University Press. https://books.google.com.my/books?id=9SHzrFhVO30C.

Fang, T., & Zhang, J. (2009). Closed-form exact solutions of MHD viscous flow over a shrinking sheet. *Communications in Nonlinear Science and Numerical Simulation, 14*(7), 2853–2857. https://doi.org/10.1016/j.cnsns.2008.10.005

Grosan, T., Pop, I., Revnic, C., & Ingham, D. B. (2009). Magnetohydrodynamic oblique stagnation-point flow. *Meccanica, 44*(5), 565. https://doi.org/10.1007/s11012-009-9196-0

Hannah, D. M. (1947). Forced flow against a rotating disc. In *Reports and memoranda (Aeronautical Research Council (Great Britain)), No. 2772*. H.M.S.O.

Hiemenz, K. (1911). Die Grenzschicht an einem in den gleichförmigen Flüssigkeitsstrom eingetauchten geraden Kreiszylinder. *Digitalisierung des Polytechnischen Journals, 326*, 321–410.

Homann, F. (1936). Der Einfluß großer Zähigkeit bei der Strömung um den Zylinder und um die Kugel. *ZAMM - Journal of Applied Mathematics and Mechanics/Zeitschrift für Angewandte Mathematik und Mechanik, 16*(3), 153–164. https://doi.org/10.1002/zamm.19360160304

Howarth, L. (1951). CXLIV. The boundary layer in three dimensional flow.—Part II. The flow near a stagnation point. *The London, Edinburgh, and Dublin Philosophical Magazine and Journal of Science, 42*(335), 1433–1440. https://doi.org/10.1080/14786445108560962

Labropulu, F., Dorrepaal, J. M., & Chandna, O. P. (1996). Oblique flow impinging on a wall with suction or blowing. *Acta Mechanica, 115*(1), 15–25. https://doi.org/10.1007/BF01187425

Liu, T. (1992). Nonorthogonal stagnation flow on the surface of a quiescent fluid — an exact solution of the Navier-Stokes equation. *Quarterly of Applied Mathematics, 50*(1), 39–47. http://www.jstor.org/stable/43637755.

Lok, Y. Y., Merkin, J. H., & Pop, I. (2015). MHD oblique stagnation-point flow towards a stretching/shrinking surface. *Meccanica, 50*(12), 2949–2961. https://doi.org/10.1007/s11012-015-0188-y

Lok, Y. Y., Merkin, J. H., & Pop, I. (2018). Axisymmetric rotational stagnation-point flow impinging on a permeable stretching/shrinking rotating disk. *European Journal of Mechanics - B: Fluids, 72*, 275–292. https://doi.org/10.1016/j.euromechflu.2018.05.013

Lok, Y. Y., & Pop, I. (2014). Stretching or shrinking sheet problem for unsteady separated stagnation-point flow. *Meccanica, 49*(6), 1479–1492. https://doi.org/10.1007/s11012-014-9932-y

Mahapatra, T. R., Dholey, S., & Gupta, A. S. (2007). Heat transfer in oblique stagnation-point flow of an incompressible viscous fluid towards a stretching surface. *Heat and Mass Transfer, 43*(8), 767–773. https://doi.org/10.1007/s00231-006-0116-8

Mahapatra, T. R., & Gupta, A. S. (2002). Heat transfer in stagnation-point flow towards a stretching sheet. *Heat and Mass Transfer, 38*(6), 517–521. https://doi.org/10.1007/s002310100215

Ma, P. K. H., & Hui, W. H. (1990). Similarity solutions of the two-dimensional unsteady boundary-layer equations. *Journal of Fluid Mechanics, 216*, 537–559. https://doi.org/10.1017/S0022112090000520

Merchant, G. J., & Davis, S. H. (1989). Modulated stagnation-point flow and steady streaming. *Journal of Fluid Mechanics, 198*, 543–555. https://doi.org/10.1017/S0022112089000248

Merkin, J. H. (1980). Mixed convection boundary layer flow on a vertical surface in a saturated porous medium. *Journal of Engineering Mathematics, 14*(4), 301–313. https://doi.org/10.1007/BF00052913

Merkin, J. H., & Mahmood, T. (1989). Mixed convection boundary layer similarity solutions: Prescribed wall heat flux. *Zeitschrift Für Angewandte Mathematik Und Physik ZAMP, 40*(1), 51–68. https://doi.org/10.1007/BF00945309

Merkin, J. H., & Pop, I. (2010). Natural convection boundary-layer flow in a porous medium with temperature-dependent boundary conditions. *Transport in Porous Media, 85*(2), 397–414. https://doi.org/10.1007/s11242-010-9569-9

Nandy, S. K., Mahapatra, T. R., & Pop, I. (2015). Unsteady separated stagnation-point flow over a moving porous plate in the presence of a variable magnetic field. *European Journal of Mechanics - B: Fluids, 53*, 229–240. https://doi.org/10.1016/j.euromechflu.2015.05.003

Pozrikidis, C. (1997). *Introduction to theoretical and computational fluid dynamics*. Oxford University Press.

Reza, M., & Gupta, A. S. (2005). Steady two-dimensional oblique stagnation-point flow towards a stretching surface. *Fluid Dynamics Research, 37*(5), 334–340. https://doi.org/10.1016/j.fluiddyn.2005.07.001

Rosca, N. C., & Pop, I. (2017). Axisymmetric rotational stagnation point flow impinging radially a permeable stretching/shrinking surface in a nanofluid using Tiwari and Das model. *Scientific Reports,* 7(1), 40299. https://doi.org/10.1038/srep40299

Sajid, M., Javed, T., & Hayat, T. (2008). MHD rotating flow of a viscous fluid over a shrinking surface. *Nonlinear Dynamics, 51*(1), 259–265. https://doi.org/10.1007/s11071-007-9208-3

Slater, L. J. (1960). *Confluent hypergeometric functions*. Cambridge University Press.

Stuart, J. T. (1959). The viscous flow near a stagnation point when the external flow has uniform vorticity. *Journal of the Aerospace Sciences, 26*(2), 124–125. https://doi.org/10.2514/8.7963

Takemitsu, N., & Matunobu, Y. (1979). Unsteady stagnation-point flow impinging obliquely on an oscillating flat plate. *Journal of the Physical Society of Japan, 47*(4), 1347–1353. https://doi.org/10.1143/JPSJ.47.1347

Tamada, K. (1979). Two-dimensional stagnation-point flow impinging obliquely on a plane wall. *Journal of the Physical Society of Japan, 46*(1), 310–311. https://doi.org/10.1143/JPSJ.46.310

Tooke, R. M., & Blyth, M. G. (2008). A note on oblique stagnation-point flow. *Physics of Fluids, 20*, 033101. https://doi.org/10.1063/1.2876070

Turkyilmazoglu, M. (2012). Three dimensional MHD stagnation flow due to a stretchable rotating disk. *International Journal of Heat and Mass Transfer, 55*(23), 6959–6965. https://doi.org/10.1016/j.ijheatmasstransfer.2012.05.089

Wang, C. Y. (2008). Stagnation flow towards a shrinking sheet. *International Journal of Non-Linear Mechanics, 43*(5), 377–382. https://doi.org/10.1016/j.ijnonlinmec.2007.12.021

Weidman, P. (2015). Axisymmetric rotational stagnation-point flow impinging on a rotating disk. *Zeitschrift für Angewandte Mathematik und Physik, 66*(6), 3425–3431. https://doi.org/10.1007/s00033-015-0587-x

Weidman, P. (2016a). Axisymmetric rotational stagnation point flow impinging on a flat liquid surface. *European Journal of Mechanics - B: Fluids, 56*, 188–191. https://doi.org/10.1016/j.euromechflu.2015.11.012

Weidman, P.. (2016b). Axisymmetric rotational stagnation point flow impinging on a radially stretching sheet. *International Journal of Non-Linear Mechanics, 82*, 1–5. https://doi.org/10.1016/j.ijnonlinmec.2016.01.016

Weidman, P. D., Kubitschek, D. G., & Davis, A. M. J. (2006). The effect of transpiration on self-similar boundary layer flow over moving surfaces. *International Journal of Engineering Science, 44*(11), 730–737. https://doi.org/10.1016/j.ijengsci.2006.04.005

Weidman, P. D., & Putkaradze, V. (2003). Axisymmetric stagnation flow obliquely impinging on a circular cylinder. *European Journal of Mechanics - B: Fluids, 22*(2), 123–131. https://doi.org/10.1016/S0997-7546(03)00019-0

CHAPTER 4

Nanofluids

4.1 Forced convection boundary layer flow past nonisothermal thin needles in nanofluids

4.1.1 Introduction

Thin needle is a body of revolution whose diameter is of the same order as the velocity or thermal boundary layer that it develops. By appropriately varying the radius of the needle, the partial differential boundary layer equations admit similarity solutions, which are more revealing than the direct numerical integration of the partial differential equations. The problem of steady forced boundary layer flow past a needle was considered by many researchers (e.g, Chen, 1987; Chen & Smith, 1978; Narain, 1972; Narain & Uberoi, 1973). Grosan and Pop (2011b) extended these studies for the nanofluid case, using the mathematical model proposed by Tiwari and Das (2007) and Oztop and Abu-Nada (2008).

4.1.2 Basic equations

The steady forced convection boundary layer flow and heat transfer of an incompressible viscous fluid over a thin needle with variable wall temperature in a nanofluid at a constant temperature T_∞ is considered. The slender paraboloid needle has the radius $\overline{r} = \overline{R}(\overline{x})$, where $\overline{x}$ and $\overline{r}$ are the axial and radial coordinates with the $\overline{x}$-axis measuring from the needle leading edge. When the thickness of the needle is larger than the dimension of the boundary layer over it, we say that the needle is thin. It is assumed that the velocity of the external flow (inviscid fluid) is $\overline{u}_e(\overline{x})$ and that the needle is subject to a variable surface temperature, $\overline{T}_w(\overline{x})$, see Fig. 4.1. Because the needle is assumed thin, the effect of its transverse curvature is of importance but the pressure gradient along the body may be neglected. Under the boundary layer approximations, and using the nanofluid model

Similarity Solutions for the Boundary Layer Flow and Heat Transfer of Viscous Fluids, Nanofluids, Porous Media, and Micropolar Fluids
ISBN 978-0-12-821188-5
https://doi.org/10.1016/B978-0-12-821188-5.00005-9

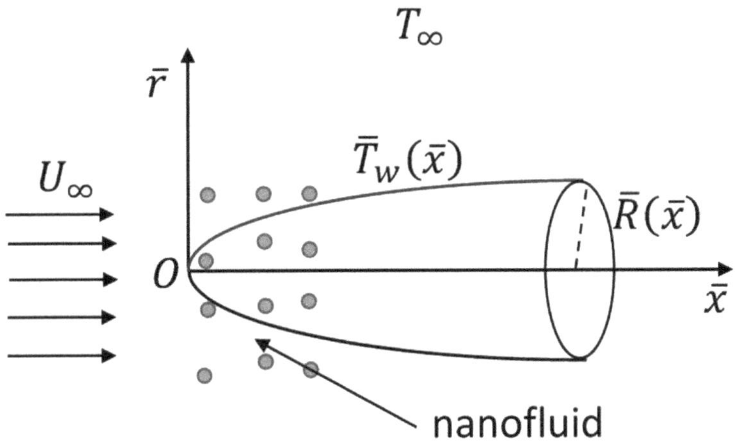

Figure 4.1 Physical models and coordinate systems.

proposed by Tiwari and Das (2007), the basic boundary layer equations written in dimensional form and in cylindrical coordinates are (see Chen & Smith, 1978).

$$\frac{\partial}{\partial \bar{x}}(\bar{r}\,\bar{u}) + \frac{\partial}{\partial \bar{r}}(\bar{r}\,\bar{v}) = 0 \tag{4.1}$$

$$\bar{u}\,\frac{\partial \bar{u}}{\partial \bar{x}} + \bar{v}\,\frac{\partial \bar{u}}{\partial \bar{r}} = \bar{u}_e\frac{d\bar{u}_e}{d\bar{x}} + \frac{\mu_{nf}}{\rho_{nf}}\frac{1}{\bar{r}}\frac{\partial}{\partial \bar{r}}\left(\bar{r}\,\frac{\partial \bar{u}}{\partial \bar{r}}\right) \tag{4.2}$$

$$\bar{u}\,\frac{\partial \overline{T}}{\partial \bar{x}} + \bar{v}\,\frac{\partial \overline{T}}{\partial \bar{r}} = \frac{\alpha_{nf}}{\bar{r}}\frac{\partial}{\partial \bar{r}}\left(\bar{r}\,\frac{\partial \overline{T}}{\partial \bar{r}}\right) \tag{4.3}$$

$$\begin{aligned} &\bar{v} = 0,\ \bar{u} = 0,\ \overline{T} = \overline{T}_w(\bar{x})\ at\ \bar{r} = \overline{R}(\bar{x}) \\ &\bar{u} = \bar{u}_e(\bar{x}),\ \overline{T} = T_\infty\ \ at\ \ \bar{r} \to \infty \end{aligned} \tag{4.4}$$

where μ_{nf} is the viscosity of the nanofluid, α_{nf} is the thermal diffusivity of the nanofluid, and ρ_{nf} is the density of the nanofluid, which are defined in Eq. (1.16) (see (Tiwari & Das, 2007) and (Oztop & Abu-Nada, 2008)).

The following dimensionless variables are introduced:

$$\begin{aligned} &x = \frac{\bar{x}}{L},\ \ r = Re^{1/2}\frac{\bar{r}}{L},\ \ R(x) = Re^{1/2}\frac{\overline{R}(\bar{x})}{L},\ \ u = \frac{\bar{u}}{U_\infty}, \\ &v = Re^{1/2}\frac{\bar{v}}{U_\infty},\ \ u_e(x) = \frac{\bar{u}_e(\bar{x})}{U_\infty},\ \ T = \frac{\overline{T} - T_\infty}{\Delta T} \end{aligned} \tag{4.5}$$

where $Re = \frac{U_\infty L}{\nu_f}$ is the Reynolds number. Substituting Eq. (4.5) into Eqs. (4.1)–(4.33), they reduce to the following dimensionless form:

$$\frac{\partial}{\partial x}(r\,u) + \frac{\partial}{\partial r}(r\,v) = 0 \tag{4.6}$$

$$u\frac{\partial u}{\partial x} + v\frac{\partial u}{\partial r} = u_e\frac{du_e}{dx} + \frac{\mu_{nf}}{\rho_{nf}}\frac{1}{r}\frac{\partial}{\partial r}\left(r\frac{\partial u}{\partial r}\right) \tag{4.7}$$

$$u\frac{\partial T}{\partial x} + v\frac{\partial T}{\partial r} = \frac{1}{Pr}\frac{\alpha_{nf}}{\alpha_f}\frac{1}{r}\frac{\partial}{\partial r}\left(r\frac{\partial T}{\partial r}\right) \tag{4.8}$$

$$\begin{aligned} &v = 0,\;\; u = 0,\;\; T = T_w(x) \;\text{ at }\; r = R(x) \\ &u = ue(x),\;\; T = 0 \;\text{ as }\; r \to \infty \end{aligned} \tag{4.9}$$

In order that Eqs. (4.7)–(4.9) become similar, it is necessary that

$$u_e(x) = x^m, \quad T_w(x) = x^n \tag{4.10}$$

where m and n are constants. We introduce now the following similarity variables:

$$\psi = x f(\eta), \quad T(x) = x^n\,\theta(\eta), \quad \eta = x^{m-1}r^2 \tag{4.11}$$

where ψ is the stream function which is defined in the usual way as $u = (1/r)\partial\psi/\partial r$ and $v = -(1/r)\partial\psi/\partial x$. The surfaces of constant $\eta = a$, where a is a nondimensional constant and is numerically small for a slender body, correspond to the surfaces of revolution. Setting $\eta = a$, Eq. $\eta = x^{m-1}r^2$ prescribes both shape and size of the body with its surface given by

$$R(x) = a^{1/2}x^{(1-m)/2} \tag{4.12}$$

Of practical interest are pointed bodies and cylinders for which we must have, from Eq. (4.12), $m \le 1$. For example, the body is a cylinder when $m = 1$, a paraboloid when $m = 0$, and a cone when $m = -1$.

Substituting Eq. (4.11) into Eqs. (4.7) and (4.8), we get the following ordinary differential equations:

$$\frac{8}{(1-\varphi)^{2.5}\left[1-\varphi + \varphi(\rho_s/\rho_f)\right]}(\eta f'')' + 4f f'' + m\left(1-4f'^2\right) = 0 \tag{4.13}$$

$$\frac{2}{Pr}\frac{k_{nf}/k_f}{(1-\varphi) + \varphi(\rho C_p)_s/(\rho C_p)_f}(\eta\theta')' + f\theta' - nf'\theta = 0 \tag{4.14}$$

subject to the boundary conditions

$$\begin{aligned} &f(a) = f'(a) = 0, \quad \theta(a) = 1 \\ &f'(\infty) = 1/2, \quad \theta(\infty) = 0 \end{aligned} \tag{4.15}$$

For a regular Newtonian fluid ($\varphi = 0$), Eqs. (4.13) and (4.14) reduce to Eqs. (4.7) and (4.13) from the paper by Chen and Smith (1978).

The physical quantities of interest are the skin friction coefficient C_f and the local Nusselt number Nu_x, which are defined as

$$C_f = \frac{\overline{\tau}_w}{\rho_f \overline{u}_e^2}, \quad Nu_x = \frac{\overline{x}\overline{q}_w}{k_f\left(\overline{T}_w - T_\infty\right)} \tag{4.16}$$

where the skin friction $\overline{\tau}_w$ and the heat transfer $\overline{q}_w$ from the surface of the needle are given by

$$\overline{\tau}_w = \mu_{nf}\left(\frac{\partial \overline{u}}{\partial \overline{r}}\right)_{\overline{r}=\overline{R}(\overline{x})}, \quad \overline{q}_w = -k_{nf}\left(\frac{\partial \overline{T}}{\partial \overline{r}}\right)_{\overline{r}=\overline{R}(\overline{x})} \tag{4.17}$$

Next, using (4.5), (4.16), and (4.17) one gets

$$Re_x^{1/2} C_f = \frac{4a^{1/2}}{(1-\varphi)^{2.5}} f''(a), \quad Re_x^{-1/2}\, Nu_x = -2a^{1/2}\frac{k_{nf}}{k_f}\theta'(a) \tag{4.18}$$

where $Re_x = \overline{u}_e\left(\overline{x}\right)\overline{x}/\nu_f$ is the local Reynolds number.

4.1.3 Results and discussion

Numerical solutions to the governing ordinary differential Eqs. (4.13) and (4.14) with the boundary conditions (4.15) were obtained using the boundary value problem solver bvp4c from Matlab. In order to assess the accuracy of these results, reported by Grosan and Pop (2011b), comparison with existing solution in literature for the special case of a Newtonian or regular fluid $\varphi = 0$ with those of Chen and Smith (1978) for $m = 0{,}0.5$ and $n = 0{,}1$ when $Pr = 0.733$ is shown in Table 4.1. Results for Cu−water and Al_2O_3−water nanofluids (the water's Prandtl number is $\Pr = 7$) when $a = 0.001{,}0.01$ and 0.1, $m = n = 0$ and the range of the solid volume fraction φ is in $0 \leq \varphi \leq 0.2$ are also given.

Tables 4.2 and 4.3 show the variation of the reduced skin friction coefficient, $f''(a)$, and of the reduced Nusselt number, $-\theta'(a)$ with a and φ

Table 4.1 Values of $f''(a)$ and $-\theta'(a)$ for $\varphi = 0$ reported by Grosan and Pop, (2011b). In brackets are the values reported by Chen and Smith (1978).

	$f''(a)$		$-\theta'(a)$ ($Pr = 0/733$)			
	$m = 0$	$m = 0.5$	$m = 0$		$m = 0.5$	
a			$n = 0$	$n = 1$	$n = 0$	$n = 1$
0.1	1.289074 (1.28881)	1.721998 (1.72178)	2.441675 (2.434)	2.920733 (2.920)	2.516082 (2.505)	3.027603 (3.026)
0.01	8.492173 (8.49244)	10.350656 (10.35056	16.306544	18.704361	16.535072	19.069678
0.001	62.161171 (62.16372)	71.667683 71.66594)	120.55034	134.17011	121.51686	135.67860

Table 4.2 Values of $f''(a)$ for $m = 0$.

		$f''(a)$	
φ	a	Cu	Al_2O_3
0.05	0.1 0.01 0.001	1.347208 8.771680 63.884384	1.293086 8.509033 62.544689
0.1	0.1 0.01 0.001	1.382008 8.935933 64.653616	1.288703 8.491924 62.373393
0.2	0.1 0.01 0.001	1.404136 9.041011 65.235057	1.266109 8.378166 61.516511

for Cu and Al_2O_3 nanoparticles. It is seen from these tables that the skin friction and Nusselt number have greater values for Cu than for Al_2O_3. This is due to the physical properties of fluid and nanoparticles (i.e., thermal conductivity of Cu is much higher than that of Al_2O_3), see Table 1.1. It can also be seen that $f''(a)$ increases with φ for Cu, while it decreases for Al_2O_3. However, the values of $-\theta'(a)$ decrease with φ for both types of nanoparticles. Table 4.2.

The variation of the dimensionless velocity and temperature profiles with the nanofluid volume fraction φ of the Cu nanoparticles and with the needle size parameter a is shown in Figs. 4.2 and 4.3. The thickness of the both velocity and temperature profiles increases with φ. Physically, the increases of φ leads to an increase of the viscosity and thermal conductivity of the nanofluid. This behavior of the thermal boundary layer is in agreement with the decrease of the heat transfer rate from the surface of the needle to the nanofluid. Figs. 4.4 and 4.5 present the velocity and temperature profiles for Cu and Al_2O_3 nanoparticles when $\varphi = 0.2$ and $a = 0.01$. These figures indicate that the thickness of the velocity boundary layer is greater for Cu, while the thickness of the temperature boundary layer is greater for Al_2O_3. However, the two temperature profiles are very close to each other.

For $a = 0.01$, the reduced Nusselt number decreases with about 8.18% (9.16%), when in the clear fluid (water, $\varphi = 0$) a fraction of nanoparticles of Cu (Al_2O_3) is added ($\varphi = 0.2$).

4.2 Axisymmetric mixed convection boundary layer flow past a vertical cylinder in a nanofluid

4.2.1 Introduction

Mixed convection flows, or combined free- and forced convection flows, occur in many technological and industrial applications and in nature. The mixed convection boundary layer flow past a vertical cylinder is a classical

Table 4.3 Values of $\theta'(a)$ for $Pr = 7$.

a	Particles	φ 0	0.05	0.1	0.2
0.1	Cu Al_2O_3	3.770504	3.682009 3.652075	3.586544 3.538902	3.389762 3.325065
	Cu Al_2O_3	22.738762	22.284916 22.183096	21.816182 21.653419	20.877668 20.654662
	Cu Al_2O_3	156.137817	153.570113 153.157168	150.977665 150.314106	145.860889 144.934874

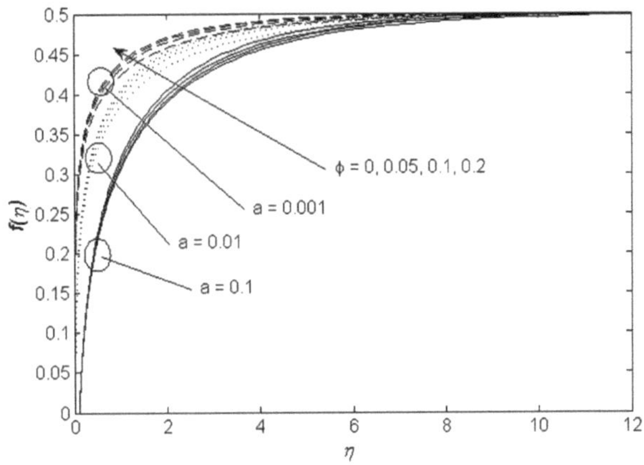

Figure 4.2 Dimensionless velocity profiles for Cu nanoparticles. *(Modified from Grosan, T., & Pop, I. (2011b). Forced convection boundary layer flow past nonisothermal thin needles in nanofluids.* Journal of Heat Transfer, 133*(5), 054503. https://doi.org/10.1115/1.4003059054503-3.)*

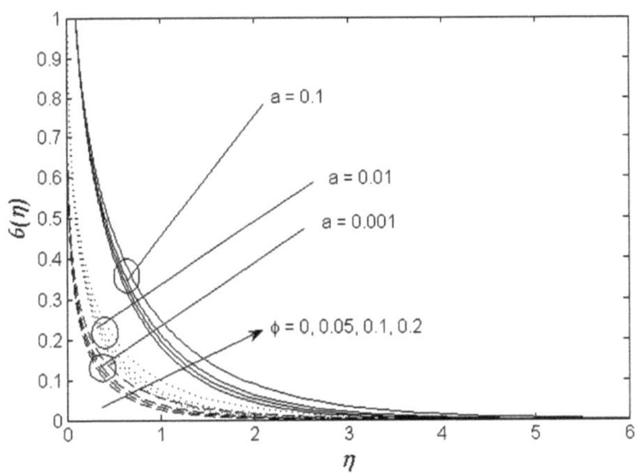

Figure 4.3 Dimensionless temperature profiles for Cu nanoparticles. *(From Grosan, T., & Pop, I. (2011b). Forced convection boundary layer flow past nonisothermal thin needles in nanofluids.* Journal of Heat Transfer, 133*(5), 054503. https://doi.org/10.1115/1.4003059054503-3.)*

problem and has been studied by many investigators. Chen and Mucoglu (1975) have investigated the effects of mixed convection over a vertical slender cylinder due to the thermal diffusion with prescribed wall temperature and the solution was obtained by using the local nonsimilarity method. Furthermore, (Mahmood & Merkin, 1988a) have solved this problem using an implicit finite difference scheme. Ishak et al. (2007)

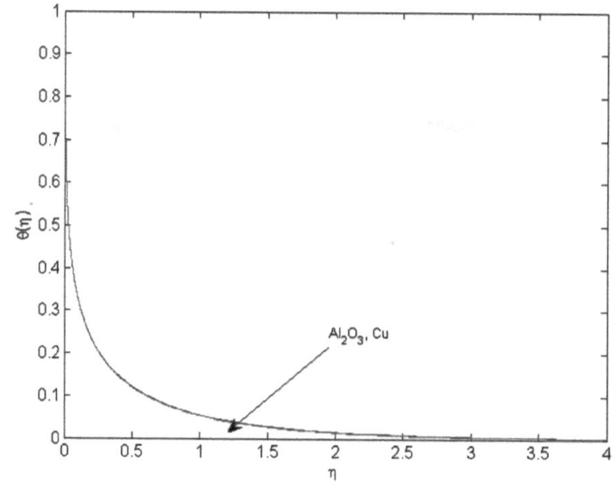

Figure 4.4 Dimensionless temperature profiles for Al_2O_3 and Cu nanoparticles when $\varphi = 0.2$, $a = 0.01$, and $Pr = 7$. *(From Grosan, T., & Pop, I. (2011b). Forced convection boundary layer flow past nonisothermal thin needles in nanofluids.* Journal of Heat Transfer, 133*(5), 054503. https://doi.org/10.1115/1.4003059054503-3.)*

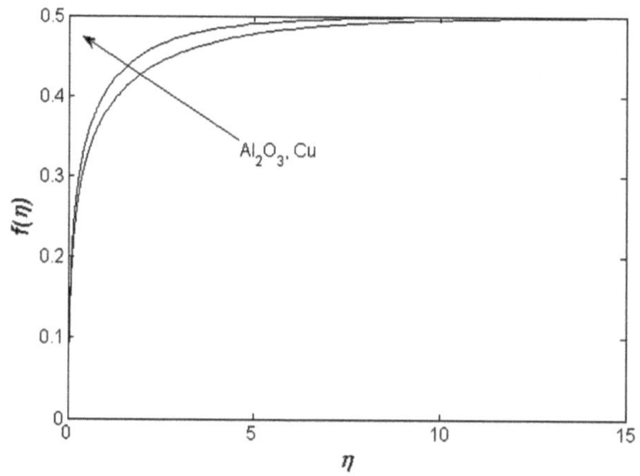

Figure 4.5 Dimensionless velocity profiles for Al_2O_3 and Cu nanoparticles when $\varphi = 0.2$, $a = 0.01$, and $Pr = 7$. *(From Grosan, T., & Pop, I. (2011b). Forced convection boundary layer flow past nonisothermal thin needles in nanofluids.* Journal of Heat Transfer, 133*(5), 054503. https://doi.org/10.1115/1.4003059054503-3.)*

analyzed the effects of injection and suction on the steady mixed convection boundary layer flows over a vertical slender cylinder with a free stream velocity and a wall surface temperature proportional to the axial distance along the surface of the cylinder. Mahmood and Merkin (1988b) studied the steady axisymmetric mixed convection boundary layer flow from a vertical cylinder when $U(x) \sim x$ and $T_w(x) \sim x$. Grosan and Pop (2011a) considered, further, the effect of the addition of the nanoparticles in the base fluid and examined the effects of the solid volume fraction parameter, mixed convection parameter, and the curvature parameter on the thermal performance of flow and heat transfer characteristics.

4.2.2 Basic equations

We consider the steady axisymmetric mixed convection boundary layer flow past a vertical circular cylinder of radius a placed in a nanofluid (Tiwari & Das, 2007), see Fig. 4.6. It is assumed that the mainstream velocity is $U(x)$ and the temperature of the ambient nanofluid is T_∞, while the temperature of the cylinder is $T_w(x)$. Under these assumptions and using the model of the nanofluid proposed by Grosan and Pop (2011a), the equations governing this problem are

$$\frac{\partial}{\partial x}(ru) + \frac{\partial}{\partial r}(rw) = 0 \tag{4.19}$$

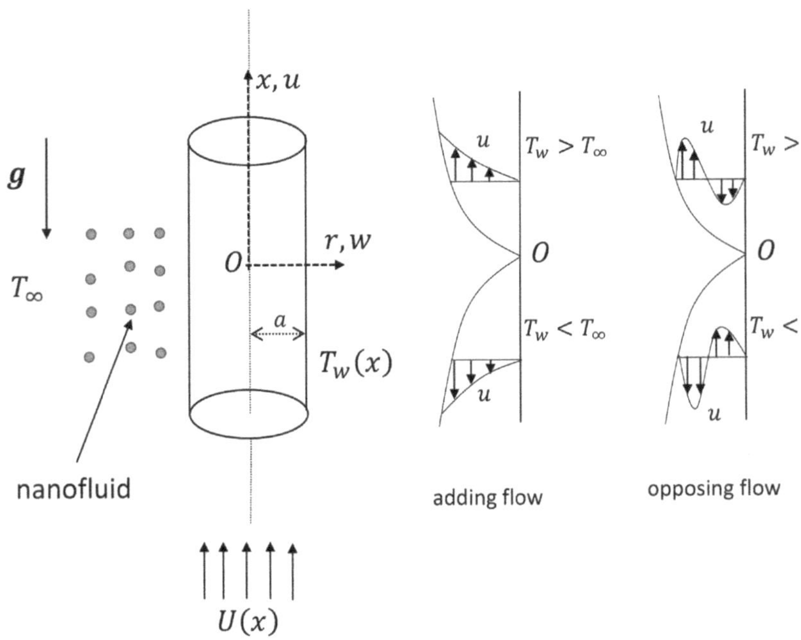

Figure 4.6 Geometry of the mixed convection on the cylinder. Adding and opposing cases of flow.

$$u\frac{\partial u}{\partial x}+w\frac{\partial u}{\partial r}=U\frac{dU}{dx}+\frac{\mu_{nf}}{\rho_{nf}}\left(\frac{\partial^2 u}{\partial r^2}+\frac{1}{r}\frac{\partial u}{\partial r}\right)$$
$$+\frac{\varphi\rho_s\beta_s+(1-\varphi)\rho_f\beta_f}{\rho_{nf}}g(T-T_\infty) \qquad (4.20)$$

$$u\frac{\partial T}{\partial x}+w\frac{\partial T}{\partial r}=\alpha_{nf}\left(\frac{\partial^2 T}{\partial r^2}+\frac{1}{r}\frac{\partial T}{\partial r}\right) \qquad (4.21)$$

$$u=w=0, \quad T=T_w(x)=T_\infty+\Delta T\left(\frac{x}{l}\right) \quad \text{at} \quad r=a$$
$$u=U(x)\rightarrow U_\infty(x/l), \quad T\rightarrow T_\infty \quad \text{as} \quad r\rightarrow\infty \qquad (4.22)$$

Here x and r are the coordinates measured in the axial and radial directions, respectively, u and w are the velocity components along x and r directions, and T is the temperature of the nanofluid. The viscosity, the thermal diffusivity, and the thermal conductivity of the nanofluid are given in Eq. (1.16) (see (Oztop & Abu-Nada, 2008) and (Brinkman, 1952)).

Mahmood and Merkin (1988b) looked for similarity solutions of Eqs. (4.19)–(4.21) subject to the boundary conditions (4.22) of the following form:

$$\psi=\left(U_\infty\nu_f a^2/l^{1/2}\right)xf(\eta),\ T-T_\infty=\Delta T(x/l)\theta(\eta),$$
$$\eta=\frac{r^2-a^2}{2\nu_f l}\left(U_\infty\nu_f l/a^2\right)^{1/2} \qquad (4.23)$$

where ψ is the stream function. Substituting (4.23) into Eqs. (4.20) and (4.21), we obtain the following ordinary differential equations (see (Grosan & Pop, 2011a)):

$$\frac{1}{(1-\varphi)^{2.5}\left(1-\varphi+\varphi\rho_s/\rho_f\right)}\left[(1+2\gamma\eta)f'''+2\gamma f''\right]+ff''+1-f'^2$$
$$+\frac{(1-\varphi)+\varphi\left(\rho_s/\rho_f\right)\left(\beta_s/\beta_f\right)}{(1-\varphi)+\varphi\left(\rho_s/\rho_f\right)}\lambda\theta=0 \qquad (4.24)$$

$$\frac{1}{Pr}\frac{k_{nf}/k_f}{(1-\varphi)+\varphi\left(\rho C_p\right)_s/\left(\rho C_p\right)_f}\left[(1+2\gamma\eta)\theta''+2\gamma\theta'\right]+f\theta'-f'\theta=0 \qquad (4.25)$$

$$f(0) = f'(0) = 0, \quad \theta(0) = 1$$
$$f'(\eta) \to 1, \quad \theta(\eta) \to 0 \quad as \quad \eta \to \infty \tag{4.26}$$

Here primes denote differentiation with respect to η, $Pr = \nu_f / \alpha_f$ is the Prandtl number, λ is the mixed convection parameter, and γ is the curvature parameter, which are defined as

$$\lambda = \frac{Gr}{Re^2}, \quad \gamma = \left(\frac{\nu_f l}{U_\infty a^2}\right)^{1/2} \tag{4.27}$$

where $Gr = g\beta_f \, \Delta T \, l^3 / \nu_f^2$ is the Grashof number. It should be noticed that $\lambda > 0$ corresponds to a heated cylinder (assisting flow), $\lambda < 0$ corresponds to a cooled cylinder (opposing flow), and $\lambda = 0$ corresponds to forced convection flow ($T_w = T_\infty$).

In this case, the skin friction coefficient and the Nusselt number are defined as

$$C_f = \frac{\tau_w}{\rho_f U_\infty^2}, \quad Nu = \frac{l q_w}{k_f \Delta T}, \quad \tau_w = \mu_{nf} \left(\frac{\partial u}{\partial r}\right)_{r=a}, \quad q_w = -k_{nf} \left(\frac{\partial T}{\partial r}\right)_{r=a} \tag{4.28}$$

In dimensionless form the skin friction coefficient and the Nusselt number become

$$Re^{1/2} C_f = \frac{\overline{x}}{(1-\varphi)^{2.5}} f''(0), \quad Re^{-1/2} Nu = \frac{k_{nf}}{k_f} \overline{x} [-\theta'(0)] \tag{4.29}$$

where $\overline{x} = x/l$.

4.2.3 Results and discussion

Grosan and Pop (2011a) solved the nonlinear ordinary differential Eqs. (4.24) and (4.25) subject to the boundary conditions (4.26) numerically, using bvp4c routine from Matlab, for some values of the mixed convection parameter λ, the volume fraction parameter φ, and the curvature parameter γ. One type of nanofluid (Cu—water) was considered, and the value of the Prandtl number was taken $Pr = 6.2$ (for water). Different values of the volume fraction parameter φ (0—regular fluid, 0.1 and 0.2) and the curvature parameter γ (0—flat plate, 1 and 5) were considered. When the volume fraction nanoparticle parameter $\varphi = 0$ (regular fluid), this problem reduces to that of viscous fluid (see (Mahmood & Merkin, 1988b). For this case, (Grosan & Pop, 2011a) have compared the variation with $\lambda < 0$(opposing flow) of the reduced skin friction f'(0) for $\gamma = 0$,1, and 5 with those reported by Mahmood and Merkin (1988b). The obtained results are illustrated in Fig. 4.7. The variation of the reduced skin friction coefficient $f''(0)$ with $\lambda < 0$ (opposing flow) for $\varphi = 0$, 0.1, and 0.2 and $\gamma = 0$, 1, and 5 is illustrated in Fig. 4.8, Fig. 4.9, and Fig. 4.10. These figures show that it is possible to get dual solutions of the similarity Eqs. (4.24) and (4.25) subject to the boundary conditions (4.26) with two branches, an upper and lower branch, respectively. These dual solutions are in the range $\lambda_{cr} < \lambda < 0$ and no solutions exist for $\lambda < \lambda_{cr} < 0$, where λ_{cr} represents the minimum values of λ for which the solution exists. The values of λ_{cr} depend on the parameters φ and γ, and, in addition, the minimum value $|\lambda_{cr}|$ increases as both of the curvature parameter γ and the volume fraction parameter φ increase. This suggests that higher values of φ increase the range of existence of the solutions of Eqs. (4.24) - (4.26). Between these two dual solutions, the first solution (upper branch solution) is stable and physically relevant, while the second solution (lower branch solution) is not stable and, therefore, it is not physically realizable.

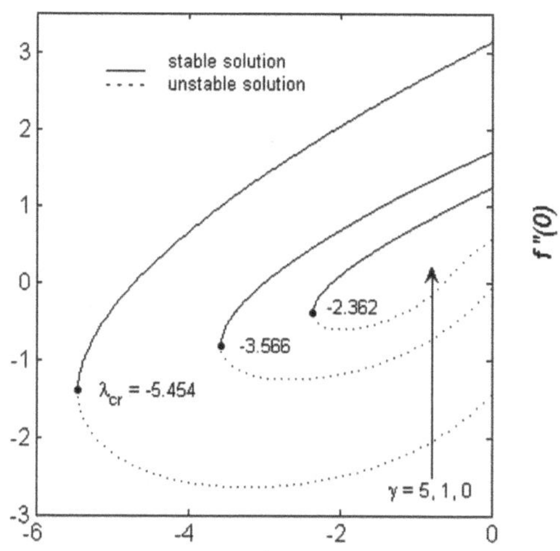

Figure 4.7 Variation of f''(0) with λ for $\gamma = 0$, 1, and 5 when $\varphi = 0$ (regular fluid) and $\lambda < 0$ (opposing flow). The results are identical to those obtained by Mahmood and Merkin (1988b). *(From Grosan, T., & Pop, I. (2011a). Axisymmetric mixed convection boundary layer flow past a vertical cylinder in a nanofluid.* International Journal of Heat and Mass Transfer, *5(15–16), 3139–3145. https://doi.org/10.1016/j.ijheatmasstransfer. 2011.04.0183143.)*

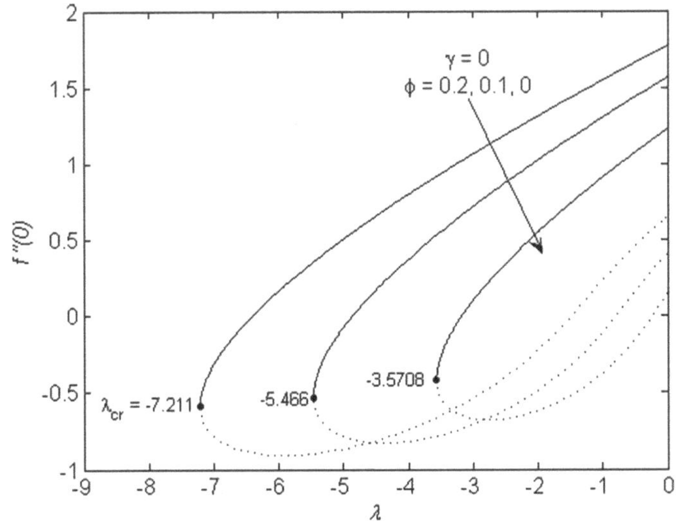

Figure 4.8 Variation of f''(0) with λ for $\gamma = 0$ and $\varphi = 0, 0.1, 0.2$ when $\lambda < 0$ (opposing flow). *(From Grosan, T., & Pop, I. (2011a). Axisymmetric mixed convection boundary layer flow past a vertical cylinder in a nanofluid.* International Journal of Heat and Mass Transfer, *5(15–16), 3139–3145. https://doi.org/10.1016/j.ijheatmasstransfer.2011.04.0183143.)*

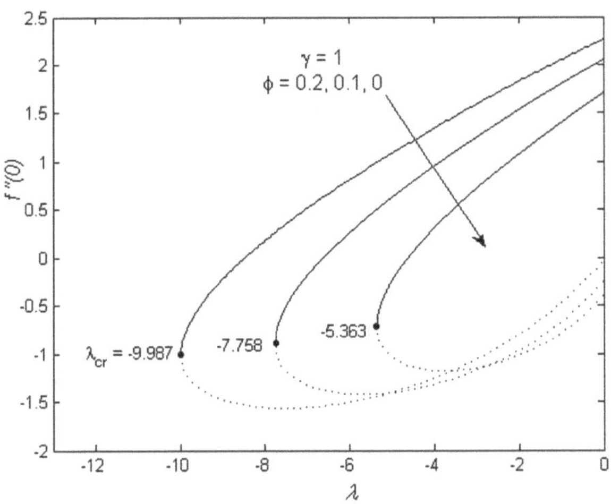

Figure 4.9 Variation of f''(0) with λ for $\gamma = 1$ and $\varphi = 0, 0.1, 0.2$ when $\lambda < 0$ (opposing flow). *(From Grosan, T., & Pop, I. (2011a). Axisymmetric mixed convection boundary layer flow past a vertical cylinder in a nanofluid.* International Journal of Heat and Mass Transfer, *5(15–16), 3139–3145. https://doi.org/10.1016/j.ijheatmasstransfer.2011.04.0183143.)*

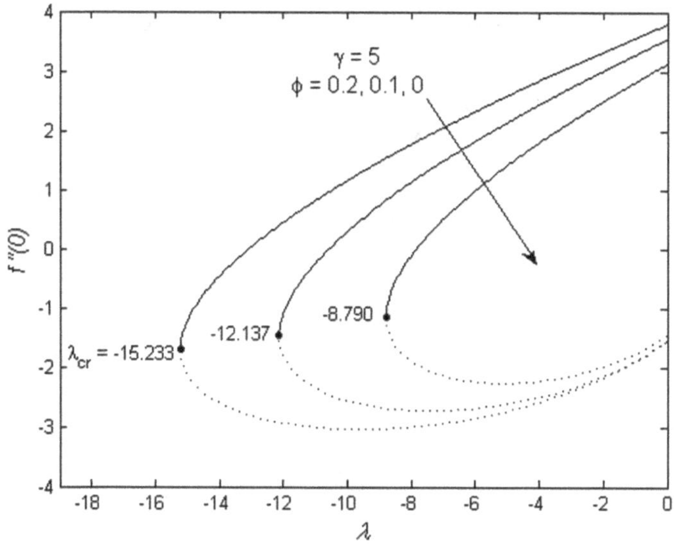

Figure 4.10 Variation of f''(0) with λ for $\gamma = 5$ and $\varphi = 0$, 0.1, 0.2 when $\lambda < 0$ (opposing flow). *(From Grosan, T., & Pop, I. (2011a). Axisymmetric mixed convection boundary layer flow past a vertical cylinder in a nanofluid.* International Journal of Heat and Mass Transfer, *5(15–16), 3139–3145. https://doi.org/10.1016/j.ijheatmasstransfer. 2011.04.0183143.)*

Figs. 4.11 and 4.12 display typical dimensionless velocity and temperature profiles for opposing flow case ($\lambda < 0$). These profiles are for Cu–water nanofluid, for both first and second solutions satisfy the far field boundary conditions asymptotically. The profiles of the lower branch of solutions have a much higher boundary layer thickness. The reversed flow near the wall is present for the lower branch of solutions (Fig. 4.11). On the other hand, Fig. 4.12 illustrates the existence of the heat generation inside the boundary layer for the lower temperature branch solution. However, this is not physically possible, while the viscous dissipation effect has not been considered in the present physical model.

For the assisting flow case ($\lambda \geq 0$) only one solution exists, this is in agreement with (Ishak et al., 2007; Mahmood & Merkin, 1988b). Furthermore, Figs. 4.13 and 4.14 present the dimensionless velocity and temperature profiles for large values of λ (free convection limit); the dimensionless velocity shows an overshoot near the wall (see, Fig. 4.13), and this is in agreement with the results reported by Mahmood and Merkin (1988b).

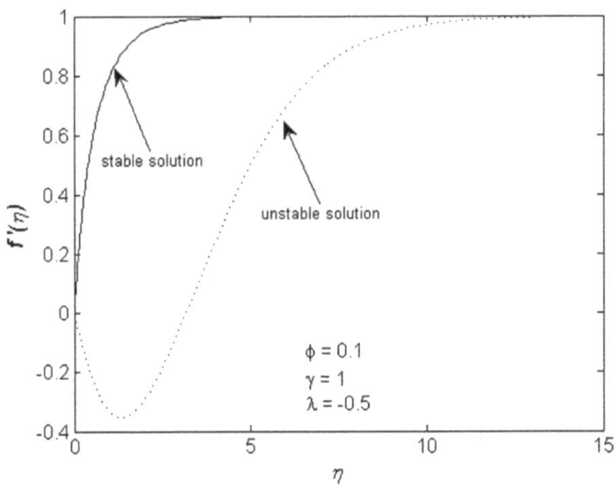

Figure 4.11 Dual solutions for the dimensionless velocity profiles. *(Modified from Grosan, T., & Pop, I. (2011a). Axisymmetric mixed convection boundary layer flow past a vertical cylinder in a nanofluid.* International Journal of Heat and Mass Transfer, *5(15–16), 3139–3145. https://doi.org/10.1016/j.ijheatmasstransfer.2011.04.0183143.)*

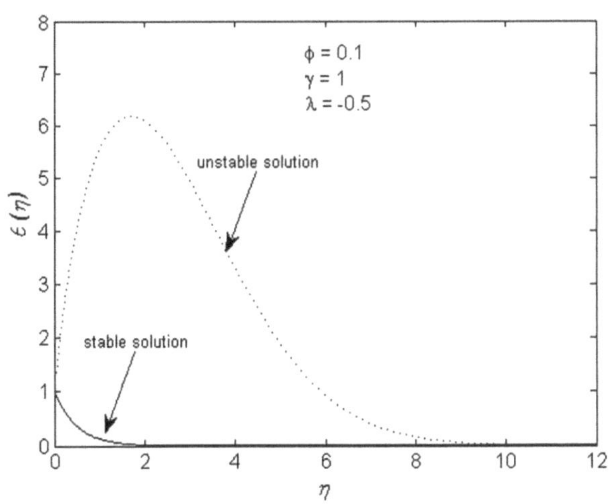

Figure 4.12 Dual solutions for the dimensionless temperature profiles. *(Modified from Grosan, T., & Pop, I. (2011a). Axisymmetric mixed convection boundary layer flow past a vertical cylinder in a nanofluid.* International Journal of Heat and Mass Transfer, *5(15–16), 3139–3145. https://doi.org/10.1016/j.ijheatmasstransfer.2011.04.0183143.)*

Variation of the reduced skin friction coefficient and reduced Nusselt number with $\lambda \geq 0$ (assisting flow), γ and φ is given in Table 4.4. It can be noticed that both skin friction coefficient and Nusselt number increase

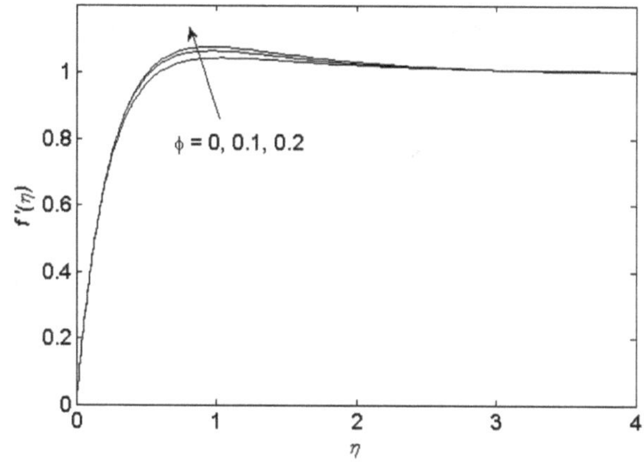

Figure 4.13 Variation of dimensionless velocity profiles with φ for $\lambda = 15$ and $\gamma = 1$. *(From Grosan, T., & Pop, I. (2011a). Axisymmetric mixed convection boundary layer flow past a vertical cylinder in a nanofluid.* International Journal of Heat and Mass Transfer, *5(15–16), 3139–3145. https://doi.org/10.1016/j.ijheatmasstransfer.2011.04.0183143.)*

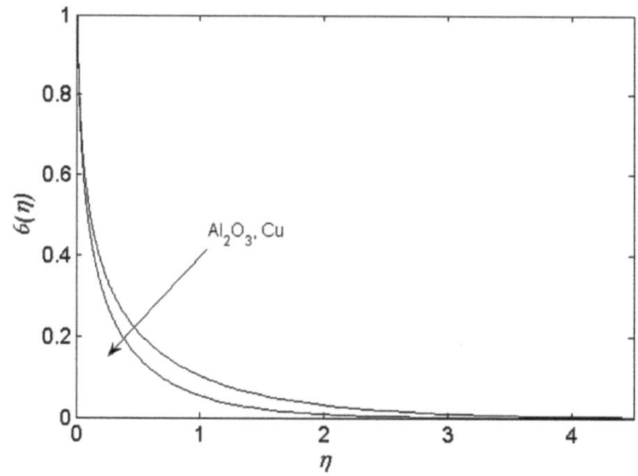

Figure 4.14 Variation of dimensionless temperature profiles with φ for $\lambda = 15$ and $\gamma = 1$. *(From Grosan, T., & Pop, I. (2011a). Axisymmetric mixed convection boundary layer flow past a vertical cylinder in a nanofluid.* International Journal of Heat and Mass Transfer, *5(15–16), 3139–3145. https://doi.org/10.1016/j.ijheatmasstransfer.2011.04.0183143.)*

when all three parameters increase. For example, the skin friction coefficient and Nusselt number increase when $\lambda = \gamma = 5$ with about 28% and 24% for $\varphi = 0.1$ and about 66% and 52% for $\varphi = 0.2$, respectively.

Table 4.4 Values related to the reduced skin friction coefficient and reduced Nusselt number for different values of the governing parameters.

		$f''(0)/(1-\varphi)^{2.5}$		$-(k_{nf}/k_f)\theta'(0)$	
λ	γ	$\varphi = 0$	$\varphi = 0.1$	$\varphi = 0$	$\varphi = 0.1$
0	0 1 5	1.232587 1.707622 3.137181	1.884324 2.512130 4.429302	1.573432 2.151733 3.837074	1.969206 2.696671 4.852504
5	0 1 5	2.560232 3.020042 4.533701	3.223206 3.836088 5.840746	1.883998 2.409351 4.049479	2.240041 2.922855 5.041794
15	0 1 5	4.687465 5.185480 6.913786	5.464257 6.109829 8.332511	2.236582 2.732082 4.346323	2.586342 3.236515 5.328565

4.3 Blasius and Sakiadis problems in nanofluids

4.3.1 Introduction

The classical forced convection boundary layer flow and heat transfer problems past a semiinfinite static flat plate (Blasius problem, 1908) and past a moving semiinfinite flat plate (Sakiadis problem, 1961) in nanofluids are theoretically studied. The problems are studied for three types of metallic or nonmetallic nanoparticles such as copper (Cu), alumina (Al_2O_3), and titania (TiO_2) in the base fluid of water with the Prandtl number $Pr = 6.2$ to investigate the effect of the solid volume fraction parameter of the nanofluids. The skin friction coefficient, Nusselt number, and the velocity and temperature profiles are presented and discussed in detail. It is found that the solid volume fraction affects the fluid flow and heat transfer characteristics.

4.3.2 Basic equations

Consider a Cartesian coordinate system (x,y), where x and y are the coordinates measured along the plate and normal to it, respectively, and assume that the flow takes place at $y \geq 0$. It is also assumed that the constant temperature of the flat plate is T_w and that of the ambient nanofluid is T_∞. Under the boundary layer approximations and the fact that this flow is one of zero pressure gradient, the basic equations can be written in the Cartesian coordinates x and y as follows:

$$\frac{\partial u}{\partial x} + \frac{\partial v}{\partial y} = 0 \tag{4.30}$$

$$u\frac{\partial u}{\partial x} + v\frac{\partial u}{\partial y} = \frac{\mu_{nf}}{\rho_{nf}}\frac{\partial^2 u}{\partial y^2} \tag{4.31}$$

$$u\frac{\partial T}{\partial x} + v\frac{\partial T}{\partial y} = \frac{k_{nf}}{\left(\rho C_p\right)_{nf}}\frac{\partial^2 T}{\partial y^2} \tag{4.32}$$

subject to the boundary conditions (see (Ahmad et al., 2011))

i) Blasius problem

$$\begin{aligned} &v = 0, \quad u = 0, \quad T = T_w \ \text{ at } \ y = 0 \\ &u \to U, \quad T \to T_\infty \ \text{ as } \ y \to \infty \end{aligned} \tag{4.33}$$

ii) Sakiadis problem

$$v = 0,\ \ u = U,\ \ T = T_w \text{ at } y = 0$$
$$u \to 0,\ \ T \to T_\infty \text{ as } y \to \infty \tag{4.34}$$

where u and v are the velocity components along the axes x and y, U is the constant velocity of the free stream (inviscid flow) or that of a moving flat plate.

A solution of the boundary value problem (4.30)–(4.34) is looked for as

$$\psi = \left(xU\nu_f\right)^{1/2} f\,(\eta),\ \theta(\eta) = (T - T_\infty)/(T_w - T_\infty),\ \eta = \left(yU/x\nu_f\right)^{1/2} \tag{4.35}$$

where ψ is the stream function, which is defined in the usual way as $u = \partial\psi/\partial y$ and $v = -\partial\psi/\partial x$. Substituting (4.35) into Eqs. (4.31) and (4.32), the following ordinary (similarity) differential equations are obtained

$$\frac{1}{(1-\varphi)^{2.5}\left(1-\varphi+\varphi\rho_s/\rho_f\right)} f''' + \frac{1}{2} f f'' = 0 \tag{4.36}$$

$$\frac{1}{Pr}\frac{1}{(1-\varphi)^{2.5}\left(1-\varphi+\varphi\left(\rho C_p\right)_s/\left(\rho C_p\right)_f\right)}\theta'' + \frac{1}{2} f\theta' = 0 \tag{4.37}$$

i) Blasius problem

$$f\,(0) = 0,\ \ f'(0) = 0,\ \ \theta(0) = 1$$
$$f'(\eta) \to 1,\ \ \theta \to 0 \text{ as } \eta \to \infty \tag{4.38}$$

i) Sakiadis problem

$$f\,(0) = 0,\ \ f'(0) = 1,\ \ \theta(0) = 1$$
$$f'(\eta) \to 0,\ \ \theta(\eta) \to 0 \text{ as } \eta \to \infty \tag{4.39}$$

where primes denote differentiation with respect to the variable η and $Pr = \nu_f/\alpha_f$ is the Prandtl number. It is worth mentioning that Eq. (4.36) reduces to that derived by (Blasius, 1908) or (Sakiadis, 1961) when $\varphi = 0$ (regular fluid).

Quantities of practical interest in this study are the skin friction coefficient C_f and the local Nusselt number Nu_x, which are defined as

$$C_f = \frac{\mu_{nf}}{\rho_f U^2}\left(\frac{\partial u}{\partial y}\right)_{y=0}, \quad Nu_x = -\frac{x k_{nf}}{k_f (T_w - T_\infty)}\left(\frac{\partial T}{\partial y}\right)_{y=0} \tag{4.40}$$

Using (4.35) and (4.40), we get

$$Re_x^{1/2} C_f = \frac{\mu_{nf}}{\mu_f} f''(0), \quad Re_x^{-1/2} Nu_x = -\frac{k_{nf}}{k_f}\theta'(0) \tag{4.41}$$

where $Re_x = Ux/\nu_f$ is the local Reynolds number.

4.3.3 Comments

Numerical solutions of the boundary value problem (4.36)–(4.39) were obtained by Ahmad et al. (2011) using the shooting method. The effects of the solid volume fraction of nanofluid φ and the Prandtl number Pr are analyzed for three different nanofluids copper (Cu), alumina (Al_2O_3), and titania (TiO_2) in the base fluid of water with the Prandtl number $Pr = 6.2$. The parameter φ is considered in the range of $0 \le \varphi \le 0.2$, see Oztop & Abu-Nada (2008). The comparison of their results, with those reported by Blasius (1908) and Sakiadis (1961) for the special case of a Newtonian or regular fluid ($\varphi = 0$) is shown in Tables 4.5 and 4.6. It is seen that the results reported by Ahmad et al. (2011) are in excellent agreement with those reported by the above-mentioned authors.

Figs. 4.15–4.18 present the variation of the skin friction coefficient and the Nusselt number with the concentration of the nanoparticles, φ. Velocity and temperature profiles for both Blasius and Sakiadis problems when copper nanoparticles are used are illustrated in Figs. 4.19 and 4.20.

Table 4.5 Values of the skin friction coefficient (Eq. 4.41) for the Blasius problem.

	Ahmad et al. (2011)			
φ	Cu	Al_2O_3	TiO_2	Blasius (1908)
0	0.3321			0.3321
0.004	0.3390	0.3357	0.3359	
0.01	0.3494	0.3412	0.3417	
0.016	0.3597	0.3468	0.3476	
0.02	0.3667	0.3506	0.3515	
0.1	0.5076	0.4316	0.4362	
0.2	0.7066	0.5545	0.5642	

Table 4.6 Values of the skin friction coefficient (Eq. 4.41) for the Sakiadis problem.

	Ahmad et al. (2011)			
φ	**Cu**	**Al_2O_3**	**TiO_2**	**Sakiadis (1961)**
0	0.4446			0.44375
0.004	0.4538	0.4494	0.4497	
0.01	0.4676	0.4568	0.4574	
0.016	0.4814	0.4643	0.4653	
0.02	0.4906	0.4693	0.4705	
0.1	0.6788	0.5778	0.5840	
0.2	0.9446	0.7428	0.7556	

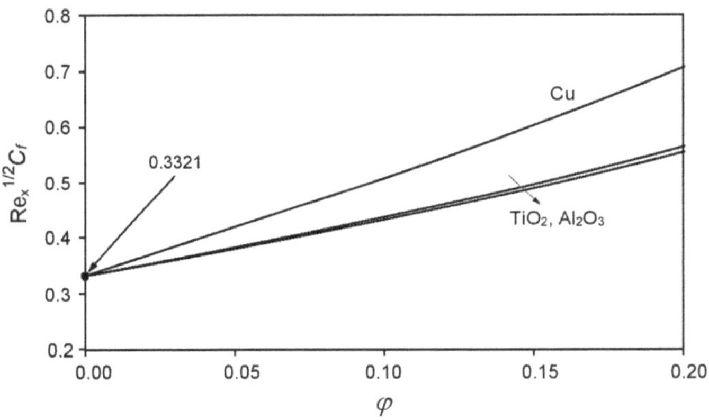

Figure 4.15 Variation of the skin friction coefficient with φ for the Blasius problem. *(From Ahmad, S., Rohni, A. M., & Pop, I. (2011). Blasius and Sakiadis problems in nanofluids.* Acta Mechanica, 218*(3–4), 195–204. https://doi.org/10.1007/s00707-010-0414-6 199.)*

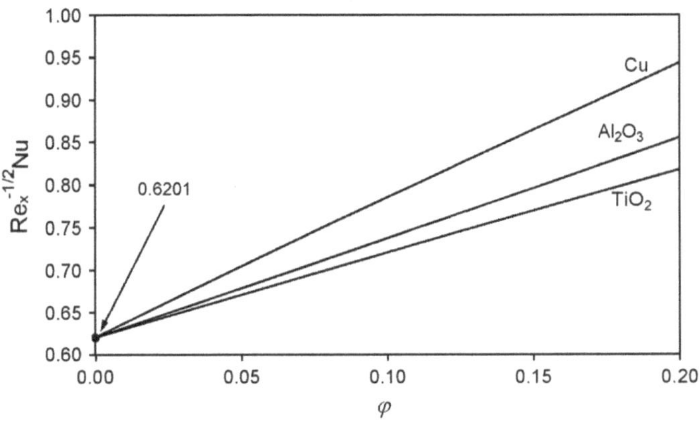

Figure 4.16 Variation of the Nusselt number with φ for the Blasius problem. *(From Ahmad, S., Rohni, A. M., & Pop, I. (2011). Blasius and Sakiadis problems in nanofluids.* Acta Mechanica, 218*(3–4), 195–204. https://doi.org/10.1007/s00707-010-0414-6 199.)*

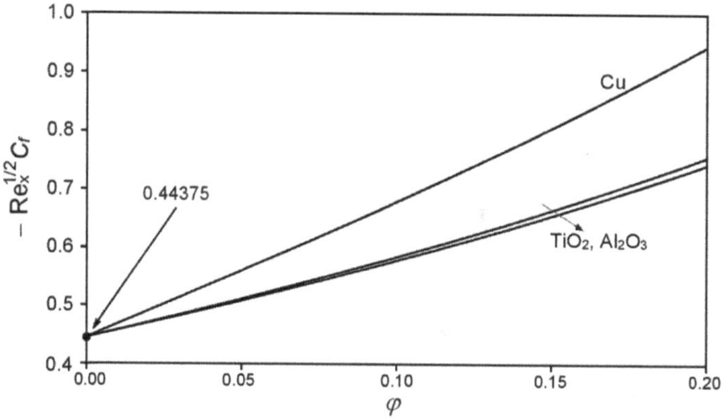

Figure 4.17 Variation of the skin friction coefficient with φ for the Sakiadis problem. *(From Ahmad, S., Rohni, A. M., & Pop, I. (2011). Blasius and Sakiadis problems in nanofluids.* Acta Mechanica, 218*(3–4), 195–204. https://doi.org/10.1007/s00707-010-0414-6 199.*)

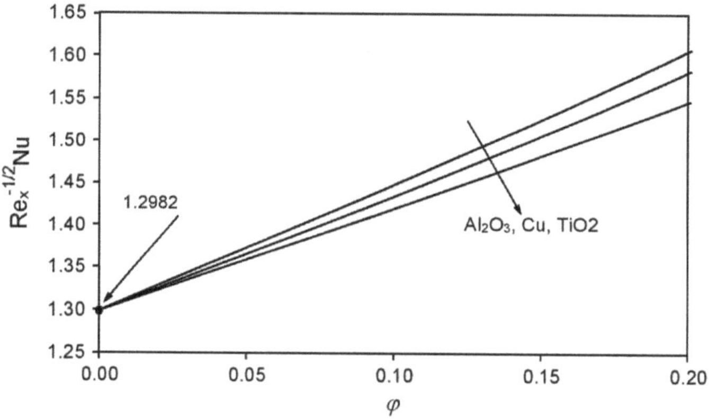

Figure 4.18 Variation of the Nusselt number with φ for the Sakiadis problem. *(From Ahmad, S., Rohni, A. M., & Pop, I. (2011). Blasius and Sakiadis problems in nanofluids.* Acta Mechanica, 218*(3–4), 195–204. https://doi.org/10.1007/s00707-010-0414-6 199.)*

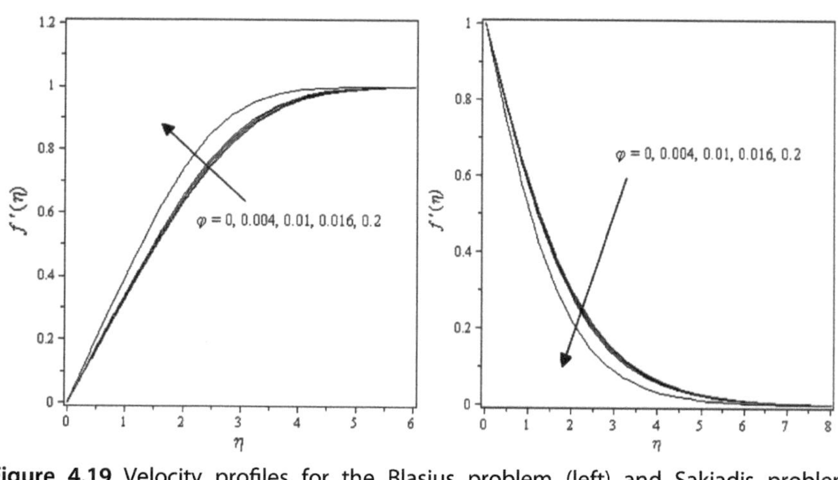

Figure 4.19 Velocity profiles for the Blasius problem (left) and Sakiadis problem (right). *(From Ahmad, S., Rohni, A. M., & Pop, I. (2011). Blasius and Sakiadis problems in nanofluids.* Acta Mechanica, 218*(3–4), 195–204. https://doi.org/10.1007/s00707-010-0414-6 199.)*

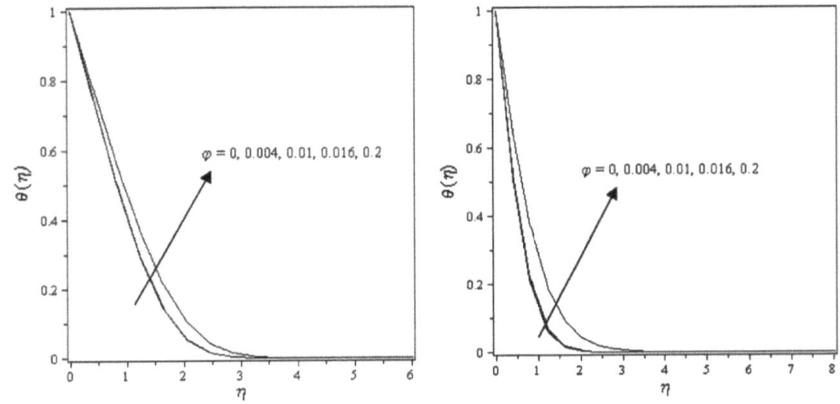

Figure 4.20 Temperature profiles for the Blasius problem (left) and Sakiadis problem (right). *(From Ahmad, S., Rohni, A. M., & Pop, I. (2011). Blasius and Sakiadis problems in nanofluids.* Acta Mechanica, *218(3–4), 195–204. https://doi.org/10.1007/s00707-010-0414-6 199.)*

Nomenclature

(x, y, z)	Cartesian coordinates (m)
u_e, $\overline{u}_e$	velocities of the external flow (m/s)
a	constant
Cp	specific heat capacity at constant pressure (J/Kg K)
g	gravitational acceleration vector (m/s^2)
Gr	Grashof number
k	thermal conductivity (W/m K)
l, L	characteristic lengths (m)
m, n	constants
Pr	Prandtl number
r, $\overline{r}$	radial coordinates
R, $\overline{R}$	radii of the needles (m)
Re	Reynolds number
T	temperature
$T\infty$	ambient temperature (K)
U	external flow velocity (m/s)
u, v, w	velocity vector components (m/s)
$U\infty$	characteristic velocity (m/s)
C_f	skin friction coefficient
$f'(\eta)$	dimensionless velocity
Gr_x	local Grashof number
Nu_x	local Nusselt number
Re_x	local Reynolds number
T_w	wall temperature (K)
$\overline{T}_w$	surface temperature (K)

$\bar{u}$, $\bar{v}$	velocity vector components (m/s)
$\bar{x}$	Cartesian coordinate (m)

Greek letters

α	thermal diffusivity (m^2/s)
β	coefficient of thermal expansion (1/K)
γ	curvature parameter
λ	mixed convection parameter
μ	dynamic viscosity (Ns/m^2)
η	dimensionless similarity variable
$\theta(\eta)$	dimensionless temperature
ν	kinematic viscosity (m^2/s)
ρ	density of fluid (kg/m^3)
φ	nanoparticle volume fraction
ψ	stream function

Subscript

e	outer free stream flow
f	base fluid
nf	nanofluid
s	solid phase
w	value at the wall/surface
∞	far field boundary conditions

Superscript

'	differentiation with respect to η

References

Ahmad, S., Rohni, A. M., & Pop, I. (2011). Blasius and Sakiadis problems in nanofluids. *Acta Mechanica, 218*(3–4), 195–204. https://doi.org/10.1007/s00707-010-0414-6

Blasius, H. (1908). Grenzschichten in Flussigkeiten mit kleiner Reibung. *Zeitschrift für angewandte Mathematik und Physik*, 1–37.

Brinkman, H. C. (1952). The viscosity of concentrated suspensions and solutions. *The Journal of Chemical Physics, 20*(4), 571. https://doi.org/10.1063/1.1700493

Chen, J. L. S. (1987). Mixed convection flow about slender bodies of revolution. *Journal of Heat Transfer, 109*(4), 1033–1036. https://doi.org/10.1115/1.3248177

Chen, T. S., & Mucoglu, A. (1975). Bouyancy effects on forced convection along a vertical cylinder. *Journal of Heat Transfer, 97*(2), 198–203. https://doi.org/10.1115/1.3450341

Chen, J. L. S., & Smith, T. N. (1978). Forced convection heat transfer from nonisothermal thin needles. *Journal of Heat Transfer, 100*(2), 358–362. https://doi.org/10.1115/1.3450809

Grosan, T., & Pop, I. (2011a). Axisymmetric mixed convection boundary layer flow past a vertical cylinder in a nanofluid. *International Journal of Heat and Mass Transfer, 5*(15–16), 3139–3145. https://doi.org/10.1016/j.ijheatmasstransfer.2011.04.018

Grosan, T., & Pop, I. (2011b). Forced convection boundary layer flow past nonisothermal thin needles in nanofluids. *Journal of Heat Transfer, 133*(5), 054503. https://doi.org/10.1115/1.4003059

Ishak, A., Nazar, R., & Pop, I. (2007). The effects of transpiration on the boundary layer flow and heat transfer over a vertical slender cylinder. *International Journal of Non-linear Mechanics, 42*(8), 1010–1017. https://doi.org/10.1016/j.ijnonlinmec.2007.05.004

Mahmood, T., & Merkin, J. H. (1988a). Mixed convection on a vertical circular cylinder. *Zeitschrift für angewandte Mathematik und Physik, 39*(2), 186–203. https://doi.org/10.1007/BF00945765

Mahmood, T., & Merkin, J. H. (1988b). Similarity solutions in axisymmetric mixed-convection boundary-layer flow. *Journal of Engineering Mathematics, 22*(1), 73–92. https://doi.org/10.1007/BF00044366

Narain, J. P. (1972). Combined forced and free-convection heat transfer from vertical thin needles in a uniform stream. *Physics of Fluids, 15*(11), 1879. https://doi.org/10.1063/1.1693798

Narain, J. P., & Uberoi, M. S. (1973). Combined forced and free-convection over thin needles. *International Journal of Heat and Mass Transfer, 16*(8), 1505–1512. https://doi.org/10.1016/0017-9310(73)90179-8

Oztop, H. F., & Abu-Nada, E. (2008). Numerical study of natural convection in partially heated rectangular enclosures filled with nanofluids. *International Journal of Heat and Fluid Flow, 29*(5), 1326–1336. https://doi.org/10.1016/j.ijheatfluidflow.2008.04.009

Sakiadis, B. C. (1961). Boundary-layer behavior on continuous solid surfaces: I. Boundary-layer equations for two-dimensional and axisymmetric flow. *AIChE Journal,* 7(1), 26–28. https://doi.org/10.1002/aic.690070108

Tiwari, R. K., & Das, M. K. (2007). Heat transfer augmentation in a two-sided lid-driven differentially heated square cavity utilizing nanofluids. *International Journal of Heat and Mass Transfer, 50*(9–10), 2002–2018. https://doi.org/10.1016/j.ijheatmasstransfer.2006.09.034

CHAPTER 5

Stretching/shrinking sheets in nanofluids and hybrid nanofluids

5.1 Flow and heat transfer over an unsteady shrinking sheet with suction in a nanofluid using Buongiorno's model

5.1.1 Introduction

There have been many studies on the mechanism that results in the enhanced heat transfer using nanofluids. Collections of papers on heat transfer in nanofluids can be found in the book by Bozhko and Suslov (2018), Das et al. (2007), Minkowycz et al. (2013), Nield and Bejan (2017), Shenoy et al. (2016), and in the review papers by Buongiorno et al. (2009), Kakaç and Pramuanjaroenkij (2009), Mahian et al. (2013), Mahian, Kolsi, Amani, Estellé, Ahmadi, Kleinstreuer, Marshall, Siavashi, et al. (2019), Mahian, Kolsi, Amani, Estellé, Ahmadi, Kleinstreuer, Marshall, Taylor, et al. (2019), etc. Rohni et al. (2013) has, for the first time, studied the unsteady flow over a continuously shrinking surface using the mathematical nanofluid model proposed by Buongiorno (2006). By introducing appropriate coordinate transformations, the number of independent variables is reduced by one or more (Sin & Chio, 2012).

5.1.2 Basis equations

Consider the two-dimensional flow over an unsteady continuously shrinking sheet with mass transfer in a nanofluid. It is assumed that the velocity of the shrinking sheet is $u_w(x,t)$ and the velocity of the mass transfer is $v_w(x,t)$, where x is the coordinate measured along the shrinking sheet and t is the time. It is also assumed that the constant surface temperature and the concentration of the sheet are T_w and C_w, while the uniform temperature and concentration far from the sheet are T_∞ and C_∞, respectively. A schematic representation of this problem is shown in Fig. 5.1.

Similarity Solutions for the Boundary Layer Flow and Heat Transfer of Viscous Fluids, Nanofluids, Porous Media, and Micropolar Fluids
ISBN 978-0-12-821188-5
https://doi.org/10.1016/B978-0-12-821188-5.00006-0

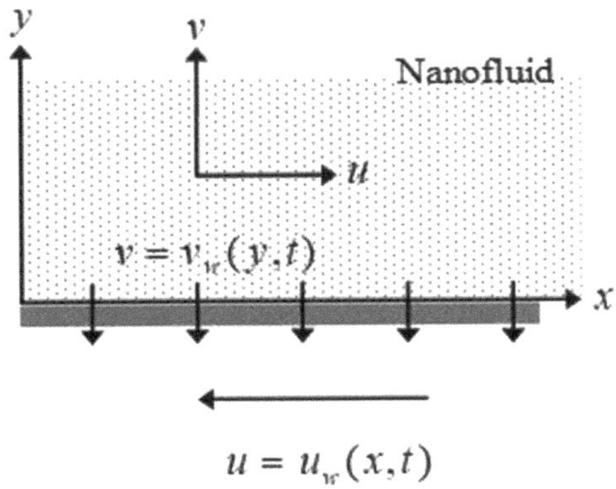

Figure 5.1 Physical model and coordinate system of the shrinking sheet.

Following the nanofluid model proposed by Buongiorno (2006), the unsteady governing two-dimensional Eqs. (1.1)–(1.4) can be written in the Cartesian coordinates x and y using boundary layer approximations as

$$\frac{\partial u}{\partial x}+\frac{\partial v}{\partial y}=0 \tag{5.1}$$

$$\frac{\partial u}{\partial t}+u\frac{\partial u}{\partial x}+v\frac{\partial u}{\partial y}=\nu\frac{\partial^2 u}{\partial y^2} \tag{5.2}$$

$$\frac{\partial T}{\partial t}+u\frac{\partial T}{\partial x}+v\frac{\partial T}{\partial y}=\alpha\frac{\partial^2 T}{\partial y^2}+\tau\left[D_B\frac{\partial C}{\partial y}\frac{\partial T}{\partial y}+\frac{D_T}{T_\infty}\left(\frac{\partial T}{\partial y}\right)^2\right] \tag{5.3}$$

$$\frac{\partial C}{\partial t}+u\frac{\partial C}{\partial x}+v\frac{\partial C}{\partial y}=D_B\frac{\partial^2 C}{\partial y^2}+\frac{D_T}{T_\infty}\frac{\partial^2 T}{\partial y^{2'}} \tag{5.4}$$

subject to the initial and boundary conditions

$$t<0:\ u=v=0,\ T=T_\infty,\ C=C_\infty \ for\ all\ x, y$$

$$t\gg 0:\ v=v_0,\ u=u_w(x,t)=-\frac{u_w(x)}{1-\gamma t'},\ T=T_w,\ C=C_w\ at\ y=0 \tag{5.5}$$

$$u\to 0,\ T\to T_\infty,\ C\to C_\infty\ as\ y\to\infty$$

where u and v are the velocity components in x- and y-directions, respectively, v_0 is the constant mass flux velocity with $v_0<0$ for suction and $v_0>0$ for injection or withdrawal of the fluid, γ is a parameter showing the unsteadiness of the physical problem, and we assume that $u_w(x)=ax$

with a positive constant. The physical meaning of the other quantities in given in the Nomenclature.

Following (Rohni et al., 2013), we consider the similarity solution of Eqs. (5.1)−(5.4) along with the boundary conditions (5.5) as

$$u = \frac{ax}{1-\gamma t} f'(\eta),\ \theta(\eta) = \frac{T - T_\infty}{T_w - T_\infty},\ \phi(\eta) = \frac{C - C_\infty}{C_w - C_\infty},\ \eta = \gamma \sqrt{\frac{a}{\nu(1-\gamma t)}} \tag{5.6}$$

where prime denotes differentiation with respect to η. The wall mass transfer velocity then becomes

$$v_w(0,t) = -\sqrt{\frac{a\nu}{1-\gamma t}}\, s \tag{5.7}$$

where s is the constant mass transfer velocity with $s > 0$ for suction and $s < 0$ for injection.

By substituting (5.6) into Eqs. (5.2)−(5.4), the following ordinary (similarity) differential equations are obtained:

$$f''' + ff'' - f'^2 - A\left(f' - \frac{\eta}{2} f''\right) = 0 \tag{5.8}$$

$$\frac{1}{Pr}\theta'' + f\theta' - A\frac{\eta}{2}\theta' + Nb\phi'\theta' + Nt\,\theta'^2 = 0 \tag{5.9}$$

$$\phi'' + Le\left(f - A\frac{\eta}{2}\right)\phi' + \frac{Nt}{Nb}\,\theta'' = 0, \tag{5.10}$$

and the boundary conditions (5.5) become

$$\begin{aligned} &f(0) = s, \quad f'(\eta) = -1, \quad \theta(\eta) = 1, \phi(0) = 1 \\ &f'(\eta) \to 0, \quad \theta(\eta) \to 0, \quad \phi(\eta) \to 0 \text{ as } \eta \to \infty \end{aligned} \tag{5.11}$$

where A is the unsteady parameter and the physical meaning of the other quantities is mentioned in the Nomenclature.

The physical quantities of interest are the skin friction coefficient C_f, the local Nusselt number Nu_x, and the local Sherwood number Sh_x which are defined by

$$C_f = \frac{\nu}{u_w^2(x)}\left(\frac{\partial u}{\partial y}\right)_{y=0}, \quad Nu_x = \frac{x}{(T_w - T_\infty)}\left(-\frac{\partial T}{\partial y}\right)_{y=0}, \quad Sh_x = \left(-\frac{\partial C}{\partial y}\right)_{y=0} \tag{5.12}$$

Using (5.6) and (5.12), we get

$$Re_x^{1/2} C_f = f''(0),\ Re_x^{-1/2} Nu_x = -\theta'(0), Re_x^{-1/2} Sh_x = -\phi'(0) \qquad (5.13)$$

with $Re_x = u_w(x)x/\nu$ being the local Reynolds number.

5.1.3 Comments

Numerical solutions of the boundary value problem (5.8–5.11) were obtained using the shooting method. The effects of the wall mass suction $s(>0)$, the unsteadiness A, the Brownian motion Nb, and the thermophoresis Nt parameters on the flow, heat transfer, and concentration characteristics are analyzed. As it is stated in the paper by Nield and Kuznetsov (2009), since most nanofluids examined to date have large values for the Lewis number Le, we are interested here mainly in the case $Le \gg 1$. We also notice that when $Nb = Nt = 0$, Eqs. (5.9) and (5.10) involve just two dependent variables, namely $\theta(\eta)$ and $\phi(\eta)$. (The boundary value problem for $\phi(\eta)$ then becomes ill-posed and is of no physical significance.)

Following (Fang & Zhang, 2009) we consider here only the case $A \leq 0$ (decelerating shrinking sheet). The results of this study are illustrated in Figs. 5.2–5.6. It is important to highlight that Fig. 5.2 is qualitatively consistent with Figure 1 in (Fang & Zhang, 2009). We notice from

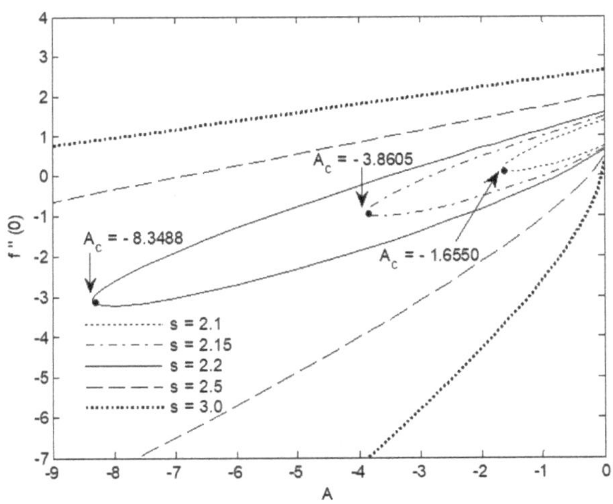

Figure 5.2 Variation of $f''(0)$ with A for some values of s. *(From Rohni, A. M., Ahmad, S., Ismail, A. I. M., & Pop, I. (2013). Flow and heat transfer over an unsteady shrinking sheet with suction in a nanofluid using Buongiorno's model.* International Communications in Heat and Mass Transfer, 43, *75–80. https://doi.org/10.1016/j.icheatmasstransfer.2013.02.001 79.)*

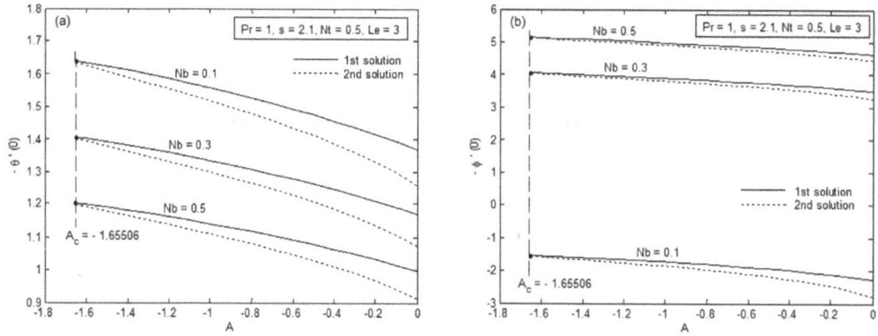

Figure 5.3 Variation of (A) -$\theta'(0)$ and (B) $\phi'(0)$ with A for different Nb when $Pr = 1$, $s = 2.1$, and $Le = 3$. *(From Rohni, A. M., Ahmad, S., Ismail, A. I. M., & Pop, I. (2013). Flow and heat transfer over an unsteady shrinking sheet with suction in a nanofluid using Buongiorno's model.* International Communications in Heat and Mass Transfer, 43, *75–80. https://doi.org/10.1016/j.icheatmasstransfer.2013.02.001 79.)*

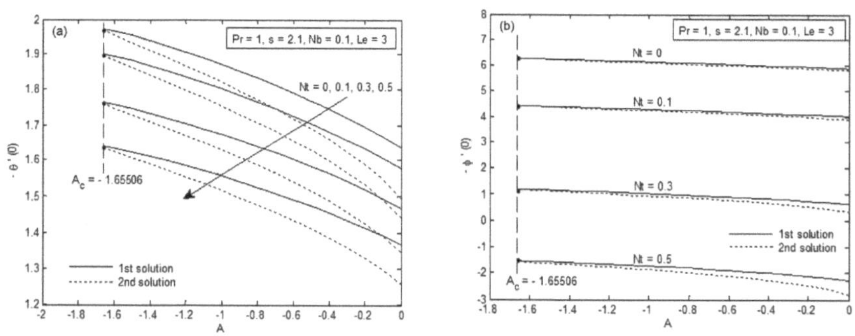

Figure 5.4 Variation of (A) -$\theta'(0)$ and (B) -$\phi'(0)$ with A for different Nt when $Pr- = 1$, $s = 2.1$, and $Le = 3$. *(From Rohni, A. M., Ahmad, S., Ismail, A. I. M., & Pop, I. (2013). Flow and heat transfer over an unsteady shrinking sheet with suction in a nanofluid using Buongiorno's model.* International Communications in Heat and Mass Transfer, 43, *75–80. https://doi.org/10.1016/j.icheatmasstransfer.2013.02.001 79.)*

Figs. 5.2–5.5 that dual solutions exist for certain chosen parameters. The two branches (first and second solutions) merge with one another at a critical point where the boundary layer solutions beyond this point are impossible to be obtained due to boundary layer separation from the surface. In Figs. 5.3 and 5.4, the variation of heat transfer rate $-\theta'(0)$and concentration rate $-\phi'(0)$ with A for different Nb and Nt, respectively, when $Pr = 1$, $s = 2.1$ and $Le = 3$ is plotted. From these figures, we can see that the increment of the value of Nb increases $-\phi'(0)$ and thus decreases the heat transfer rate. However, the increment of the value of Nt decreases both the heat transfer rate and concentration rate.

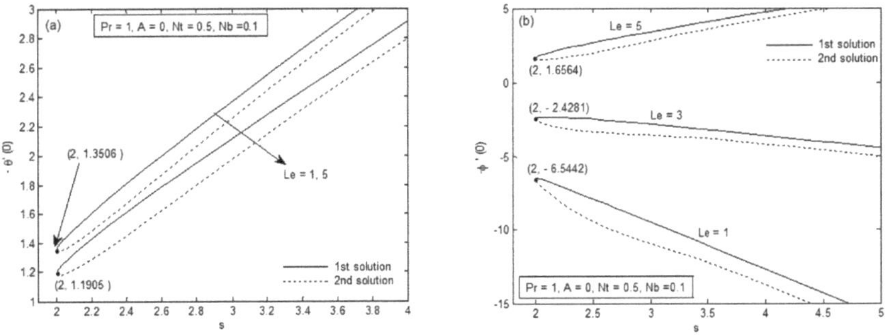

Figure 5.5 Variation of (A) -$\theta'(0)$ and (B) -$\phi'(0)$ with s for different *Le* when *Pr-* = 1, *Nb* = 0.1, *Nt* = 0.5, and *A* = 0 (steady case). *(From Rohni, A. M., Ahmad, S., Ismail, A. I. M., & Pop, I. (2013). Flow and heat transfer over an unsteady shrinking sheet with suction in a nanofluid using Buongiorno's model.* International Communications in Heat and Mass Transfer, *43, 75–80. https://doi.org/10.1016/j.icheatmasstransfer.2013.02.001 79.)*

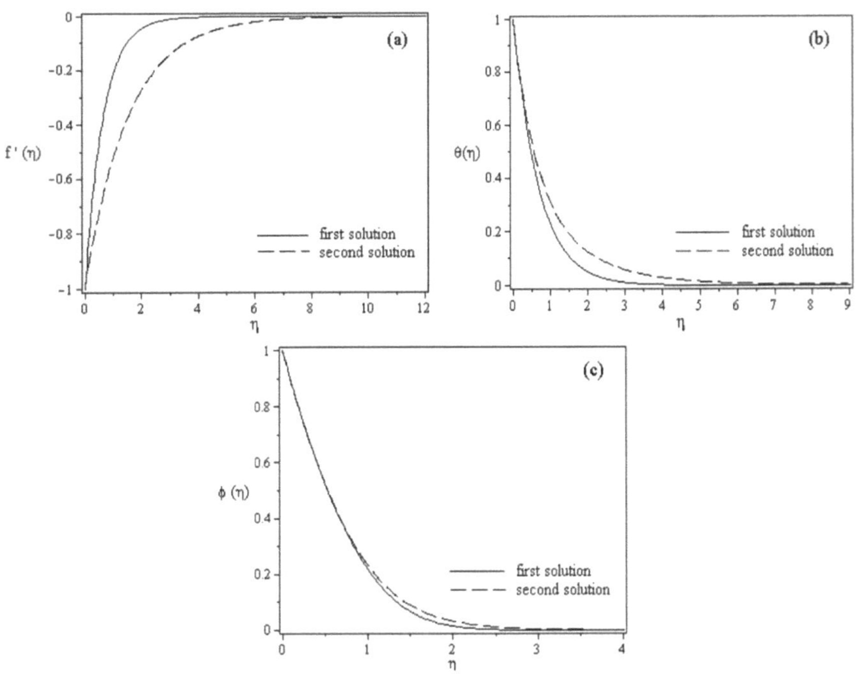

Figure 5.6 Plots of (A) velocity $f'(\eta)$; (B) temperature $\theta(\eta)$; (C) concentration $\phi(\eta)$ for *Pr-* = 1, *s* = 2.2, *Nb* = 0.1, *Nt* = 0.5, *Le* = 3, and *A* = 0 (steady flow). *(From Rohni, A. M., Ahmad, S., Ismail, A. I. M., & Pop, I. (2013). Flow and heat transfer over an unsteady shrinking sheet with suction in a nanofluid using Buongiorno's model.* International Communications in Heat and Mass Transfer, *43, 75–80. https://doi.org/10.1016/j.icheatmasstransfer.2013.02.001 79.)*

5.2 Axisymmetric rotational stagnation-point flow impinging radially a permeable stretching/shrinking surface in a nanofluid

5.2.1 Introduction

The classical problem of steady two-dimensional stagnation-point flows has been first analyzed exactly by Hiemenz (1911). The result is an exact solution for the flow directed perpendicular to an infinite flat plate. The axisymmetric stagnation flow toward a plate was solved by Homann (1936). Howarth and Bairstow (1938) and Davey (1961) extended the two-dimensional and axisymmetric flows to the three-dimensional case, which is based on boundary layer approximation in the direction normal to the plane. We mention also here the paper by Naganthran et al. (2016) on the unsteady stagnation-point flow and heat transfer of a special third-grade fluid past a permeable stretching/shrinking sheet. This class of solutions, describing the fluid flow near the stagnation region, exists on all solid bodies moving in a fluid. The stagnation region encounters the highest pressure, the highest heat transfer, and the highest rates of mass deposition. Further it should be mentioned that (Agrawal, 1957) has discovered a new axisymmetric stagnation-point flow, obtaining also an exact solution of the Navier—Stokes equations. In contrast to the irrotational outer flow of (Homann, 1936), this flow is rotational in the far field. Agrawal (1957) derived his solution using spherical coordinates. Very recently, (Weidman, 2016) has extended Agrawal's (1957) paper to the case of normal impingement of the rotational stagnation-point flow to a radially stretching sheet. Sample velocity profiles along both solution branches have been presented. A linear temporal stability analysis reveals that solutions along the upper branch are stable, while those on the lower branch are unstable. Wang (1991) examined the liquid film behavior over an unsteady stretching sheet and observed the growth of the uncommon flow owing to shrinking. The investigation of an exact solution of the Navier—Stokes equations over a shrinking sheet was conducted by Miklavčič and Wang (2006). They discovered that the vorticity within a boundary layer is not bounded, and a steady flow exists by applying an adequate suction at the boundary. As stated in Fang and Zhang (2009), a shrinking sheet flow demonstrates different physical phenomena compared to the stretching sheet flow. Goldstein (1965) has pointed out that the new type of shrinking sheet flow is essentially a backward flow and it shows physical phenomena quite distinct from the stretching flow case. A very good collection of published papers on stretching sheets can be found in the review paper by Wang (2011).

Rosca and Pop (2017) have extended the paper by Weidman (2016) on the axisymmetric rotational stagnation-point flow impinging on a radially stretching sheet in a classical viscous fluid to water-based cooper (Cu), alumina (Al_2O_3), and titanium (TiO_2) nanofluids by using the mathematical nanofluid model proposed by Tiwari and Das (2007). In addition, the case of a shrinking sheet with a stability analysis for the multiple (dual) solutions has been considered.

5.2.2 Basic equations

The problem is formulated using cylindrical coordinates (r, ϑ, z), measured in the axial and radial directions, respectively, the flow being axisymmetric $((\partial/\partial\vartheta = 0))$ about the z-axis and also symmetric to the $z = 0$ plane. The stagnation line is at $z = 0$ and the domain of flow is in the upper half plane as shown in Fig. 5.7. It is assumed that the velocity of the ambient fluid (inviscid fluid) is $u_e(r, z) = 2arz$, where a is a parameter, which measures the strength of the stagnation-point flow having units $(\mathrm{LT})^{-1}$ (see (Weidman, 2016)). It is also assumed that the constant temperature of the stretching/shrinking surface is T_w, while the temperature and nanoparticle volume fraction of the ambient fluid T_∞ and C_∞, respectively, are constant. Further, it is assumed that the mass flux velocity is $w = w_0$, where $w_0 < 0$ is

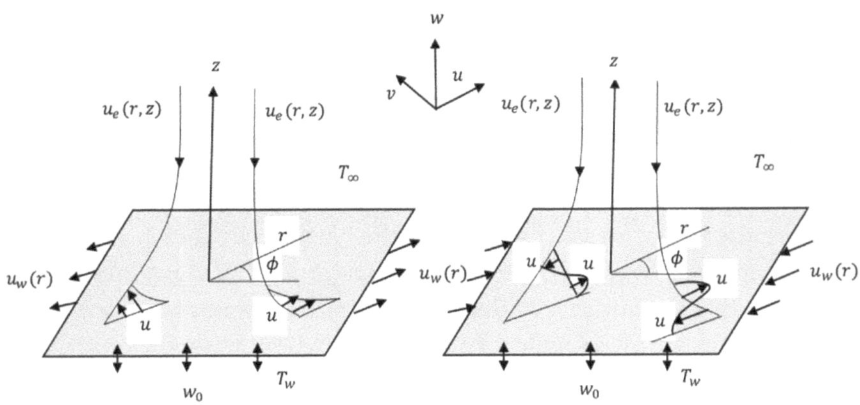

Figure 5.7 Geometry of the problem and coordinate system: (A) stretching surface; (B) shrinking surface. *(From Rosca, N. C., & Pop, I. (2017). Axisymmetric rotational stagnation point flow impinging radially a permeable stretching/shrinking surface in a nanofluid using Tiwari and Das model.* Scientific Reports, *7(1), 40299. https://doi.org/10.1038/srep40299 9.)*

for suction and $w_0 > 0$ is for injection or withdrawal of the fluid. Under these conditions, the unsteady governing Eqs. (2.1)–(2.4) can be written in the cylindrical coordinates (r,z) as (see (Rosca & Pop, 2017))

$$\frac{\partial}{\partial r}(ru)+\frac{\partial}{\partial z}(rw)=0 \tag{5.14}$$

$$\frac{\partial u}{\partial t}+u\frac{\partial u}{\partial r}+w\frac{\partial u}{\partial z}=-\frac{1}{\rho_{nf}}\frac{\partial p}{\partial r}+\frac{\mu_{nf}}{\rho_{nf}}\left(\frac{\partial^2 u}{\partial r^2}+\frac{1}{r}\frac{\partial u}{\partial r}+\frac{\partial^2 u}{\partial z^2}-\frac{u}{r^2}\right) \tag{5.15}$$

$$\frac{\partial w}{\partial t}+u\frac{\partial w}{\partial r}+w\frac{\partial w}{\partial z}=-\frac{1}{\rho_{nf}}\frac{\partial p}{\partial r}+\frac{\mu_{nf}}{\rho_{nf}}\left(\frac{\partial^2 w}{\partial r^2}+\frac{1}{r}\frac{\partial w}{\partial r}+\frac{\partial^2 w}{\partial z^2}\right) \tag{5.16}$$

$$\frac{\partial T}{\partial t}+u\frac{\partial T}{\partial r}+w\frac{\partial T}{\partial z}=\frac{k_{nf}}{\left(\rho C_p\right)_{nf}}\left(\frac{\partial^2 T}{\partial r^2}+\frac{1}{r}\frac{\partial T}{\partial r}+\frac{\partial^2 T}{\partial z^2}\right) \tag{5.17}$$

along with the initial and boundary conditions

$$\begin{gathered} t<0: u=0, w=0, T=T_\infty \text{ for any } r,\ z \\ t \gg 0: u_w(r)=U_w(r)\varepsilon, w=w_0,\ T=T_w \text{ at } z=0 \\ \frac{\partial u}{\partial z}\rightarrow\frac{\partial u_e}{\partial z},\ T\rightarrow T_\infty \text{ as } z\rightarrow\infty \end{gathered} \tag{5.18}$$

where $U_w(r)=a^{2/3}\alpha_f^{1/3}r$ is the velocity of the stretching/shrinking surface, α_f is the thermal diffusivity of the base fluid, and the physical meaning of the other quantities are mentioned in the Nomenclature.

5.2.3 Steady-state case

We introduce for this case the following similarity variables (Weidman, 2016).

$$\begin{gathered} u(r,\eta)=a^{2/3}\alpha_f^{1/3}rf'(\eta), \quad w(\eta)=-2a^{1/3}\alpha_f^{1/3}f(\eta) \\ \theta(\eta)=(T-T_\infty)/(T_w-T_\infty), \quad \eta=\left(a/\alpha_f\right)^{1/3}z \end{gathered} \tag{5.19}$$

where prime denotes differentiation with respect to η. Using (5.19), Eq. (5.14) is satisfied automatically, and since there is no longitudinal pressure gradient, Eqs. (5.15)–(5.17) reduce to the following ordinary differential equations:

$$\frac{Pr}{(1-\varphi)^{2.5}\left(1-\varphi+\varphi\rho_s/\rho_f\right)}f''' + 2ff'' - f'^2 = 0 \tag{5.20}$$

$$\frac{k_{nf}/k_f}{(1-\varphi)+\varphi\left(\rho c_p\right)_s/\left(\rho c_p\right)_f}\theta'' + 2f\theta' = 0 \tag{5.21}$$

and the boundary conditions (5.18) become

$$\begin{aligned} &f(0) = s, \quad f'(0) = \varepsilon, \ \theta(0) = 1 \\ &f''(\eta)\to 2, \ \theta(\eta)\to 0 \ as \ \eta\to\infty \end{aligned} \tag{5.22}$$

Here $s = -w_0\Big/\left(2a^{1/3}\alpha_f^{2/3}\right)$ is the dimensionless mass flux parameter, with $s > 0$ for suction and $s < 0$ for injection, respectively, and $Pr = \nu_f/\alpha_f$ is the Prandtl number. The physical quantities of practical interest are the local skin friction coefficients C_f and the local Nusselt number Nu_z, which are defined as

$$C_f = \frac{\mu_{nf}}{\rho_f U^2{}_w}\left(\frac{\partial u}{\partial z}\right)_{z=0}, \ Nu_z = \frac{k_{nf}\left(\alpha_f/a\right)^{1/3}}{k_f\left(T_w - T_\infty\right)}\left(-\frac{\partial T}{\partial z}\right)_{z=0} \tag{5.23}$$

Substituting (5.19) into (5.23), we obtain

$$(Pe_r/Pr)C_f = \frac{1}{(1-\varphi)^{2.5}}f''(0), \quad Nu_z = \frac{k_{nf}}{k_f}\left[-\theta'(0)\right] \tag{5.24}$$

where $Pe_r = U_w(r)\left(\alpha_f/a\right)^{1/3}\Big/\alpha_f$ is the local Péclet number. In the case of $\varphi = s = 0$ and $Pr = 1$, the boundary value problem for $f(\eta)$ reduces to that of (Weidman, 2016).

5.2.4 Stability analysis

We test the fact that the dual solutions of the boundary value problem (5.20–5.22) are stable or unstable by performing a stability analysis, that is similar with that presented in Section 2.1.5. Thus, we have, finally, to solve the following linear eigenvalue problem

$$\frac{Pr}{(1-\varphi)^{2.5}\left(1-\varphi+\varphi\rho_s/\rho_f\right)}F_0''' + 2f_0F_0'' + \left(\gamma - 2f_0'\right)F_0' + 2f_0''F_0 = 0 \tag{5.25}$$

$$\frac{k_{nf}/k_f}{(1-\varphi)+\varphi(\rho C_p)_s/(\rho C_p)_f}G_0''+2f_0G_0'+2\,\theta_0'\,F_0\;+\gamma\,G_0=0 \quad (5.26)$$

along with the boundary conditions (5.18) which become

$$\begin{aligned}&F_0(0)=0,\;F_0'(0)=\;0,\;G_0(0)=0\\&F_0''(\eta)\to 0,\;G_0(\eta)\to 0\text{ as }\eta\to\infty\end{aligned} \quad (5.27)$$

According to (Harris et al., 2009), the set of possible eigenvalues can be determined by relaxing a boundary condition on $F_0(\eta)$ or $G_0(\eta)$. For the present problem, we relax the boundary condition $F_0''(\eta)\to 0$ as $\eta\to\infty$ and for a fixed value of γ we solve the system of Eqs. (5.25) and (5.26) subject to (5.27) along with the new boundary condition $F_0''(0)\;=2$.

5.2.5 Comments

Rosca and Pop (2017) has solved numerically, the boundary value problem (5.20−5.22) for several values of the suction $s(>0)$, volume fraction of the nanofluid φ, and stretching/shrinking ε parameters when the Prandtl number Pr is fixed at $Pr=7$, using the bvp4c function from Matlab software. The values of $\varphi\;=0,\;0.1$ and 0.2 have been taken as in the paper by Oztop and Abu-Nada (2008). The tutorial and examples of solving boundary value problems with bvp4c can be found in the book by (Shampine et al., 2003). The asymptotic boundary conditions in (5.22) at $\eta\to\infty$ are replaced by those at a large but finite value of η where no considerable variation in velocity, temperature, etc., occur as is usually the standard practice in the boundary layer analysis. In order to validate the obtained results, (Rosca & Pop, 2017), (right) have compared the reduced skin friction coefficient $f''(0)$, see Fig. 5.8, when $\varphi=0,\;Pr=1$ and $s=0$

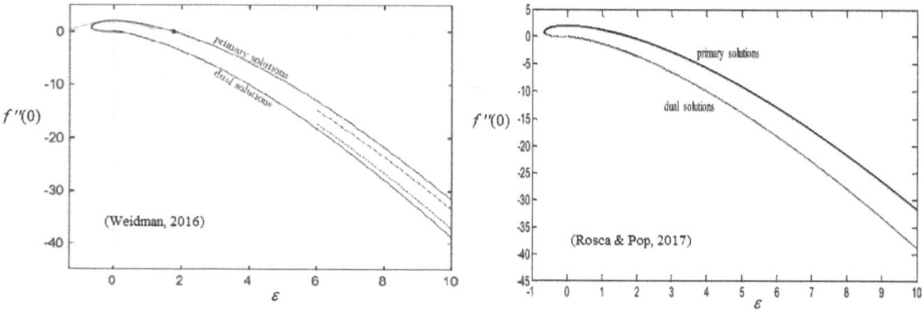

Figure 5.8 Comparisons of $f''(0)$ when $\varphi=0$, $Pr=1$, and $s=0$. *(From Rosca, N. C., & Pop, I. (2017). Axisymmetric rotational stagnation point flow impinging radially a permeable stretching/shrinking surface in a nanofluid using Tiwari and Das model.* Scientific Reports, *7(1), 40299. https://doi.org/10.1038/srep40299 9.)*

with that reported by Weidman (2016) (left). It can be seen from this figure a favorable agreement with the above-mentioned paper.

The obtained numerical results are displayed in terms of the skin friction coefficient $(Pe_r/Pr)\ C_f$, local Nusselt number Nu_z, dimensionless velocity $f'(\eta)$, and temperature profiles $\theta(\eta)$, for different values of the parameters s, ε, and φ with $Pr = 7$ (water). These results are shown in Figs. 5.9 and 5.10. We observe from these figures that the boundary value problem (5.20 and 5.21) admits multiple (double) solutions, i.e., one upper branch solution and one lower branch solution, respectively. In order to ascertain which of the double solutions is expected to be stable, we find the eigenvalues γ in (5.29) by solving numerically Eqs. (5.20) and (5.21) subject to the boundary condition (5.22) using the bvp4c routine. From the performed stability analysis it follows that the upper branch solutions are stable and physically realizable, while the lower branch solutions are not stable and hence physically not realizable. The smallest eigenvalues γ for several values of φ and ε when $s = 1$, $Pr = 7$ (water) are given in Table 5.1. It is worth mentioning that the smallest eigenvalues γ are positive for the upper branch solutions, while the lower branch solutions have negative values of γ, which correspond to initial growth of disturbances, hence the flow is

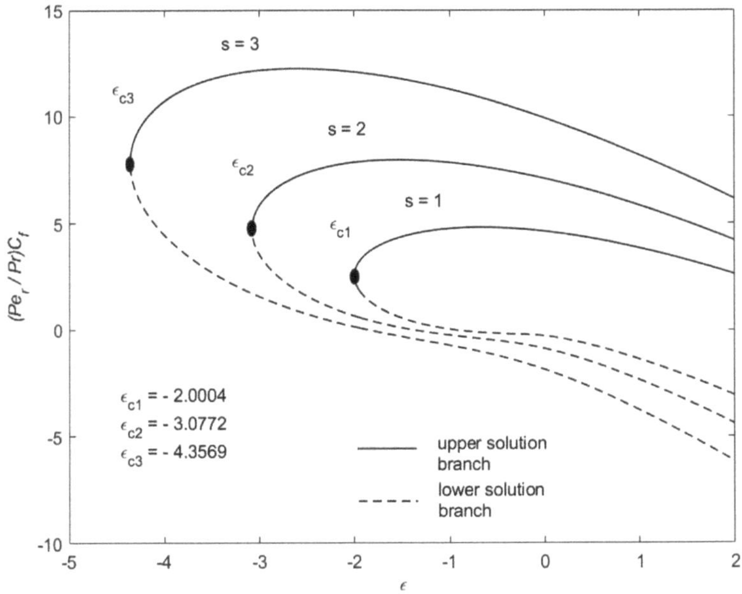

Figure 5.9 Variation of the skin friction coefficient as a function of ε for several values of s when $\varphi = 0.1$ and $Pr = 7$. *(From Rosca, N. C., & Pop, I. (2017). Axisymmetric rotational stagnation point flow impinging radially a permeable stretching/shrinking surface in a nanofluid using Tiwari and Das model.* Scientific Reports, *7(1), 40299. https://doi.org/10.1038/srep40299 9.)*

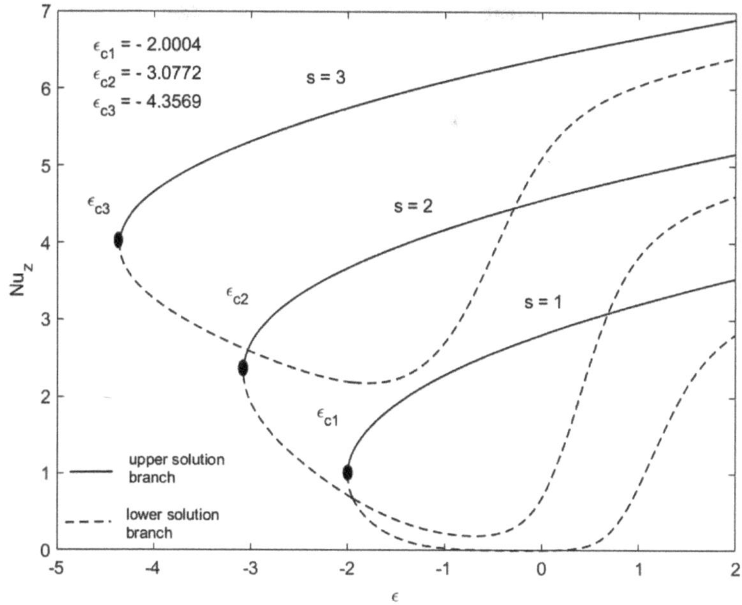

Figure 5.10 Variation of the local Nusselt number as a function of ε for several values of s when $\varphi = 0$ and $Pr = 7$. *(From Rosca, N. C., & Pop, I. (2017). Axisymmetric rotational stagnation point flow impinging radially a permeable stretching/shrinking surface in a nanofluid using Tiwari and Das model.* Scientific Reports, *7(1), 40299. https://doi.org/10.1038/srep40299 9.)*

Table 5.1 Smallest eigenvalues \gamma for several values of φ and ε when $s = 1$, $Pr = 7$ (water).

		γ	
φ	ε	**Upper branch**	**Lower branch**
0.05	1	5.7226	−1.9896
	−1	3.5125	−1.9237
	−1.5	2.5078	−1.7722
	−1.9	1.1226	−0.9798
	−2	0.2245	−0.2185
0.1	1	5.6122	−1.9841
	−1	3.4547	−1.9130
	−1.5	2.4662	−1.7572
	−1.9	1.0889	−0.9559
	−2	0.0704	−0.0698
0.2	1	5.5457	−1.9818
	−1	3.4227	−1.9079
	−1.5	2.4459	−1.7508
	−1.9	1.08	−0.9502
	−2	0.0245	−0.0245

unstable. Although only the upper branch solutions are stable, the lower branch solution is also of mathematical interest as the boundary value problem (5.20−5.22) is concerned. The numerical computations shown in Figs. 5.9 and 5.10 for $(Pe_r/Pr)\ C_f$ and Nu_z suggest that for these values of φ and Pr, the dual solutions exist for both stretching and shrinking sheets when $s = 1$: $\varepsilon \gg \varepsilon_{c1} = -2.0004$; $s = 2: \varepsilon \gg \varepsilon_{c2} = -3.0772$; $s = 3: \varepsilon \gg \varepsilon_{c3} = -4.3589$. However, for $\varepsilon < \varepsilon_{c1}$, $\varepsilon < \varepsilon_{c2}$ and $\varepsilon < \varepsilon_{c3}$ when $s = 1$, 2 and 3, respectively, solutions of Eqs. (5.20) and (5.21) subject to the boundary conditions (5.22) do not exist and the full Navier−Stokes and energy equations should be solved. It is also seen from these figures that for the upper and lower branch solutions, both skin friction coefficient and local Nusselt number increase with the suction parameter as $\varepsilon < 0$ (shrinking parameter) is fixed. Here ε_i (<0), $i = 1,2,3$ are the critical values of ε (<0) starting from which the boundary value problem (5.20−5.22) has at least one solution.

The plots of the velocity $f'(\eta)$ and temperature $\theta(\eta)$ profiles with η for various values of φ when $s = 1$, $Pr = 7$, and $\varepsilon = -1.75$ are presented in Figs. 5.11 and 5.12. The dual solutions are also obtained for both velocity

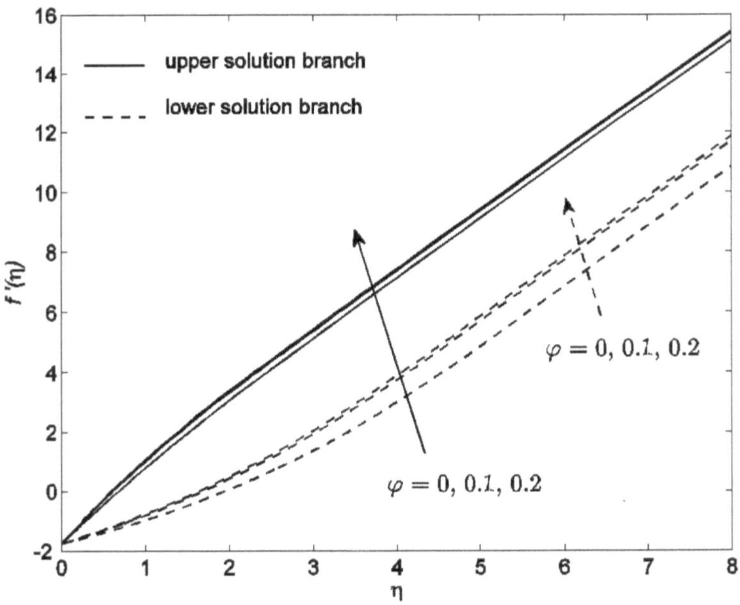

Figure 5.11 Dimensionless velocity profiles $f'(\eta)$ for several values of φ when $s = 1$, $Pr = 7$, and $\varepsilon = -1.75$ (shrinking sheet). *(From Rosca, N. C., & Pop, I. (2017). Axisymmetric rotational stagnation point flow impinging radially a permeable stretching/shrinking surface in a nanofluid using Tiwari and Das model.* Scientific Reports, *7(1), 40299. https://doi.org/10.1038/srep40299 9.)*

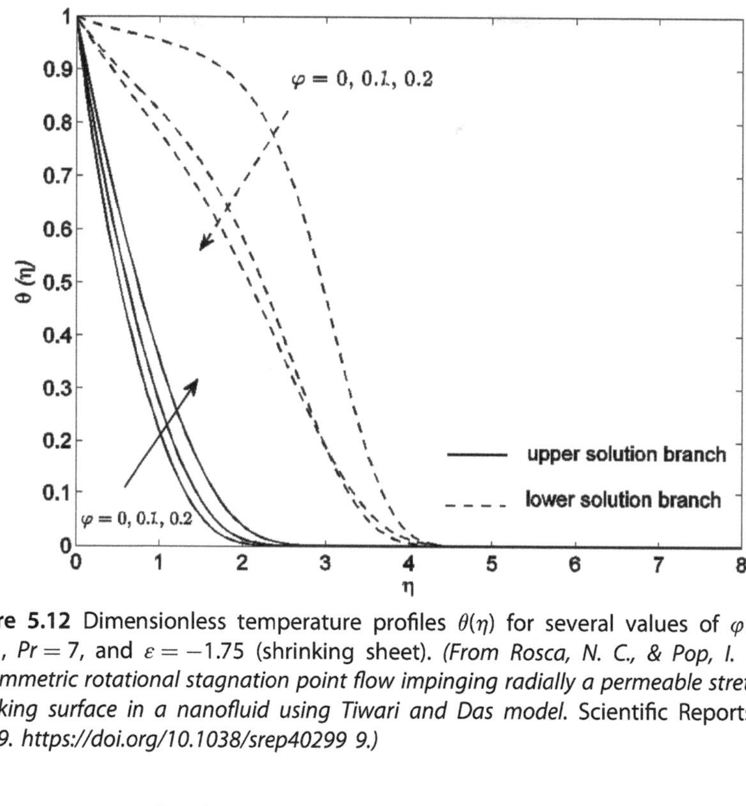

Figure 5.12 Dimensionless temperature profiles $\theta(\eta)$ for several values of φ when $s=1$, $Pr=7$, and $\varepsilon=-1.75$ (shrinking sheet). *(From Rosca, N. C., & Pop, I. (2017). Axisymmetric rotational stagnation point flow impinging radially a permeable stretching/shrinking surface in a nanofluid using Tiwari and Das model.* Scientific Reports, *7(1), 40299. https://doi.org/10.1038/srep40299 9.)*

and temperature distributions. It is seen that the lower branch solutions for these profiles exhibit a larger boundary layer thickness compared to the upper branch solutions. The enhancement of the value of nanofluid parameter φ increases the velocity profile for both upper and lower branch solutions, while it decreases the temperature profiles for lower branch solution and increases for upper branch solution. The fluid velocity is damped faster for a higher value of the parameter φ. It is noticed from Fig. 5.12 that the thermal boundary layer is blown away from the sheet with φ in the case of upper branch solution.

5.3 Flow and heat transfer over a permeable biaxial stretching/shrinking sheet in a nanofluid

5.3.1 Introduction

Wang (1984) studied the three-dimensional (3D) flow due to biaxial stretching of a flat surface, including the axisymmetric stretching surface. A similarity transform reduces the Navier–Stokes equations into a set of nonlinear ordinary differential equations which are integrated numerically by a shooting method. Extended work of (Wang, 1984) to include unsteady stretching added an energy equation for heat transfer. We consider the flow of a fluid film, which adheres to a biaxial stretching surface using perturbation and numerical integration. For all the above sources, the fluid far from the surface does not move laterally. The problem leads to a rare exact similarity solution of the Navier–Stokes equations.

This section considers the uniform flow of a nanofluid over a biaxial stretching surface as it has been solved by Grosan and Pop (2020).

5.3.2 Basic equations

The three-dimensional stagnation-point flow of a viscous nanofluid past a permeable biaxial stretching/shrinking flat surface is considered. A locally orthogonal set of Cartesian coordinates (x,y,z) is chosen with the origin in the plane of the stretching/shrinking sheet. The x- and y-coordinates are in the plane of the shrinking sheet, while the coordinate z is measured in the perpendicular direction to the shrinking surface as shown in Fig. 5.13. It is assumed that the flat surface of the sheet is stretching/shrinking continuously in the both x- and y-directions with the velocities $u_w(x,t)$ and $v_w(x,t)$, respectively. It is assumed that far from the surface at $z \to \infty$ (inviscid flow), there is a uniform flow of velocity $u_e = U$. Further, we assume that the constant mass flux velocity is $w = w_0$, where $w_0 < 0$ is for suction and $w_0 > 0$ is for injection or withdrawal of the fluid.

Under the above conditions, the boundary layer equations are (see (Wang, 2015).)

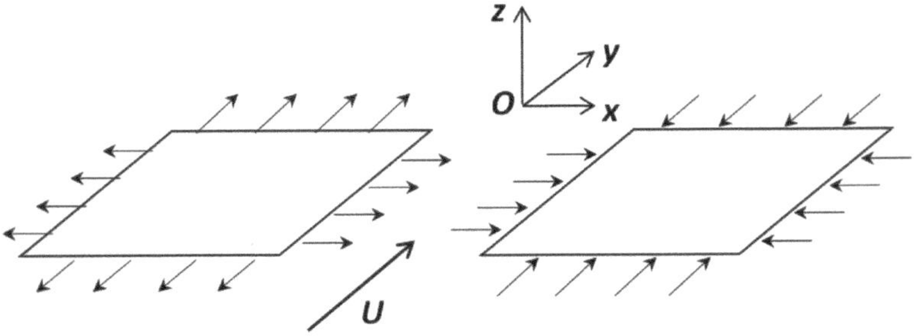

Figure 5.13 Physical model and coordinate system when the sheet is stretching/shrinking. *(From Grosan, T., & Pop, I. (2020). Flow and heat transfer over a permeable biaxial stretching/shrinking sheet in a nanofluid.* Neural Computing and Applications, *32(9), 4575–4582. https://doi.org/10.1007/s00521-018-3770-0 4576.)*

$$\frac{\partial u}{\partial x}+\frac{\partial v}{\partial y}=0 \tag{5.28}$$

$$\frac{\partial u}{\partial t}+u\frac{\partial u}{\partial x}+v\frac{\partial u}{\partial y}+w\frac{\partial u}{\partial z}=-\frac{1}{\rho}\frac{\partial p}{\partial x}+\nu\nabla^2 u \tag{5.29}$$

$$\frac{\partial v}{\partial t}+u\frac{\partial v}{\partial x}+v\frac{\partial v}{\partial y}+w\frac{\partial v}{\partial z}=-\frac{1}{\rho}\frac{\partial p}{\partial y}+\nu\nabla^2 v \tag{5.30}$$

$$\frac{\partial w}{\partial t}+u\frac{\partial w}{\partial x}+v\frac{\partial w}{\partial y}+w\frac{\partial w}{\partial z}=-\frac{1}{\rho}\frac{\partial p}{\partial z}+\nu\nabla^2 w \tag{5.31}$$

$$\begin{aligned}\frac{\partial T}{\partial t}+u\frac{\partial T}{\partial x}+v\frac{\partial T}{\partial y}+w\frac{\partial T}{\partial z}&=\alpha\nabla^2 T+\delta\left\{D_B\left(\frac{\partial C}{\partial x}\frac{\partial T}{\partial x}+\frac{\partial C}{\partial y}\frac{\partial T}{\partial y}+\frac{\partial C}{\partial z}\frac{\partial T}{\partial z}\right)\right.\\&\left.+\frac{D_T}{T_\infty}\left[\left(\frac{\partial T}{\partial x}\right)^2+\left(\frac{\partial T}{\partial y}\right)^2+\left(\frac{\partial T}{\partial z}\right)^2\right]\right\}\end{aligned} \tag{5.32}$$

$$\frac{\partial C}{\partial t}+u\frac{\partial C}{\partial x}+v\frac{\partial C}{\partial y}+w\frac{\partial C}{\partial z}=D_B\nabla^2 C+\frac{D_T}{T_\infty}\nabla^2 T \tag{5.33}$$

along with the initial and boundary conditions (see (Kuznetsov & Nield, 2013))

$$\begin{aligned}&t<0:\ u(x,y,t)=0,\ v(x,y,t)=0,\ w(x,y,t)=0,\ T=T_\infty, C=C_\infty,\ \text{for any } x,y,z\\&t\geq 0:\ u=u_w(x,t)=a\lambda x,\ v=v_w(y,t)=a\lambda y,\ w=w_w=w_0\ T=T_w,\\&\qquad D_B\frac{\partial C}{\partial z}+\frac{D_T}{T_\infty}\frac{\partial T}{\partial z}=0 \text{ at } z=0\\&u=u_e\to U,\quad v=v_e\to 0,\ w=w_e\to 0,\ T\to T_\infty,\ C\to C_\infty \text{ at } z\to\infty\end{aligned} \tag{5.34}$$

where the physical meaning of the quantities in these equations is mentioned in the Nomenclature. The boundary condition $D_B\partial C/\partial z+(D_T/T_\infty)\partial T/\partial z=0$ at $z=0$ in (5.34) is a statement that, the normal flux of nanoparticles is zero at the boundary when the thermophoresis effect is considered (Kuznetsov & Nield, 2013).

5.3.3 Solution for the steady-state case ($\partial/\partial t = 0$)

We introduce now the following similarity variables (see (Wang, 2015)):

$$u = ax f'(\eta) + U\, h(\eta), \quad w = -2\sqrt{av}\, f(\eta)$$
$$\theta(\eta) = (T - T_\infty)/(T_w - T_\infty), \phi(\eta) = (C - C_\infty)/(C_w - C_\infty),$$
$$\eta = \sqrt{a/\nu} \tag{5.35}$$

Substituting the similarity variables (5.35) into Eqs. (5.28)−(5.33), it is found that these equations are reduced to the following ordinary (similarity) differential equations:

$$f''' + 2f f'' - 2f'^2 = 0 \tag{5.36}$$

$$h'' + 2fh' - hf' = 0 \tag{5.37}$$

$$\frac{1}{Pr}\theta'' + 2 f\theta' + Nb\phi'\, \theta' + Nt\, \theta''^2 = 0 \tag{5.38}$$

$$\phi'' + 2Sc f\, \phi' + \frac{Nt}{Nb}\theta'' = 0 \tag{5.39}$$

subject to the boundary conditions

$$\begin{aligned} &f(0) = s, f'(0) = \varepsilon, h(0) = 0, \theta(0) = 1, Nb\phi'(0) + Nt\theta'(0) = 0 \\ &f'(\eta) \to 0, h(\eta) \to 1, \theta(\eta) \to 0, \phi(\eta) \to 0 \text{ as } \eta \to \infty \end{aligned} \tag{5.40}$$

The physical quantities of practical interest is the shear stress τ_{wx} and the heat flux from the surface of the sheet q_w, which are given by

$$\begin{aligned} \tau_{wx} &= \rho\nu\left(\frac{\partial u}{\partial z}\right)_{z=0} = \rho\sqrt{av}[ax f''(0) + Uh'(0)] \\ q_w &= k\left(\frac{\partial T}{\partial z}\right)_{z=0} = k(T_w - T_\infty)[-\theta'(0)] \end{aligned} \tag{5.41}$$

5.3.4 Flow stability

We test the fact that the dual solutions of the boundary value problem (5.36−5.40) are stable or unstable by performing a stability analysis, that is similar with that presented in Section 5.2.4. Thus, we have, finally, to solve the following eigenvalue problem:

$$F_0''' + 2f_0 F_0'' + (\gamma - 2f_0')F_0' + 2f_0'' F_0' = 0 \tag{5.42}$$

$$H_0'' + 2f_0H_0' + (\gamma - f_0')H_0 - h_0F_0' + 2h_0'F_0 = 0 \tag{5.43}$$

$$\frac{1}{Pr}M_0'' + (2f_0 + Nb\,\phi_0' + 2Nt\theta_0')M_0' + \gamma\, M_0 + 2\theta_0'F_0 + Nb\theta_0'P_0' = 0 \tag{5.44}$$

$$P_0'' + 2Sc\,f_0P_0' + \gamma\, ScP_0 + 2\phi_0'F_0 + \frac{Nt}{Nb}M_0'' = 0 \tag{5.45}$$

subject to the boundary conditions

$$F_0(0) = 0,\ F_0'(0) = 0,\ H(0) = 0,\ \ M(0) = 0,\ NbP_0'(0) + NtM_0'(0) = 0$$
$$F_0'(0) \to 0, \quad H_0(\eta) \to 0, \quad M_0(\eta) \to 0, \quad P_0(\eta) \to 0 \text{ as } \eta \to \infty \tag{5.46}$$

It should be mentioned that for particular values of the parameters ε, s, Pr, Sc, Nb, and Nt, the corresponding steady flow solution $f_0(\eta)$, $h_0(\eta)$, $\theta_0(\eta)$ and $\phi_0(\eta)$, stability is determined by the smallest eigenvalue γ. According to (Harris et al., 2009), the range of possible eigenvalues can be determined by relaxing a boundary condition on $F_0(\eta)$, $H_0(\eta)$, $M_0(\eta)$ *or* $P_0(\eta)$. For the present problem, we relax the condition that $F_0'(\eta) \to 0$ as $\eta \to \infty$ and for a fixed value of γ we solve the boundary value problem (5.36–5.40) along with the new normalizing boundary condition $F_0''(0) = 1$.

5.3.5 Comments

The ordinary differential Eqs. (5.36)–(5.39), subject to the boundary conditions (5.40), have been solved numerically by Grosan and Pop (2020) for several values of the parameters: s, ε, and Sc when Pr is fixed at $Pr = 1$, using the bvp4c function from Matlab software. The values of $Nb = 0.5$ and $Nt = 0.3$ have been taken as in the paper by Kuznetsov and Nield (2013). The asymptotic boundary conditions in (5.40) at $\eta \to \infty$ are replaced by those at a large but finite value of η where no considerable variation in velocity, temperature, etc., occurs. A comparison of the results reported by Grosan and Pop (2020) and Wang (2008) is given in Table 5.2

Table 5 2 Comparison results for $\varepsilon = 1$ and $Pr = 7$.

	$f''(0)$	$g''(0)$	$h''(0)$	$\theta'(0)$
Wang (2008)	−1.1737	−1.1737	0.6653	−2.723
Grosan and Pop (2020)	−1.737207	−1.737207	0.665102	−2.721923

Table 5 3 Smallest eigenvalues γ for several values of $\varepsilon < 0$ (shrinking sheet).

Case	γ_{upper}	γ_{lower}
$s = 0.5$, $\varepsilon = -0.14$	0.51385	−0.04206
$s = 1$, $\varepsilon = -0.55$	0.23435	−0.17588
$s = 2$, $\varepsilon = -2.2$	0.93276	−0.69842

for $\varepsilon = 1$ (stretching sheet) and $Pr = 7$.The smallest eigenvalues γ for several values of s and ε (<0) are given in Table 5.3. It is worth mentioning that the smallest eigenvalues γ are positive for the upper branch solutions, while the lower branch solutions have negative values of γ, which correspond to initial growth of disturbances, hence the flow is unstable. Although only the upper branch solutions are stable, the lower branch solution is also of mathematical interest as the boundary value problem (5.36−5.40) is concerned.

Figs. 5.14 and 5.15 show the reduced skin friction coefficient $f''(0)$ and the reduced Nusselt number $\theta'(0)$ for several values of s and Sc when $Pr = 1$ (water) and ε (<0, shrinking sheet). We observe from these figures that the boundary value problem (5.41−5.44) admits multiple (double) solutions, i.e., one upper branch solution and one lower branch solution, respectively. From the performed stability analysis it follows that the upper branch solutions are stable and physically realizable, while the lower branch solutions are not stable and hence physically not realizable. These figures suggest also that for $f''(0)$ dual solutions of the boundary value problem (5.41−5.44) exist for $s = 0.5$: $\varepsilon \gg \varepsilon_{cr2} = -0.1452 \leq \varepsilon_{cr2} = -0.1222$; for

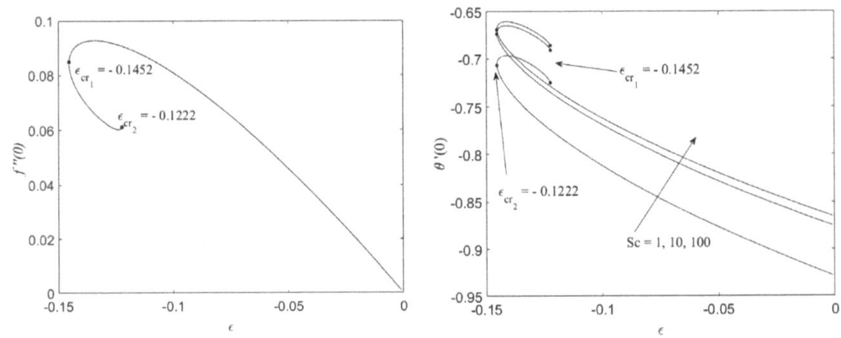

Figure 5.14 Dual solution domain for $s = 0.5$: $f''(0)$ (left) and $\theta'(0)$ (right). *(From Grosan, T., & Pop, I. (2020). Flow and heat transfer over a permeable biaxial stretching/shrinking sheet in a nanofluid.* Neural Computing and Applications, 32*(9), 4575−4582. https://doi.org/10.1007/s00521-018-3770-0 4581.)*

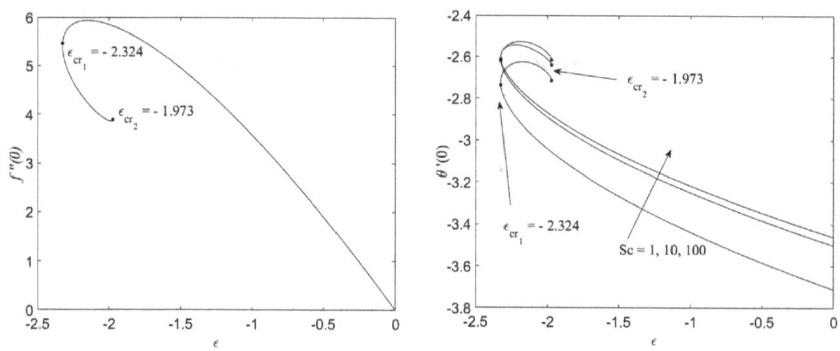

Figure 5.15 Dual solution domain for $s = 2$: $f''(0)$ (left) and $\theta'(0)$ (right). *(From Grosan, T., & Pop, I. (2020). Flow and heat transfer over a permeable biaxial stretching/shrinking sheet in a nanofluid.* Neural Computing and Applications, *32(9), 4575–4582. https://doi.org/10.1007/s00521-018-3770-0 4576.)*

$s = 1$: $\varepsilon \geq \varepsilon_{cr1} = -0.581 \leq \varepsilon_{cr2} = -0.492$; for $s = 2$: $\varepsilon \geq \varepsilon_{cr1} = -2.324 \leq \varepsilon_{cr2} = -1.973$. Also, for $\theta'(0)$ dual solutions exist when $\varepsilon \geq \varepsilon_{cr2} = -0.1222 < \varepsilon_{cr1} = -0.1452$; $\varepsilon > \varepsilon_{cr2} = -0.581 \leq \varepsilon_{cr1} = -0.4927$ when $Sc = 1$, 10, and 100. For $\varepsilon > \varepsilon_{cr2}$ a unique solution exist of the boundary value problem (5.41, 5.44). However, for $\varepsilon < \varepsilon_{cr1}$ solutions of this boundary value problem do not exist and the full Navier–Stokes, energy and concentration equations should be solved. Here ε_{cr1} is the critical values of $\varepsilon < 0$ starting from which the boundary value problem (5.41, 5.44) has at least one solution and ε_{cr2} is the point where solutions stop. It is also seen from these figures that for the upper and lower branch solutions, the reduced skin friction increases with the increase of the suction parameter *s*. Also, the reduced heat transfer increases with the increase of the Schmidt number *Sc*.

First (full line) and second (broken line) solutions for $f'(\eta)$, $h(\eta)$, $\theta(\eta)$ *and* $\phi(\eta)$ and $\varepsilon = -2.2$ for $s = 2$, $Pr = 1$, $Nb = 0.5$, $Nt = 0.3$ and $Sc = 1$ are illustrated in Fig. 5.16. It is evident from these figures that the far field boundary conditions (5.44) are approached asymptotically. Thus, it supports the numerical results obtained for the present problem. It can be also noticed from Fig. 5.16 that for all velocity, temperature, and concentration profiles, the upper branch solution displays a thinner boundary layer thickness compared to the lower branch solution.

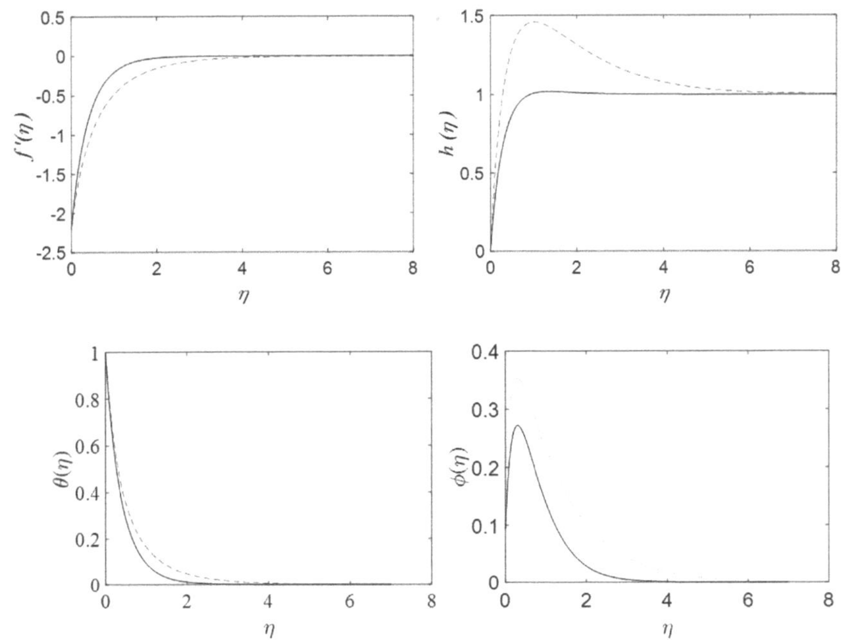

Figure 5.16 First (full line) and second (broken line) solutions for $f'(\eta)$, $h(\eta)$, $\theta(\eta)$, and $\phi(\eta)$ for $\varepsilon = -2.2$, $s = 2$, $Pr = 1$, $Nb = 0.5$, $Nt = 0.3$, and $Sc = 1$. *(From Grosan, T., & Pop, I. (2020). Flow and heat transfer over a permeable biaxial stretching/shrinking sheet in a nanofluid.* Neural Computing and Applications, *32(9), 4575–4582. https://doi.org/10.1007/s00521-018-3770-0 4581.)*

5.4 Numerical solutions of nonalignment stagnation-point flow and heat transfer of a nanofluid over a stretching/shrinking surface in a nanofluid

5.4.1 Introduction

The study of stagnation-point flow of a viscous or non-Newtonian fluid is important in several engineering and industrial applications. Knowledge of the stagnation flow assists in the design of thrust bearings and radial diffusers, drag reduction, and transpiration cooling, as well as thermal oil recovery. In addition, stagnation-point flow is often studied with a stretching and shrinking surface. Wang (2008) has considered this problem and found the nonalignment effect on the flow. According to this author, the nonalignment of the stagnation flow and the stretching/shrinking sheet can destroy the symmetry and complicate the fluid flow. The paper by Wang (2008) has been extended by many authors, such as (Fang et al., 2009), who studied the unsteady stagnation-point flow and solved the problem analytically using the homotopy analysis method. Meanwhile, (Ahmad et al., 2011; Bhattacharyya & Pop, 2011; Bhattacharyya et al., 2012; Lok et al., 2011; Mahapatra & Nandy, 2013) solved the stagnation-point flow problem past a stretching/shrinking surface by adding the effects of the thermal radiation and magnetic field.

5.4.2 Mathematical mode

Following (Hamid et al., 2016), a steady, two-dimensional nanofluid near the stagnation point past a permeable stretching/shrinking sheet as shown in Fig. 5.17 is considered. Here the Cartesian coordinates x and y are measured along and normal to the stretching/shrinking surface, respectively. In this problem, the stretching axis and the stagnation point are not aligned. The wall of the sheet is assumed to have the velocities for the stretching/shrinking, $u_w(x) = b(x + c)$, where $b > 0$ (stretching rate), $b < 0$ (shrinking rate), and c refers to the location of the stretching/shrinking origin. Also, the fluid velocities of inviscid nanofluid are denoted as $u_e(x) = ax$ and $w_e(z) = -az$, where $a>0$ is the strength of the stagnation flow.

Under the standard boundary layer approximations, the governing equations are ((Miklavčič & Wang, 2006) (Kuznetsov & Nield, 2013))

$$\frac{\partial u}{\partial x} + \frac{\partial v}{\partial y} + \frac{\partial w}{\partial z} = 0 \tag{5.47}$$

$$u\frac{\partial u}{\partial x} + v\frac{\partial u}{\partial y} + w\frac{\partial u}{\partial z} = u_e\frac{du_e}{dx} + \nu\frac{\partial^2 u}{\partial z^2} \tag{5.48}$$

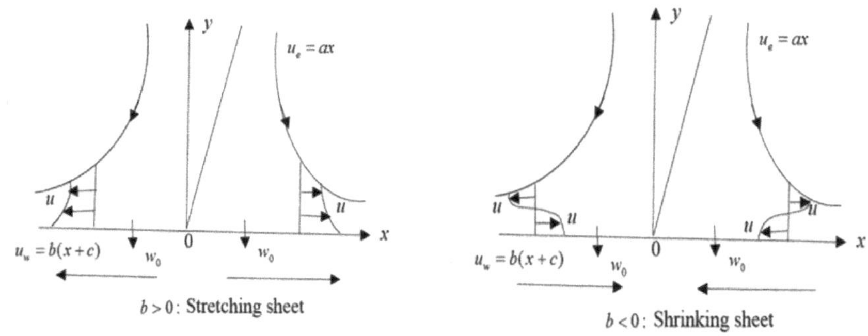

Figure 5.17 Physical model and coordinate system. *(From Hamid, R. A., Nazar, R., & Pop, I. (2016). The non-alignment stagnation-point flow toward a permeable stretching/shrinking sheet in a nanofluid using Buongiorno's model: A revised model.* Zeitschrift für Naturforschung a, *71(1), 81–89. https://doi.org/10.1515/zna-2015-0287 86.)*

$$u\frac{\partial w}{\partial x}+v\frac{\partial w}{\partial y}+w\frac{\partial w}{\partial z}=w_e\frac{dw_e}{dz}+\nu\frac{\partial^2 w}{\partial z^2} \tag{5.49}$$

$$u\frac{\partial T}{\partial x}+v\frac{\partial T}{\partial y}+w\frac{\partial T}{\partial z}=\alpha\frac{\partial^2 T}{\partial z^2}+\tau\left[D_B\frac{\partial C}{\partial z}\frac{\partial T}{\partial z}+\frac{D_T}{T_\infty}\left(\frac{\partial T}{\partial z}\right)^2\right] \tag{5.50}$$

$$u\frac{\partial C}{\partial x}+v\frac{\partial C}{\partial y}+w\frac{\partial C}{\partial z}=D_B\frac{\partial^2 C}{\partial z^2}+\frac{D_T}{T_\infty}\frac{\partial^2 T}{\partial z^2} \tag{5.51}$$

subject to the boundary conditions

$$u=u_w(x)=b(x+c),\ v=0,\ w=w_0,\ T=T_w,\ D_B\frac{\partial C}{\partial z}+\frac{D_T}{T_\infty}\frac{\partial T}{\partial z}=0$$

at $z=0$

$$u\to u_e(x)=ax,\ w\to w_e(z)=-az,\ T\to T_\infty,\ C\to C_\infty \text{ as } z\to\infty \tag{5.52}$$

Here u and v are the velocity components of the nanofluid along the x and y axes, T is the temperature of the nanofluid, C is the nanoparticle volume fraction, and the physical meaning of the other quantities is mentioned in the Nomenclature.

According to (Wang, 2008), the following variables are introduced:

$$u=axf'(\eta)+bcg(\eta),\quad v=0,\quad w=-\sqrt{a\alpha}f(\eta)$$
$$\theta(\eta)=\frac{T-T_\infty}{\Delta T},\quad \phi(\eta)=\frac{C-C_\infty}{\Delta C},\quad \eta=z\sqrt{\frac{a}{\alpha}} \tag{5.53}$$

so that

$$w_w(x)=w_0=-\sqrt{a\alpha}s \tag{5.54}$$

with $s>0$ for suction and $s<0$ for injection, respectively. Then, $\Delta T=T_w-T_\infty$ and $\Delta C=C_w-C_\infty$.

Substituting the similarity variables (5.53) into Eqs. (5.48)−(5.51), it is found that they are reduced to the following ordinary (similarity) differential equations:

$$Pr\,f'''+ff''+1-f'^2=0 \tag{5.55}$$

$$Pr\,g''+fg'-f'g=0 \tag{5.56}$$

$$\theta''+f\theta'+Nb\,\phi'\theta'+Nt\,\theta'^2=0 \tag{5.57}$$

$$\phi''+Le\,f\,\phi'+\frac{Nt}{Nb}\theta''=0 \tag{5.58}$$

subject to the boundary conditions

$$\begin{gathered} f(0) = s, \quad f'(0) = \varepsilon, \quad g(0) = 1, \quad \theta(0) = 1 \\ Nb\, \phi'(0) + Nt\, \theta'(0) = 0 \\ f'(\eta) \to 1, \quad g(\eta) \to 0, \quad \theta(\eta) \to 0, \quad \phi(\eta) \to 0 \text{ as } \eta \to \infty \end{gathered} \tag{5.59}$$

where $\varepsilon = b/a$ is the constant stretching/shrinking parameter where $\varepsilon > 0$ corresponds to the stretching sheet, $\varepsilon < 0$ corresponds to the shrinking sheet, and $\varepsilon = 0$ is for static sheet. The dimensionless constants *Pr*, *Le*, *Nb*, and *Nt* denote the Prandtl number, the nanofluid Lewis number, the Brownian motion parameter, and the thermophoresis parameter, which are defined as

$$Pr = \frac{\nu}{\alpha}, \quad Le = \frac{\alpha}{D_B}, \quad Nb = \frac{\tau D_B \Delta C}{\alpha}, \quad Nt = \frac{\tau D_T \Delta T}{\alpha} \tag{5.60}$$

It is to be noticed that Eqs. (5.55) and (5.56) have been studied by Wang (2008) for an impermeable sheet ($s = 0$) and a viscous fluid.

The physical quantities of interest are the skin friction coefficient C_f and the local Nusselt number Nu_x, which are defined as

$$C_f = \frac{\mu}{\rho u_e^2(x)} \left(\frac{\partial u}{\partial z} \right)_{z=0}, \quad Nu_x = -\frac{x}{(T_w - T_\infty)} \left(\frac{\partial T}{\partial z} \right)_{z=0} \tag{5.61}$$

Using (5.53) and (5.61) we get

$$Re_x^{1/2} C_f = f''(0) + \frac{bc}{u_e(x)} g'(0), \quad Re_x^{-1/2} Nu_x = -\theta'(0) \tag{5.62}$$

where $Re_x = u_x(x)x/\nu$ is the local Reynolds number.

For two-dimensional flow, the dimensionless streamlines ψ can be defined as

$$\psi = xf(\eta) + \lambda c \int_0^\eta g(s)\, ds \tag{5.63}$$

where $\psi = \overline{\psi} \big/ \sqrt{a\alpha}$ with $\overline{\psi}$ defined as $u = \partial \overline{\psi} \big/ \partial z$ and $w = -\partial \overline{\psi} \big/ \partial x$.

5.4.3 Comments

The ordinary differential Eqs. (5.55)–(5.59) were solved numerically by (Hamid et al., 2016) using the *bvp4c* function in Matlab software. The parameters used in this study are Prandtl number $Pr = 10$ (water-based nanofluid), Lewis number $Le = 10$, suction parameter ($s > 0$) is chosen for $s = 10, 13, 15$, the stretching sheet ($\varepsilon > 0$) takes the value $\varepsilon = 3$ and the shrinking sheet ($\varepsilon < 0$) takes the value $\varepsilon = -3$, $Nb = 0.5$, and $Nt = 0.5$.The results ($f''(0)$ and $g'(0)$) obtained by Hamid et al. (2016) are compared with those reported by Wang (2008), where $Nb = Nt = Le = 0$ (classical viscous fluid) and $Pr = 1$ as shown in Table 5.4.

Furthermore, Figs. 5.18 and 5.19 show that the boundary value problem (5.55) to (5.59) admits multiple (dual), an upper and a lower branch solutions. The smallest eigenvalues for some values of s and ε are given in Table 5.5. The lower branch solution is found to exist in the shrinking region where the critical values ε_a are for $s = 10$, ε_b for $s = 13$, and ε_c for $s = 15$ as reported in the previous mentioned figures. The values for the critical numbers ε_a, ε_b, and ε_c are shown in Table 5.6. It is observed that $\varepsilon_a > \varepsilon_b > \varepsilon_c$.

The dimensionless streamlines ψ, defined by relation (5.63) for the stretching and shrinking sheet, is illustrated in Fig. 5.20 when the stagnation flow and the sheet is not aligned ($c = 0.5$) and using different values of the suction parameter s. One can see that in both cases, the suction parameter causes the flows being drag into the center. It is obvious that the nonalignment destroys the symmetry of the flow. On the other hand, for both stretching and shrinking sheets, the fluid suction causes the fluid to move into the same direction and we can see that the flow gets smoother into the downstream. Further, it is also noted that the higher value of

Table 5 4 Values of $f''(0)$ and $g'(0)$ when $Pr = 1$. Results in square brackets [ ] are the second (dual) solutions.

	(Wang, 2008)		(Hamid et al., 2016)	
ε	$f''(0)$	$g'(0)$	$f''(0)$	$g'(0)$
−0.5	1.49567	−0.50145	1.49567	−0.501448
−0.75	1.4893	−0.29376	1.489298	−0.293763
−1	1.32882 [0]	0 [∞]	1.328817 [0]	−0.69756 × 10-7 [∞]
−1.15	1.08223 [0.116702]	0.297995 [0.276345]	1.082231 [0.116702]	0.297995 [2.763446]
0.1	1.14656	−0.86345	1.146561	−0.863452
0.2	1.05113	−0.9133	1.05113	−0.913303
0.5	0.7133	−1.05239	0.713295	−1.051458
1	0	−1.25331	0	−1.253314
5	−10.26475	−2.3381	−10.264749	−2.338099

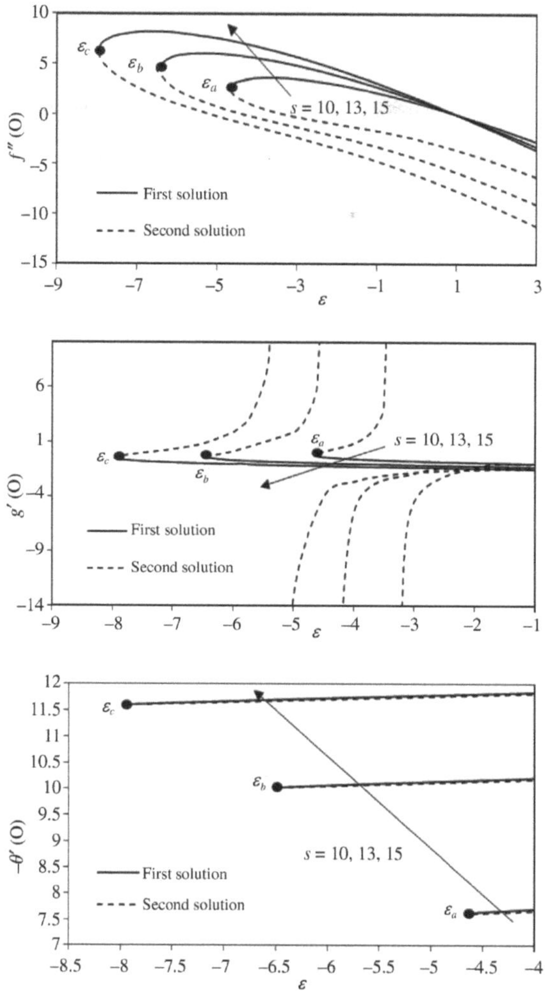

Figure 5.18 Variations of $f''(0)$, $g'(0)$, and $-\theta'(0)$ with ε when $Pr = 10$, $Le = 10$, and $Nb = Nt = 0.5$. *(From Hamid, R. A., Nazar, R., & Pop, I. (2016). The non-alignment stagnation-point flow toward a permeable stretching/shrinking sheet in a nanofluid using Buongiorno's model: A revised model.* Zeitschrift für Naturforschung a, *71(1), 81–89. https://doi.org/10.1515/zna-2015-0287 86.)*

suction causes the flow to be more closely spaced. It is known that density of the streamlines is proportional to the velocity of the flow. As the streamlines are close together, the velocities are higher. Hence, the streamlines in Fig. 5.20 provide evidence that strong fluid suction will increase the velocity of the flow.

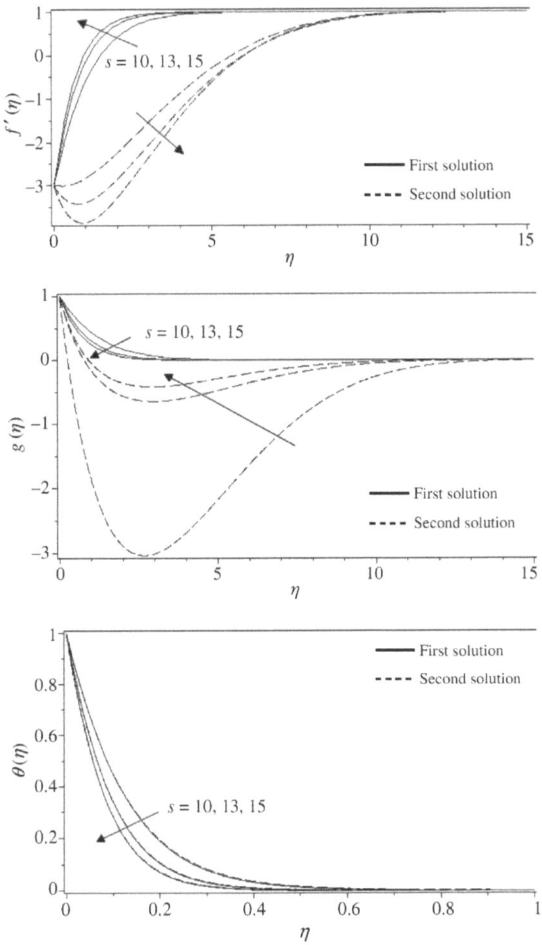

Figure 5.19 Velocity profiles $f'(\eta)$, nonalignment function profiles $g(\eta)$ and temperature profiles $\theta(\eta)$ for ε when $Pr = 10$ and $Le = 10$. *(From Hamid, R. A., Nazar, R., & Pop, I. (2016). The non-alignment stagnation-point flow toward a permeable stretching/shrinking sheet in a nanofluid using Buongiorno's model: A revised model.* Zeitschrift Für Naturforschung a, 71*(1), 81–89. https://doi.org/10.1515/zna-2015-0287 86.)*

5.5 Flow and heat transfer along a permeable stretching/shrinking curved surface in a hybrid nanofluid

5.5.1 Introduction

The flow of a viscous fluid past a curved surface has received great importance. Sajid et al. (2010) studied the linear stretching on a curved

Table 5 5 Smallest eigenvalues for some values of s and ε when $Pr = 10$.

s	ε	First solution	Second solution
10	−4.6	0.8531	−0.8098
	−4.65	0.2139	−0.2111
	−4.653	0.0785	−0.0785
13	−6.5	0.2939	−0.2866
	−6.505	0.0774	−0.0642
	−6.5052	0.0529	−0.0506
15	−7.9	1.1465	−1.0932
	−7.96	0.4282	−0.4195
	−7.969	0.1346	−0.1303

Table 5.6 The critical values of ε

ε_a ($s = 10$)	ε_b ($s = 13$)	ε_c ($s = 15$)
−4.6534	−6.5052	−7.9699

surface by a numerical approach, showing that the boundary layer thickness is greater for a curved surface as compared to a flat surface. They further indicated that the drag force is reduced on a curved surface as compared with the flat surface and the pressure variation is not negligible. The viscous flow over a power law stretching curved surface was investigated numerically by Sanni et al. (2017). The authors showed that a slight variation in the curvature causes a significant increase in the velocity and skin friction coefficient, and the nonlinearity of the stretching velocity also increases the skin friction. Finally, we mention here the paper by Okechi et al. (2017) who have studied the flow of viscous fluid along an exponentially stretching curved surface.

It has been observed that no study has taken place which blends together curved surface and nanofluids, and hybrid nanofluids. The objective of this section is to study the stretching/shrinking of a curved surface in a hybrid nanofluid. Analytic solution for the nonlinear equations is highly improbable and hence numerical solutions are presented.

5.5.2 Basic equations

Following (Waini et al., 2019) we consider the steady two-dimensional boundary layer flow of a hybrid nanofluid over a permeable stretching/shrinking curved surface coiled in a circle of radius R about the curvilinear coordinates (r, S), as it is shown in Fig. 5.21A. The large values of R correspond to a slightly curved surface. The geometry for the stretching

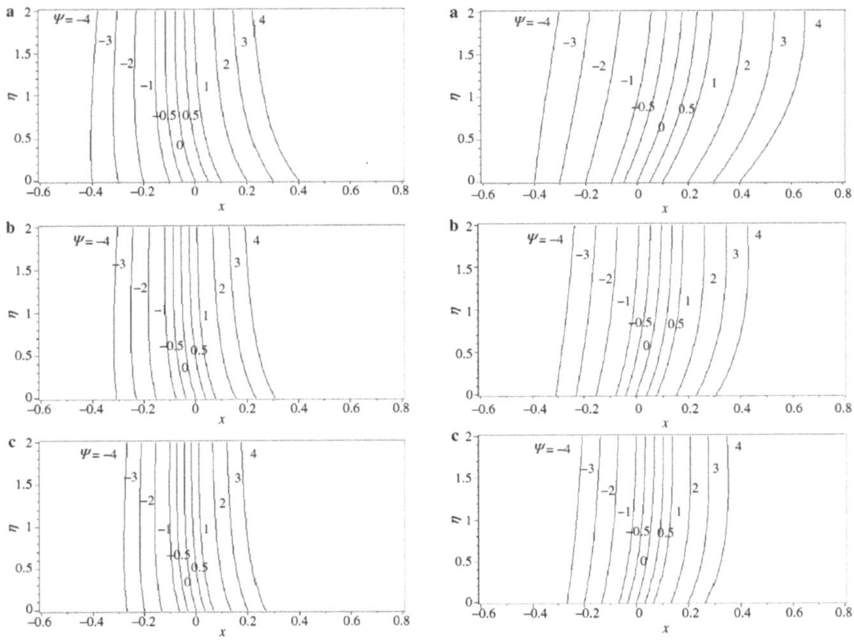

Figure 5.20 Streamlines for two dimensional stretching sheet when $\lambda = 3$ (left) and for two-dimensional shrinking sheet when $\lambda = -3$ (right) and $c = 0.5$ for different values of s: (A) $s = 10$, (B) $s = 13$; and (C) $s = 15$ when $Pr = 10$. *(From Hamid, R. A., Nazar, R., & Pop, I. (2016). The non-alignment stagnation-point flow toward a permeable stretching/shrinking sheet in a nanofluid using Buongiorno's model: A revised model.* Zeitschrift Für Naturforschung a, *71(1), 81–89. https://doi.org/10.1515/zna-2015-0287 86.)*

surface is shown in Fig. 5.21B, while that for the shrinking surface is shown in Fig. 5.21C. However, there is no free stream velocity in both cases.

It is assumed that the surface is stretched/shrunk with the velocity $u_w(s)$ along the s direction, and that the constant mass flux velocity is v_w. The constant temperature of the surface is T_w, while the constant temperature of the ambient hybrid nanofluid is T_∞. Considering the hybrid nanofluid, it is assumed that the size of nanoparticles is uniform, and the effect of the agglomeration of nanoparticles on the thermophysical properties is neglected because the nanofluids are synthesized as a stable mixture of the base fluid and nanoparticles. Under these assumptions along with the boundary layer approximations, Eqs. (1.8)–(1.10) can be written in the Cartesian coordinates (r, S) as

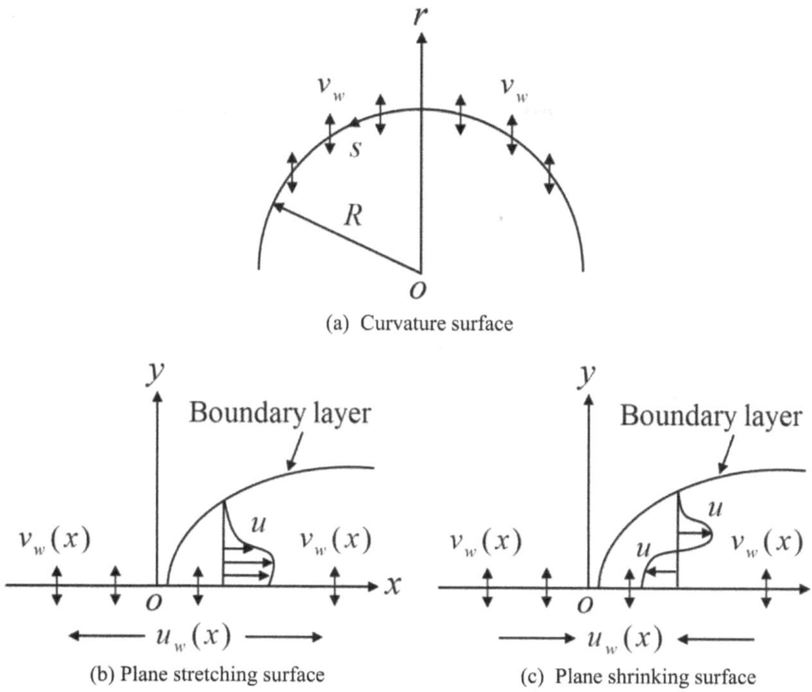

Figure 5.21 Physical model and coordinate system: (A) curvature surface; (B) plane stretching surface; (C) plane shrinking surface. *(From Waini, I., Ishak, A., & Pop, I. (2019). Flow and heat transfer along a permeable stretching/shrinking curved surface in a hybrid nanofluid.* Physica Scripta, *94, 105219. https://doi.org/10.1088/1402-4896/ab0fd5.)*

$$\frac{\partial}{\partial r}[(r+R)v]+R\,\frac{\partial u}{\partial S}=0 \tag{5.64}$$

$$\frac{u^2}{r+R}=\frac{1}{\rho_{hnf}}\frac{\partial p}{\partial r} \tag{5.65}$$

$$v\frac{\partial u}{\partial r}+\frac{R}{r+R}u\frac{\partial u}{\partial S}+\frac{uv}{r+R}=-\frac{1}{\rho_{hnf}}\frac{R}{r+R}\frac{\partial p}{\partial S}+\frac{\mu_{hnf}}{\rho_{hnf}}\left(\frac{\partial^2 u}{\partial r^2}+\frac{1}{r+R}\frac{\partial u}{\partial r}-\frac{u}{(r+R)^2}\right) \tag{5.66}$$

$$v\frac{\partial T}{\partial r}+\frac{R}{r+R}u\frac{\partial T}{\partial S}=\frac{k_{hnf}}{(\rho C_p)_{hnf}}\left(\frac{\partial^2 T}{\partial r^2}+\frac{1}{r+R}\frac{\partial T}{\partial r}\right) \tag{5.67}$$

subject to the boundary conditions

$$v = v_w,\ u = u_w(S)\varepsilon,\ T = T_w \text{ at } r = 0$$
$$u \to 0,\ \frac{\partial u}{\partial r} \to 0,\ T \to T_\infty \quad \text{at } r \to \infty \tag{5.68}$$

where the physical meaning of the quantities in these equations is mentioned in the Nomenclature, and we assume that $u_w(S) = aS$ is a positive constant.

5.5.3 Solution

In order to solve Eqs. (5.64)–(5.67) along with the boundary conditions (5.68), we introduce the following similarity variables (see (Waini et al., 2019)).

$$u = a\,Sf'(\eta),\quad v = -\frac{R}{r+R}\sqrt{a\nu_f}\,f(\eta),\quad p = p_0 + \rho_f a^2 S^2 P(\eta),$$
$$\theta(\eta) = \frac{T - T_\infty}{T_w - T_\infty},\quad \eta = r\sqrt{a/\nu_f} \tag{5.69}$$

where prime denotes differentiation with respect to η. Thus, we take

$$v_w = -\sqrt{a\nu_f}\,s \tag{5.70}$$

Substituting (5.69) into Eqs. (5.64)–(5.67), the following ordinary differential equations (similarity) are obtained:

$$p' = \frac{\rho_{hnf}}{\rho_f}\frac{f'^2}{\eta + K} \tag{5.71}$$

$$\frac{2K}{\eta + K}\frac{\rho_f}{\rho_{hnf}}P = \frac{\mu_{hnf}/\mu_f}{\rho_{hnf}/\rho_f}\left[f''' + \frac{f''}{\eta + K} + \frac{f'}{(\eta + K)^2}\right] +$$
$$\frac{k}{(\eta + K)^2}ff' - \frac{K}{\eta + K}f'^2 + \frac{K}{\eta + K}f'' \tag{5.72}$$

where $K = R\sqrt{a/\nu_f}$ is the constant curvature parameter. Eliminating the pressure P between these equations, one obtains

$$A\left[f^{iV}+\frac{2f'''}{\eta+K}-\frac{f''}{(\eta+K)^2}+\frac{f'}{(\eta+K)^3}+\frac{K}{\eta+K}(ff'''-f'f'')\right]+$$
$$\frac{K}{(\eta+K)^2}(ff''-f'^2)-\frac{K}{(\eta+K)^3}ff'=0$$
$$A=\frac{\mu_{hnf}/\mu_f}{\rho_{hnf}/\rho_f} \tag{5.73}$$

$$B\left(\theta''+\frac{\theta'}{\eta+K}\right)+\frac{K}{\eta+K}f\theta'=0$$
$$B=\frac{1}{Pr}\frac{k_{hnf}/k_f}{(\rho C_p)_{hnf}/(\rho C_p)_f} \tag{5.74}$$

and the boundary conditions (5.68) become

$$\begin{aligned}&f(0)=s,\ f'(0)=\varepsilon,\quad \theta(0)=1\\&f'(\eta)\to 0,\quad f''(\eta)\to 0,\quad \theta(\eta)\to 0 \text{ as } \eta\to\infty\end{aligned} \tag{5.75}$$

The quantity of physical interest in this problem are the skin friction coefficient C_f and the local Nusselt number Nu_s, which are defined as

$$C_f=\frac{\mu_{hnf}/\mu_f}{\rho_f u_w^2(S)}\left(\frac{\partial u}{\partial r}-\frac{u}{r+R}\right)_{r=0}\qquad Nu_S=\frac{S\,k_{hnf}/k_f}{T_w-T_\infty}\left(-\frac{\partial T}{\partial r}\right)_{r=0} \tag{5.76}$$

Using (5.69) and (5.76), we get

$$Re_S^{1/2}C_f=\frac{\mu_{hnf}}{\mu_f}\left(f''(0)-\frac{\lambda}{K}\right),\quad Re_S^{1/2}\,Nu_S=-\frac{k_{hnf}}{k_f}\theta'(0) \tag{5.77}$$

where $Re_S=u_w(S)\ S/\nu_f$ is the local Reynolds number.

In the case of the plane stretching/shrinking surface ($K\to\infty$) in a regular fluid ($\varphi_1=\varphi_2=0$), the skin friction coefficient in Eq. (5.77) reduces to

$$Re_S^{1/2}C_f=f''(0) \tag{5.78}$$

It is worth mentioning that by taking $K\to\infty$ and $\varphi_1=\varphi_2=0$ (regular fluid), Eq. (5.55) in the absence of pressure gradient reduces to the classical problem of flat stretching surface as discussed by Crane (1970). Moreover, for $K \to \infty$ when $\beta = 0$ (steady case) and $\varphi_1 = \varphi_2 = 0$ (regular fluid) in the absence of pressure gradient over a shrinking surface, Eq. (5.55) can be reduced to Eq. (6) from the paper by (Fang & Zhang, 2009).

5.5.4 Stability analysis

We test the fact that the dual solutions of the boundary value problem (5.73–5.75) are stable or unstable by performing a stability analysis, that is similar with that presented in Section 2.1.5. Thus, we have, finally, to solve the following eigenvalue problem:

$$A\left[F^{iv} + \frac{2}{\eta + K}F''' - \frac{1}{(\eta + K)^2}F'' - \frac{1}{(\eta + K)^3}F' + \frac{K}{\eta + K}(fF''' - f'F'' - f''F' + f'''F)\right] + \frac{K}{(\eta + K)^2}(fF'' - 2f'F' + f'''F)$$

$$-\frac{1}{(\eta + K)^3}(fF' + f'F) + \frac{\gamma}{(\eta + K)}F' + \gamma F' = 0 \tag{5.79}$$

$$B\left(G'' + \frac{1}{\eta + K}G'\right) + \frac{K}{\eta + K}(fG' + \theta' F) + \gamma\, G = 0 \tag{5.80}$$

along with the boundary conditions

$$\begin{aligned} &F(0) = 0,\ F'(0) = 0,\ G(0) = 0 \\ &F'(\eta) \to 0, F''(\eta) \to 0, \quad G(\eta) \to 0 \text{ as } \eta \to \infty \end{aligned} \tag{5.81}$$

For particular set of values Pr, s, K, and ε, the stability of the steady boundary value problem (5.79–5.81) is analyzed by using the smallest eigenvalues γ.

5.5.5 Comments

The resulting nonlinear ordinary differential equations, Eqs. (5.56) and (5.57) along with the boundary conditions (5.58), were solved by Waini et al. (2019), using the boundary value problem solver, bvp4c, available in the Matlab software. The details of this method can be found in (Shampine et al., 2003). We mention that (Aly, 2016; Saqib et al., 2018a, 2018b, Saqib, Ali, Khan, Sheikh, & Khan, 2019, Saqib, Ali, Khan, Sheikh, & Shafie, 2019) used the Atangana–Baleanu fractional derivative without singular and the nonlocal kernel in the mathematical formulation to get the time fractional governing equations subject to physical initial and boundary conditions. The numerical results obtained are validated in Tables 5.7 and 5.8. Table 5.7 displays the values of $-Re_S^{1/2}C_f$ for several values of K when $\varphi_1 = \varphi_2 = 0$ (regular fluid), $s = 0$ and $\varepsilon = 1$ (stretching case). The numerical results are obtained for the case of Prandtl number, $Pr- = 6.2$ (water), which is fixed throughout this problem (see (Khanafer et al., 2003; Oztop & Abu-Nada, 2008)). It is observed that the present results are in a good agreement with the solutions obtained by (Abbas et al., 2013; Rosca et al., 2015; Pop et al., 2016). Table 5.8 shows the values of $Re_S^{1/2}C_f$ for the shrinking case ($\varepsilon < 0$) when $\varphi 1 = \varphi 2 = 0$ (regular fluid), $s = 0$ and $K \rightarrow \infty$. The present numerical solutions of $Re_S^{1/2}C_f$ are compared with the

Table 5 7 Values of $-Re_S^{1/2}C_f$ for several values of K when $\varphi_1 = \varphi_2 = 0$ (regular fluid), $s = 0$ and $\varepsilon = 1$ (stretching case).

K	Abbas et al. (2013)	Rosca and Pop (2015)	Pop et al. (2016)	Waini et al. (2019)
5	1.15763	1.15076	1.15763	1.15077
10	1.07349	1.07172	1.07349	1.07173
20	1.03561	1.03501	1.03561	1.03501
30	1.02353	1.02315	1.02353	1.02316
40	1.01759	1.01729	1.01759	1.0173
50	1.01405	1.0138	1.01405	1.01381
100	1.00704	1.00687	1.00704	1.00687
200	1.00356	1.00342	1.00356	1.00343
1000	1.00079	1.00068	1.00079	1.00068

Table 5 8 Values of $Re_S^{1/2}C_f$ for $\varepsilon < 0$ (shrinking case) when $\varphi_1 = \varphi_2 = 0$ (regular fluid), $s = 0$ and $K \rightarrow \infty$

	Rosca and Pop (2015)		Pop et al. (2016)		Waini et al. (2019)	
ε	First solution	Second solution	First solution	Second solution	First solution	Second solution
−0.5	0.85355	0.14644	0.85355	0.14645	0.85355	0.14645
−0.6	0.97947	0.22052	0.97947	0.22053	0.97947	0.22053
−0.7	1.0834	0.31659	1.08341	0.31659	1.08341	0.31659
−0.75	1.125	0.375	1.125	0.375	1.125	0.375
−0.8	1.15777	0.44222	1.15777	0.44223	1.15777	0.44223
−0.9	1.1846	0.61539	1.18461	0.6154	1.18461	0.6154
−0.95	1.16242	0.73757	1.16243	0.73757	1.16243	0.73757
−0.99	1.089	0.891	1.089	0.891	1.089	0.891

analytical solutions obtained by Rosca et al. (2015) and Pop et al. (2016), which show a good agreement.

The effects of several parameters on the skin friction coefficient $Re_S^{1/2}C_f$ and the local Nusselt number $Re_S^{-1/2}Nu_S$ as well as the velocity $f'(\eta)$ and temperature $\theta(\eta)$ profiles are presented in graphical form through Figs. 5.22–5.25. Figs. 5.22 and 5.23 show the variation of $Re_S^{1/2}C_f$ and

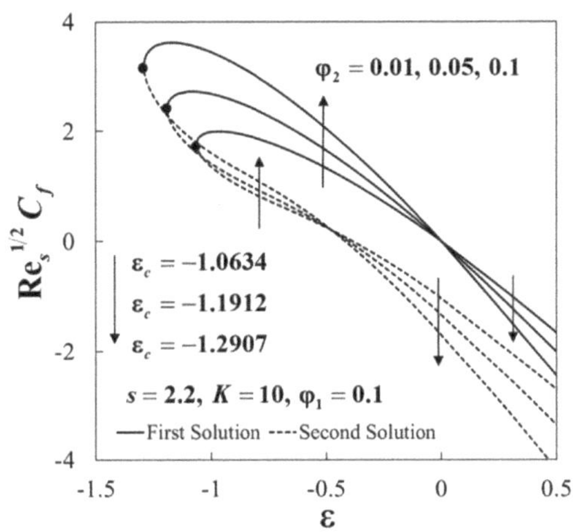

Figure 5.22 Variation of $Re_S^{1/2}C_f$ with ε for various values of φ_2. *(From Waini, I., Ishak, A., & Pop, I. (2019). Flow and heat transfer along a permeable stretching/shrinking curved surface in a hybrid nanofluid.* Physica Scripta, 94, *105219. https://doi.org/10.1088/1402-4896/ab0fd5.)*

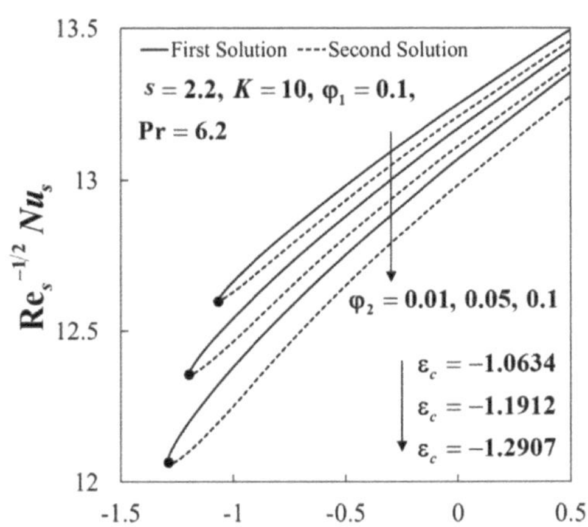

Figure 5.23 Variation of $Re_S^{1/2}Nu_s$ with ε for various values of φ_2. *(From Waini, I., Ishak, A., & Pop, I. (2019). Flow and heat transfer along a permeable stretching/shrinking curved surface in a hybrid nanofluid.* Physica Scripta, 94, *105219. https://doi.org/10.1088/1402-4896/ab0fd5.)*

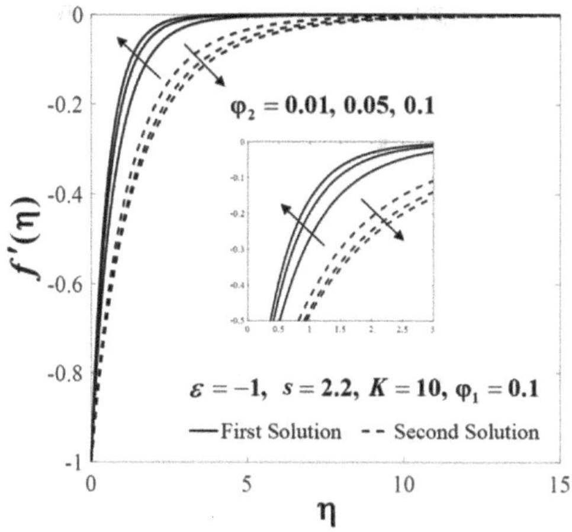

Figure 5.24 The velocity profiles $f'(\eta)$ for various values of φ_2. *(From Waini, I., Ishak, A., & Pop, I. (2019). Flow and heat transfer along a permeable stretching/shrinking curved surface in a hybrid nanofluid.* Physica Scripta, 94, *105219. https://doi.org/10.1088/1402-4896/ab0fd5.)*

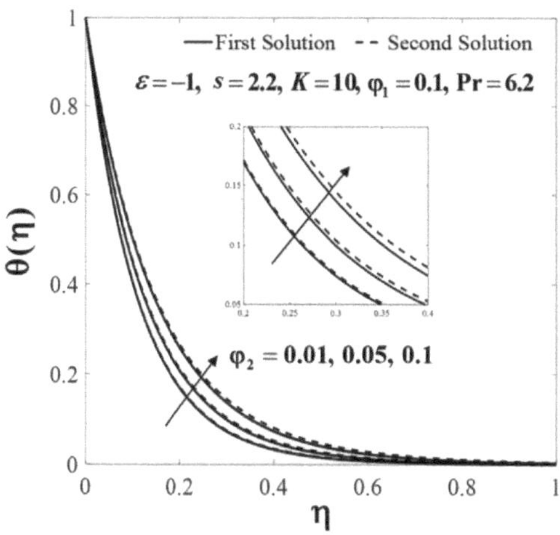

Figure 5.25 The temperature profiles $\theta(\eta)$ for various values of φ_2. *(From Waini, I., Ishak, A., & Pop, I. (2019). Flow and heat transfer along a permeable stretching/shrinking curved surface in a hybrid nanofluid.* Physica Scripta, 94, *105219. https://doi.org/10.1088/1402-4896/ab0fd5.)*

$Re_S^{-1/2}\,Nu_S$ with ε for various values of φ_2 when $s = 2.2$, $K = 10$, $\varphi 1 = 0.1$, and $Pr = 6.2$. From these figures, it can be observed that dual solutions occur for both cases of stretching ($\varepsilon > 0$) and shrinking ($\varepsilon < 0$) surface. This observation is different for the case of the plane surface

($K \to \infty$) where no dual solutions exist for the stretching surface ($\varepsilon > 0$). The dual solutions only exist when the value of ε is equal to a certain critical value ε_c, that is the value where the first (upper branch) solution meets the second (lower branch) solution. Due to the separation of the boundary layer from the surface, no similarity solutions exist beyond this critical value ε_c, and the solutions based upon the boundary layer approximations do not exist. We observe that with the increase of φ_2 and K, the solution domain expands with ε_c moving slightly to the left. Based on the computations made by Waini et al. (2019), the critical value of ε for $\varphi_2 = 0.01, 0.05$, and 0.1 are $\varepsilon_c = -1.0634, -1.1912$, and -1.2907 as shown in Figs. 5.25 and 5.26, respectively. In the same figures, dual behavior is observed for the values of $Re_S^{1/2} C_f$ in a certain range of ε, while the values of $Re_S^{-1/2} Nu_S$ decrease with the increasing values of φ_2 for both branches. Meanwhile, the critical value of ε for $K = 5$, 10, and 20 are $\varepsilon_c = -1.1175$, -1.2907, and -1.3816 as shown in Figs. 5.22 and 5.23. We also notice that both values of $Re_S^{1/2} C_f$ and $Re_S^{-1/2} Nu_S$ increase for the upper branch, while decrease for the lower branch with the increasing K.

The velocity profiles $f'(\eta)$ and the temperature profiles $\theta(\eta)$ for various values of the solid volume fractions φ_2 when $\varepsilon = -1$, $s = 2.2$, $K = 10$, $\varphi_1 = 0.1$, and $Pr- = 6.2$ are illustrated in Fig. 5.25. These figures show that there exist two different profiles for a particular value of φ_2. The increment of the velocity and the temperature for the upper branch is observed,

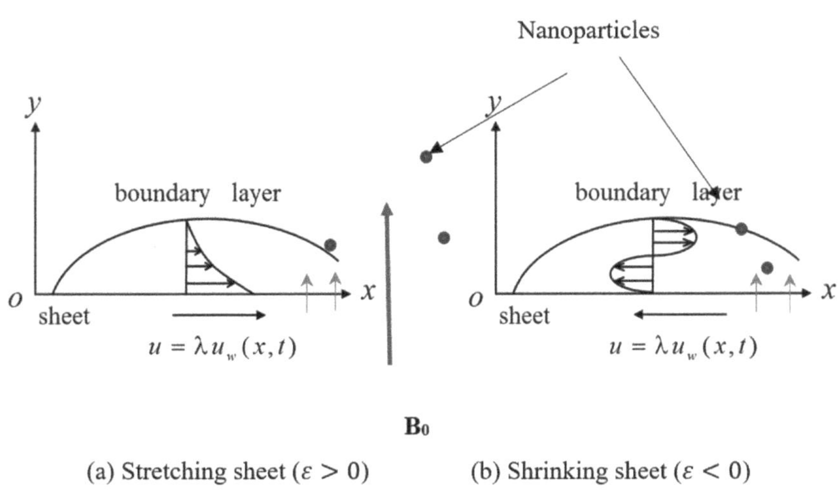

Figure 5.26 Physical models and coordinate systems.

meanwhile the velocity decreases but the temperature increases for the lower branch solution with the increasing φ_2.

5.6 MHD flow and heat transfer over a permeable stretching/shrinking sheet in a hybrid nanofluid with a convective boundary condition

5.6.1 Basic equations

Let us consider the steady two-dimensional magnetohydrodynamic (MHD) boundary layer flow and heat transfer of a hybrid nanofluid over a permeable flat plate as shown in Fig. 5.26, where x and y axes are the Cartesian coordinates with the x-axis measured along the plate and y-axis is normal to it, the plate being located at $y = 0$. It is assumed that the velocity of the stretching/shrinking surface is $u_w(x)$ and the velocity of mass flux is v_0. A uniform transverse magnetic field of strength B_0 is applied parallel to the y-axis. It is also assumed that the hybrid nanofluid is electrically conducting and the magnetic Reynolds number is small so that the induced magnetic field is neglected. No electric field is assumed to exist. The temperature of the surface is due to convective heating process which is characterized by T_f and a heat transfer coefficient h, while the constant temperature of the ambient hybrid nanofluid is T_∞.

The governing equations of the conservations of the hybrid nanofluids' mass, momentum, and energy under the boundary layer approximation are written as (see (Devi & Davi, 2017))

$$\frac{\partial u}{\partial x} + \frac{\partial v}{\partial y} = 0 \tag{5.82}$$

$$u\frac{\partial u}{\partial x} + v\frac{\partial u}{\partial y} = \frac{\mu_{hnf}}{\rho_{hnf}}\frac{\partial^2 u}{\partial y^2} - \frac{\sigma_{hnf} B_0^2}{\rho_{hnf}} u \tag{5.83}$$

$$u\frac{\partial T}{\partial x} + v\frac{\partial T}{\partial y} = \frac{k_{hnf}}{\left(\rho C_p\right)_{hnf}}\frac{\partial^2 T}{\partial y^2} \tag{5.84}$$

subject to

$$\begin{aligned} v = v_0,\, u = u_w(x)\varepsilon,\ -k_{hnf}\frac{\partial T}{\partial y} = h_f\left(T_f - T\right) \text{ at } y = 0 \\ u \to 0, \quad T \to T_\infty \end{aligned} \tag{5.85}$$

where the physical meaning of the quantities mentioned here is given in Nomenclature, and we assume that $U_w(x) = ax$, where a is a positive constant.

In order to solve the boundary value problem (5.82–5.85), we introduce the following similarity variables:

$$u = U_w(x) f'(\eta), \quad v = -\sqrt{a\nu_f} f(\eta),$$
$$\theta(\eta) = \frac{T - T_\infty}{T_f - T_\infty}, \quad \eta = y\sqrt{a/\nu_f} \tag{5.86}$$

Thus,

$$v_0 = -\sqrt{a\nu_f}\, s \tag{5.87}$$

Invoking the similarity variables (5.86), Eqs. (5.82)–(5.84) along with the boundary conditions (5.85) are transformed into the following ordinary (similarity) differential equations:

$$\beta_2 f''' + \beta_1 (ff'' - f'^2) - \beta_3 M f' = 0 \tag{5.88}$$

$$\beta_6 \theta'' + \beta_4 f \theta' = 0 \tag{5.89}$$

subject to the boundary conditions

$$f(0) = s, \quad f'(0) = \varepsilon, \ -\beta_5 \theta'(0) = Bi\,[1 - \theta(0)]$$
$$f'(\eta) \to 0, \quad \theta(\eta) \to 0 \text{ as } \eta \to \infty \tag{5.90}$$

where $M = \frac{\sigma_f B_0^2}{a\rho_f}$ is the constant magnetic parameter and $\beta_i, i = 1$ to 6 are defined as

$$\beta_1 = \frac{\rho_{hhf}}{\rho_f}, \quad \beta_2 = \frac{\mu_{hnf}}{\mu_f}, \quad \beta_3 = \frac{\sigma_{hnf}}{\sigma_f}, \quad \beta_4 = \frac{(\rho C_p)_{hnf}}{(\rho C_p)_f},$$
$$\beta_5 = \frac{k_{hnf}}{k_f}, \quad \beta_6 = \frac{1}{Pr}\frac{\beta_5}{\beta_4} \tag{5.91}$$

It is seen that for $Bi>>1$ (large), Eq. (5.90) reduces to $\theta(0) = 1$ (isothermal stretching/shrinking surfaces). It is worth mentioning that when $\varphi_1 = \varphi_2 = 0$ and $\varepsilon = -1$(shrinking sheet), Eq. (5.88) reduces to Eq. (6) from Fang and Zhang (2009). Further, at $\varphi_2 = 0$ and $\varepsilon = \pm 1$ (stretching/shrinking sheet) Eqs. (5.88) and (5.89) matches with Eqs. (11) and (12) in (Aly, 2016).

The physical quantities of interest are the skin friction coefficient C_f and the local Nusselt number Nux, which are defined as

$$C_f = \frac{\mu_{hnf}}{\rho_f u_w^2(x)}\left(-\frac{\partial u}{\partial y}\right)_{y=0}, \quad Nu_x = \frac{x k_{hnf}}{k_{ff}\left(T_f - T_\infty\right)}\left(-\frac{\partial T}{\partial y}\right)_{y=0} \tag{5.92}$$

On using (5.86) and (5.92), we get

$$Sr = \sqrt{Re_x} C_f = -\beta_1 f''(0), \quad Nur = \frac{Nu}{\sqrt{Re_x}} = -\beta_5\, \theta'(0) \tag{5.93}$$

5.6.2 Exact analytical solutions

5.6.2.1 Exact solution of the dimensionless stream function *f(η)*

On considering the boundary conditions (5.90), *f(η)* can be then obtained as

$$f(\eta) = s + \frac{\varepsilon}{\alpha_1}\left(1 - e^{-\alpha_1 \eta}\right) \tag{5.94}$$

where $\alpha 1$ is a constant to be determined and has to be positive to give a physical meaning. Therefore, by substituting (5.94) in Eq. (5.88), $\alpha 1$ can be obtained as follows:

$$\alpha_1 = \frac{1}{2\beta_2}\left[\beta_1 s \pm \sqrt{\beta_1^2 (s^2 + 4\beta_2(\beta_1 \varepsilon + \beta_3 M)}\,\right] \tag{5.95}$$

Hence, from (5.93), the reduced skin friction coefficient (*Sr*) is obtained as

$$Sr = \beta_1 \alpha_1 \varepsilon \tag{5.96}$$

5.6.2.2 Exact solution of the dimensionless temperature *θ(η)*

On substituting (5.94) into Eq. (5.89), we have

$$\frac{\theta''}{\theta'} = -\frac{1}{\beta_6}\left[S + \frac{\varepsilon}{\alpha_1}\left(1 - e^{-\alpha_1 \eta}\right)\right] \tag{5.97}$$

By integrating this equation twice, applying the boundary conditions (5.90) for θ and θ', we get

$$\theta(\eta) = \theta(0)\, \frac{\Gamma(\gamma_1; 0; \gamma_2 e^{-\alpha_1 \eta})}{\Gamma(\gamma_1; 0; \gamma_2)} \tag{5.98}$$

where Γ is the generalized incomplete gamma function and $\gamma 1$ and $\gamma 2$ are constants defined as

$$\gamma_1 = \frac{\alpha_1 s + \varepsilon}{\beta_6 \alpha_1^2} > 0, \quad \gamma_2 = \frac{\varepsilon}{\beta_6 \alpha_1^2} \tag{5.99}$$

By substituting (5.98) in the third relation of (5.90), one can obtain $\theta(0)$ and hence Nur from Eq. (5.93) as

$$\theta(0) = \frac{1}{1 + \beta_5 \beta_7}, \quad Nur = Bi \left(1 + \frac{1}{\beta_5 \beta_7}\right)^{-1} \tag{5.100}$$

where

$$\beta_7 = \frac{\alpha_1 e^{-\gamma_2} (\gamma_2)^{\gamma_1}}{Bi\, \Gamma(\gamma_1; 0; \gamma_2)} \tag{5.101}$$

5.6.2.3 Conditions to get the flow and temperature solutions (case of the stretching sheet, ε > 0)

In this case, on studying the discriminant of Eq. (5.95) for any values of M and s, one can note that only the positive sign of the root makes α_1 positive. This means that there exists a unique solution of the boundary value problem (5.88–5.90) for any combination of the considered parameters.

5.6.3 Comments

Table 5.9 presents the comparison between the results reported by Aly and Pop (2019) with those of (Aly, 2016) when $\varphi_2 = 0$, $M = 0$, $Pr = 6.2$, $s = 0$ and $\varepsilon = 1$ (stretching sheet).

Fig. 5.27 shows the regions of no and unique solutions of the temperature profiles $\theta(\eta)$ for M as a function of s for stretching sheet ($\varepsilon > 0$) when $\varphi_1 = 0.1$; $\varphi_2 = 0.04$; (a) $\varepsilon = 1$ and (b) $\varepsilon = 0.5,1,2,3$. Further, it is seen from this figure that the regions of obtaining no and unique solutions decrease and increase, respectively, with an increase of ε. On taking these results into account, the effects of various physical parameters over a

Table 5.9 Comparison results of - $f''(0)$ and -$\theta'(0)$.

	- $f''(0)$		- $\theta'(0)$	
φ_1	(Aly and Pop, 2019)	(Aly, 2016)	(Aly and Pop, 2019)	(Aly, 2016)
0.05 0.10	1.005377436	1.0053774	1.622462563	1.6224626
0.15 0.20	0.998771987	0.9987720	1.491697885	1.4916979
	0.981844545	0.9818445	1.375427827	1.3754278
	0.955918805	0.9559188	1.271181062	1.2711811

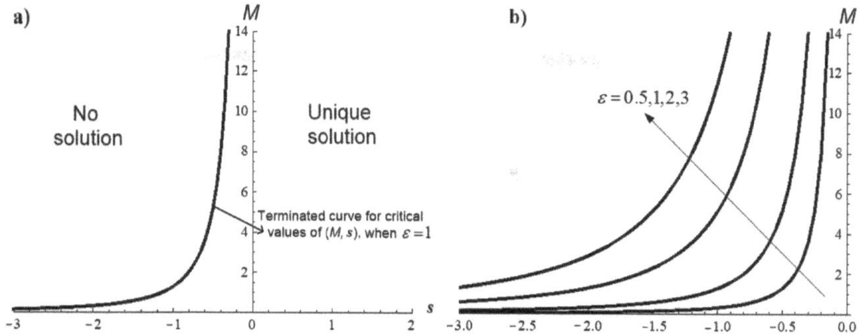

Figure 5.27 Regions of no and unique solutions of temperature for M as a function of s, for stretching sheet when $\varphi_1 = 0.1$, $\varphi_2 = 0.04$; (A) $\varepsilon = 1$ and (B) $\varepsilon = 0.5, 1, 2, 3$. *(From Aly, E. H., & Pop, I. (2019). MHD flow and heat transfer over a permeable stretching/shrinking sheet in a hybrid nanofluid with a convective boundary condition.* International Journal of Numerical Methods for Heat & Fluid Flow, *29(9), 3012–3038. https://doi.org/10.1108/HFF-12-2018-0794 3027.)*

stretching sheet can be studied to compare the flow of hybrid nanofluid versus nanofluid. This is done in the paper.

Fig. 5.28 illustrates the dual solution profiles of the reduced skin friction coefficient $|Sr|$ in absolute sense, of hybrid nanofluid ($Cu - Al_2O_3$/water),

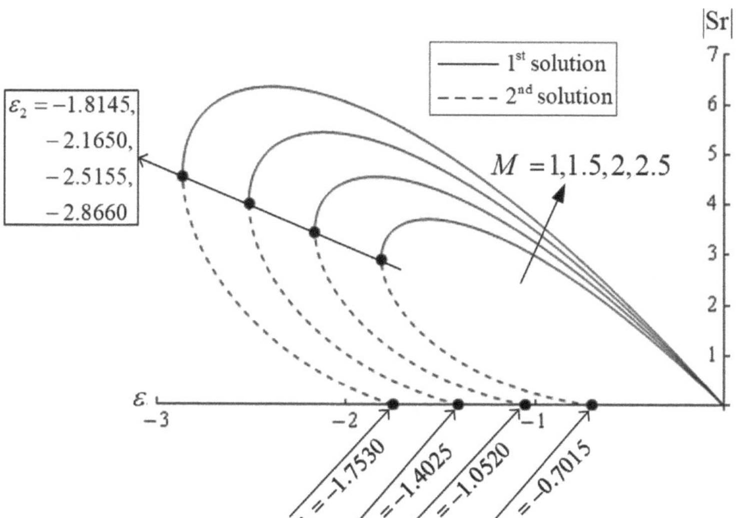

Figure 5.28 Dual solution profiles of the reduced skin friction coefficient $|Sr|$ of hybrid nanofluid, given by Eq. (5.79), as a function of $\varepsilon < 0$ (shrinking sheet) for different values of M when $s = 2.24$. *(From Aly, E. H., & Pop, I. (2019). MHD flow and heat transfer over a permeable stretching/shrinking sheet in a hybrid nanofluid with a convective boundary condition.* International Journal of Numerical Methods for Heat & Fluid Flow, *29(9), 3012–3038. https://doi.org/10.1108/HFF-12-2018-0794 3027.)*

given by Eq. (5.96), as a function of $\varepsilon < 0$ (shrinking sheet) for different values of M when $s = 2$. From this figure, one can see that there is no physical solution when $\varepsilon < \varepsilon_2$; two equal solutions when $\varepsilon = \varepsilon_2$, where the first and second solutions bifurcate; dual solution in the range $\varepsilon_2 < \varepsilon \leq \varepsilon_1$ and a unique solution when $\varepsilon_1 < \varepsilon \leq 0$. On the other hand, Fig. 5.28 shows that an increase in the value of the magnetic parameter M leads to an increase of $|Sr|$. The presence of magnetic field produces a resistive Lorentz force toward the flow which causes the slowing movement of the fluid and thus increase $|Sr|$. Moreover, Fig. 5.29 indicates the regions of no and unique or identical physical solutions of the flow/temperature combination for hybrid nanofluid ($Cu-Al_2O_3$/water) and nanofluid (Al_2O_3/water) when $\varphi_1 = 0.1$, $\varphi_2 = 0.04$, and $\varepsilon = -1$ (shrinking sheet).

Finally, we mention that (Aly & Pop, 2020) have numerically studied, the steady MHD stagnation-point flow and heat transfer over a stretching/

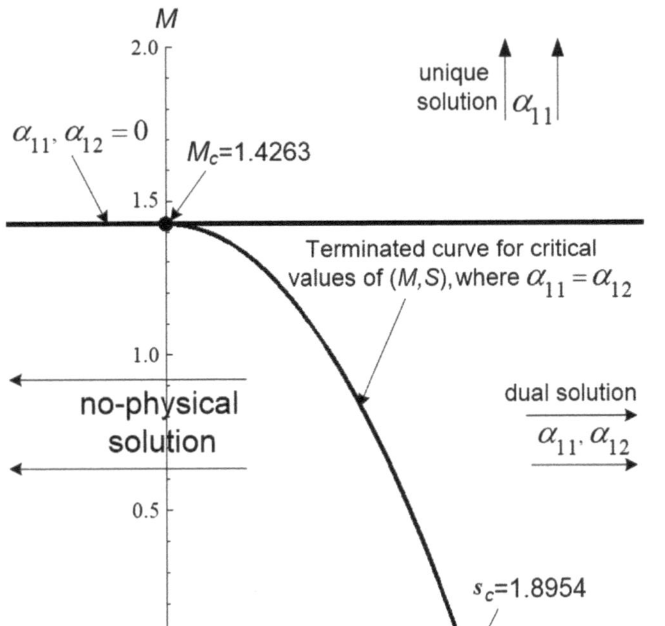

Figure 5.29 Regions of no, unique, and dual solutions of M as a function of s when $\varepsilon = -1$ (shrinking sheet), $\varphi_1 = 0.1$ and $\varphi_2 = 0.04$. *(From Aly, E. H., & Pop, I. (2019). MHD flow and heat transfer over a permeable stretching/shrinking sheet in a hybrid nanofluid with a convective boundary condition.* International Journal of Numerical Methods for Heat & Fluid Flow, *29(9), 3012–3038. https://doi.org/10.1108/HFF-12-2018-0794 3027.)*

shrinking surface in a hybrid nanofluid with partial slip and viscous dissipation. The governing equations of the problem were turned into ODEs by using appropriate transformations. The plots of most important parameters were presented for both surfaces. It was found that on increasing the magnetic parameter, the hybrid nanofluid is better as a heater rather than nanofluid; however, it is better as a cooler on rising Eckert number, the stretching and slip parameters. For shrinking sheet, critical regions of unique/dual solution have been introduced. Further, stability of the unique/dual solution was studied where three regions of stability were noticed. It was deduced that the velocity and temperature are sensitively available close to the terminated line in the dual solution region. It was also proved that their behavior is absolutely different over the three regions of stability.

Nomenclature

a	positive constant (Sections 5.1, 5.5, and 5.6)
a	parameter measuring the strength of the stagnation-point flow $(LT)^{-1}$ (Section 5.2)
A	unsteadiness parameter
b	stretching/shrinking rate
B_0	uniform transverse magnetic field
Bi	Biot number
c	constant, refers to the location of the stretching/shrinking origin
C	the nanoparticle volume fraction
C_f	skin friction coefficient
C_w	constant surface concentration of the sheet
C_∞	nanoparticle volume fraction of the ambient fluid
D_B	inertial resistance coefficient (m^2/s)
D_T	thermophoretic diffusion coefficient (m^2/s)
$f'(\eta)$, $g(\eta)$, $h(\eta)$	dimensionless velocity functions
k	thermal conductivity (W/mK)
K	constant curvature parameter
Le	local Lewis number
M	magnetic parameter
Nb	Brownian motion parameter
Nt	thermophoresis parameter
Nur	reduced Nusselt number
Nu_x, Nu_z, Nu_S	local Nusselt numbers
P	pressure
Pe	Péclet number
Pr	Prandtl number
q_w	heat flux

R	radius (m)
(r, S)	curvilinear coordinates
(r, ϑ, z)	cylindrical coordinates
Re_x, Re_S	local Reynolds numbers
s	constant mass transfer velocity (suction/injection parameter)
Sc	Schmidt number
Sh_x	local Sherwood number
Sr	reduced skin friction coefficient
t	dimensional time (s)
T	dimensional fluid temperature (K)
T_w	temperature of the stretching/shrinking surface (K)
T_∞	temperature of the ambient fluid (K)
u, v	velocity components in x- and y-directions (m/s)
u_e, U	velocity of the ambient fluid (m/s)
u_w, U_w, v_w	velocity of the stretching/shrinking sheet (m/s)
v_0	constant mass flux velocity ($m\ s^{-1}$)
v_w	wall mass transfer velocity (Section 5.1)
w, w_0	mass flux velocity in z-direction

Greek letters

α	thermal diffusivity (m^2/s)
$\alpha 1$	constant (Section 5.6)
β_i	constants (Section 5.6)
γ	parameter related to the unsteadiness of the physical problem (Section 5.1)
γ	the smallest eigenvalue (Section 5.1)
γ_1, γ_2	constants (Section 5.1)
ε	stretching/shrinking parameter
$\theta(\eta)$	dimensionless temperature profile
η	dimensionless similarity variable
τ_{wx}	shear stress
ν	kinematic viscosity (m^2/s)
ρ	density of fluid (kg/m^3)
φ	volume fraction of the nanofluid
$\phi(\eta)$	nanoparticles' concentration
ψ	stream function

Subscript

c	critical point
e, ∞	condition outside of the boundary layer
f	fluid phase
hnf	hybrid nanofluid
nf	nanofluid
w	condition at the wall

Superscript

′ differentiation with respect to η

References

Abbas, Z., Naveed, M., & Sajid, M. (2013). Heat transfer analysis for stretching flow over a curved surface with magnetic field. *Journal of Engineering and Thermophysics, 22*(4), 337–345. https://doi.org/10.1134/S1810232813040061

Agrawal, H. L. (1957). A new exact solution of the equations of viscous motion with axial symmetry. *Quarterly Journal of Mechanics & Applied Mathematics, 10*(1), 42–44. https://doi.org/10.1093/qjmam/10.1.42

Ahmad, K., Nazar, R., & Pop, I. (2011). Boundary layer flow and heat transfer of a micropolar fluid near the stagnation point on a stretching vertical surface with prescribed skin friction. *International Journal of Minerals, Metallurgy, and Materials, 18*(4), 502–507. https://doi.org/10.1007/s12613-011-0469-y

Aly, E. H. (2016). Existence of the multiple exact solutions for nanofluid flow over a stretching/shrinking sheet embedded in a porous medium at the presence of magnetic field with electrical conductivity and thermal radiation effects. *Powder Technology, 301*, 760–781. https://doi.org/10.1016/j.powtec.2016.06.024

Aly, E. H., & Pop, I. (2019). MHD flow and heat transfer over a permeable stretching/shrinking sheet in a hybrid nanofluid with a convective boundary condition. *International Journal of Numerical Methods for Heat and Fluid Flow, 29*(9), 3012–3038. https://doi.org/10.1108/HFF-12-2018-0794

Aly, E. H., & Pop, I. (2020). MHD flow and heat transfer near stagnation point over a stretching/shrinking surface with partial slip and viscous dissipation: Hybrid nanofluid versus nanofluid. *Powder Technology, 367*, 192–205. https://doi.org/10.1016/j.powtec.2020.03.030

Bhattacharyya, K., Mukhopadhyay, S., Layek, G. C., & Pop, I. (2012). Effects of thermal radiation on micropolar fluid flow and heat transfer over a porous shrinking sheet. *International Journal of Heat and Mass Transfer, 55*(11–12), 2945–2952. https://doi.org/10.1016/j.ijheatmasstransfer.2012.01.051

Bhattacharyya, K., & Pop, I. (2011). MHD boundary layer flow due to an exponentially shrinking sheet. *Magnetohydrodynamics, 47*(4), 337–344. https://doi.org/10.22364/mhd

Bozhko, A. A., & Suslov, S. A. (2018). *Convection in ferro-nanofluids: Experiments and theory* (Vol. 40). Springer International Publishing. https://doi.org/10.1007/978-3-319-94427-2

Buongiorno, J. (2006). Convective transport in nanofluids. *Journal of Heat Transfer*. https://doi.org/10.1115/1.2150834

Buongiorno, J., Venerus, D. C., Prabhat, N., McKrell, T., Townsend, J., Christianson, R., Tolmachev, Y. V., Keblinski, P., Hu, L. W., Alvarado, J. L., Bang, I. C., Bishnoi, S. W., Bonetti, M., Botz, F., Cecere, A., Chang, Y., Chen, G., Chen, H., Chung, S. J., … Zhou, S. Q. (2009). A benchmark study on the thermal conductivity of nanofluids. *Journal of Applied Physics, 106*, 094312. https://doi.org/10.1063/1.3245330

Crane, L. J. (1970). Flow past a stretching plate. *Zeitschrift für Angewandte Mathematik und Physik, 21*, 645–647. https://doi.org/10.1007/BF01587695

Das, S. K., Choi, S. U. S., Yu, W., & Pradeep, T. (2007). *Nanofluids*. John Wiley & Sons, Inc.. https://doi.org/10.1002/9780470180693

Davey, A. (1961). Boundary-layer flow at a saddle point of attachment. *Journal of Fluid Mechanics, 10*(04), 593. https://doi.org/10.1017/S0022112061000391

Devi, S. S. U., & Davi, S. P. A. (2017). Heat transfer enhancement of Cu-Al_2 O_3 water hybrid nanofluid flow over a stretching sheet. *Journal of the Nigerian Mathematical Society, 36*(2), 419–433.

Fang, T., & Zhang, J. (2009). Closed-form exact solutions of MHD viscous flow over a shrinking sheet. *Communications in Nonlinear Science and Numerical Simulation, 14*(7), 2853–2857. https://doi.org/10.1016/j.cnsns.2008.10.005

Fang, T. G., Zhang, J., & Yao, S. S. (2009). Viscous flow over an unsteady shrinking sheet with mass transfer. *Chinese Physics Letters, 26*(1), 014703. https://doi.org/10.1088/0256-307X/26/1/014703

Goldstein, S. (1965). On backward boundary layers and flow in converging passages. *Journal of Fluid Mechanics, 21*(01), 33. https://doi.org/10.1017/S0022112065000034

Grosan, T., & Pop, I. (2020). Flow and heat transfer over a permeable biaxial stretching/shrinking sheet in a nanofluid. *Neural Computing and Applications, 32*(9), 4575–4582. https://doi.org/10.1007/s00521-018-3770-0

Hamid, R. A., Nazar, R., & Pop, I. (2016). The non-alignment stagnation-point flow towards a permeable stretching/shrinking sheet in a nanofluid using Buongiorno's model: A revised model. *Zeitschrift für Naturforschung A, 71*(1), 81–89. https://doi.org/10.1515/zna-2015-0287

Harris, S. D., Ingham, D. B., & Pop, I. (2009). Mixed convection boundary-layer flow near the stagnation point on a vertical surface in a porous medium: Brinkman model with slip. *Transport in Porous Media,* 77(2), 267–285. https://doi.org/10.1007/s11242-008-9309-6

Hiemenz, K. (1911). Die Grenzschicht an einem in der gleichförmigen Flüssigkeitsstrom eingetauchten geraden Kreiszylinder. *Dinglers Polytechnisches Journal, 326*, 321–324.

Homann, F. (1936). Der Einfluß großer Zähigkeit bei der Strömung um den Zylinder und um die Kugel. *Zeitschrift für Angewandte Mathematik und Mechanik, 16*(3), 153–164. https://doi.org/10.1002/zamm.19360160304

Howarth, L., & Bairstow, L. (1938). On the solution of the laminar boundary layer equations. *Proceedings of the Royal Society A: Mathematical, Physical and Engineering Sciences, 164*(919), 547–579. https://doi.org/10.1098/rspa.1938.0037

Kakaç, S., & Pramuanjaroenkij, A. (2009). Review of convective heat transfer enhancement with nanofluids. *International Journal of Heat and Mass Transfer, 52*(13–14), 3187–3196. https://doi.org/10.1016/j.ijheatmasstransfer.2009.02.006

Khanafer, K., Vafai, K., & Lightstone, M. (2003). Buoyancy-driven heat transfer enhancement in a two-dimensional enclosure utilizing nanofluids. *International Journal of Heat and Mass Transfer, 46*(19), 3639–3653. https://doi.org/10.1016/S0017-9310(03)00156-X

Kuznetsov, A. V., & Nield, D. A. (2013). The Cheng–Minkowycz problem for natural convective boundary layer flow in a porous medium saturated by a nanofluid: A revised model. *International Journal of Heat and Mass Transfer, 65*, 682–685. https://doi.org/10.1016/j.ijheatmasstransfer.2013.06.054

Lok, Y. Y., Ishak, A., & Pop, I. (2011). MHD stagnation-point flow towards a shrinking sheet. *International Journal of Numerical Methods for Heat and Fluid Flow, 21*(1), 61–72. https://doi.org/10.1108/09615531111095076

Mahapatra, T. R., & Nandy, S. K. (2013). Momentum and heat transfer in MHD axisymmetric stagnation-point flow over a shrinking sheet. *Journal of Applied Fluid Mechanics, 6*(1), 121–129.

Mahian, O., Kianifar, A., Kalogirou, S. A., Pop, I., & Wongwises, S. (2013). A review of the applications of nanofluids in solar energy. *International Journal of Heat and Mass Transfer, 57*(2), 582–594. https://doi.org/10.1016/j.ijheatmasstransfer.2012.10.037

Mahian, O., Kolsi, L., Amani, M., Estellé, P., Ahmadi, G., Kleinstreuer, C., Marshall, J. S., Siavashi, M., Taylor, R. A., Niazmand, H., Wongwises, S., Hayat, T., Kolanjiyil, A.,

Kasaeian, A., & Pop, I. (2019). Recent advances in modeling and simulation of nanofluid flows-part I: Fundamentals and theory. *Physics Reports, 790*, 1–48. https://doi.org/10.1016/j.physrep.2018.11.004

Mahian, O., Kolsi, L., Amani, M., Estellé, P., Ahmadi, G., Kleinstreuer, C., Marshall, J. S., Taylor, R. A., Abu-Nada, E., Rashidi, S., Niazmand, H., Wongwises, S., Hayat, T., Kasaeian, A., & Pop, I. (2019). Recent advances in modeling and simulation of nanofluid flows—part II: Applications. *Physics Reports, 791*, 1–59. https://doi.org/10.1016/j.physrep.2018.11.003

Miklavčič, M., & Wang, C. Y. (2006). Viscous flow due to a shrinking sheet. *Quarterly of Applied Mathematics, 64*(2), 283–290. https://doi.org/10.1090/S0033-569X-06-01002-5

Minkowycz, W. J., Sparrow, E. M., & Abraham, J. P. (2013). Nanoparticle heat transfer and fluid flow. In *Nanoparticle heat transfer and fluid flow*. CRC Press. https://doi.org/10.1201/b12983

Naganthran, K., Nazar, R., & Pop, I. (2016). Unsteady stagnation-point flow and heat transfer of a special third grade fluid past a permeable stretching/shrinking sheet. *Scientific Reports, 6*(1), 24632. https://doi.org/10.1038/srep24632

Nield, D. A., & Bejan, A. (2017). Convection in porous media. In *Convection in porous Media* (5th ed.). Springer International Publishing. https://doi.org/10.1007/978-3-319-49562-0

Nield, D. A., & Kuznetsov, A. V. (2009). The Cheng-Minkowycz problem for natural convective boundary-layer flow in a porous medium saturated by a nanofluid. *International Journal of Heat and Mass Transfer, 52*(25–26), 5792–5795. https://doi.org/10.1016/j.ijheatmasstransfer.2009.07.024

Okechi, N. F., Jalil, M., & Asghar, S. (2017). Flow of viscous fluid along an exponentially stretching curved surface. *Results in Physics*, 7, 2851–2854. https://doi.org/10.1016/j.rinp.2017.07.059

Oztop, H. F., & Abu-Nada, E. (2008). Numerical study of natural convection in partially heated rectangular enclosures filled with nanofluids. *International Journal of Heat and Fluid Flow, 29*(5), 1326–1336. https://doi.org/10.1016/j.ijheatfluidflow.2008.04.009

Pop, I., Rosca, N. C., & Rosca, A. V. (2016). Additional results for the problem of MHD boundary-layer flow past a stretching/shrinking surface. *International Journal of Numerical Methods for Heat and Fluid Flow, 26*(7), 2283–2294. https://doi.org/10.1108/HFF-06-2015-0223

Rohni, A. M., Ahmad, S., Ismail, A. I. M., & Pop, I. (2013). Flow and heat transfer over an unsteady shrinking sheet with suction in a nanofluid using Buongiorno's model. *International Communications in Heat and Mass Transfer, 43*, 75–80. https://doi.org/10.1016/j.icheatmasstransfer.2013.02.001

Rosca, N. C., & Pop, I. (2017a). Axisymmetric rotational stagnation point flow impinging radially a permeable stretching/shrinking surface in a nanofluid using Tiwari and Das model. *Scientific Reports,* 7(1), 40299. https://doi.org/10.1038/srep40299

Rosca, N. C., & Pop, I. (2017b). A numerical study of the axisymmetric rotational stagnation point flow impinging radially a permeable stretching/shrinking surface in a nanofluid. *International Journal of Numerical Methods for Heat and Fluid Flow,* 27(11), 2415–2432. https://doi.org/10.1108/HFF-11-2016-0464

Rosca, A. V., Rosca, N. C., & Pop, I. (2015). Mixed convection heat and mass transfer from a vertical surface embedded in a porous medium. *Transport in Porous Media, 109*(2), 279–295. https://doi.org/10.1007/s11242-015-0517-6

Sajid, M., Ali, N., Javed, T., & Abbas, Z. (2010). Stretching a curved surface in a viscous fluid. *Chinese Physics Letters,* 27(2), 024703. https://doi.org/10.1088/0256-307X/27/2/024703

Sanni, K. M., Asghar, S., Jalil, M., & Okechi, N. F. (2017). Flow of viscous fluid along a nonlinearly stretching curved surface. *Results in Physics, 7*, 1–4. https://doi.org/10.1016/j.rinp.2016.11.058

Saqib, M., Ali, F., Khan, I., Sheikh, N. A., & Khan, A. (2019). Entropy generation in different types of fractionalized nanofluids. *Arabian Journal for Science and Engineering, 44*(1), 531–540. https://doi.org/10.1007/s13369-018-3342-8

Saqib, M., Ali, F., Khan, I., Sheikh, N. A., & Shafie, S. B. (2019). Convection in ethylene glycol-based molybdenum disulfide nanofluid. *Journal of Thermal Analysis and Calorimetry, 135*(1), 523–532. https://doi.org/10.1007/s10973-018-7054-9

Saqib, M., Khan, I., & Shafie, S. (2018a). Application of Atangana–Baleanu fractional derivative to MHD channel flow of CMC-based-CNT's nanofluid through a porous medium. *Chaos, Solitons & Fractals, 116*, 79–85. https://doi.org/10.1016/j.chaos.2018.09.007

Saqib, M., Khan, I., & Shafie, S. (2018b). Natural convection channel flow of CMC-based CNTs nanofluid. *The European Physical Journal Plus, 133*(12), 549. https://doi.org/10.1140/epjp/i2018-12340-3

Shampine, L. F., Gladwell, I., & Thompson, S. (2003). Solving ODEs with MATLAB. In *Solving ODEs with MATLAB*. Cambridge University Press. https://doi.org/10.1017/CBO9780511615542

Shenoy, A., Sheremet, M., & Pop, I. (2016). Convective flow and heat transfer from wavy surfaces. In *Convective flow and heat transfer from wavy surfaces: Viscous fluids, porous media, and nanofluids*. CRC Press. https://doi.org/10.1201/9781315367637

Sin, V. K., & Chio, C. K. (2012). Unsteady reversed stagnation-point flow over a flat plate. *ISRN Applied Mathematics, 2012*, 1–13. https://doi.org/10.5402/2012/430432

Tiwari, R. K., & Das, M. K. (2007). Heat transfer augmentation in a two-sided lid-driven differentially heated square cavity utilizing nanofluids. *International Journal of Heat and Mass Transfer, 50*(9–10), 2002–2018. https://doi.org/10.1016/j.ijheatmasstransfer.2006.09.034

Waini, I., Ishak, A., & Pop, I. (2019). Flow and heat transfer along a permeable stretching/shrinking curved surface in a hybrid nanofluid. *Physica Scripta, 94*, 105219. https://doi.org/10.1088/1402-4896/ab0fd5

Wang, C. Y. (1984). The three-dimensional flow due to a stretching flat surface. *Physics of Fluids, 27*(8), 1915. https://doi.org/10.1063/1.864868

Wang, C. Y. (1991). Flow across a stretching surface. *Journal of Applied Mechanics, 58*(2), 579–580. https://doi.org/10.1115/1.2897224

Wang, C. Y. (2008). Stagnation flow towards a shrinking sheet. *International Journal of Non-linear Mechanics, 43*(5), 377–382. https://doi.org/10.1016/j.ijnonlinmec.2007.12.021

Wang, C. Y. (2011). Review of similarity stretching exact solutions of the Navier–Stokes equations. *European Journal of Mechanics - B: Fluids, 30*(5), 475–479. https://doi.org/10.1016/j.euromechflu.2011.05.006

Wang, C. Y. (2015). Uniform flow over a Bi-axial stretching surface. *Journal of Fluids Engineering, 137*(8), 084502. https://doi.org/10.1115/1.4029447

Weidman, P. (2016). Axisymmetric rotational stagnation point flow impinging on a radially stretching sheet. *International Journal of Non-linear Mechanics, 82*, 1–5. https://doi.org/10.1016/j.ijnonlinmec.2016.01.016

CHAPTER 6

Mixed convection flow in porous medium

6.1 Introduction

Convective flow and heat transfer in a porous medium has received much attention because of their possible applications in many branches of science and technology. Practical importance of such applications is, for example, in geothermal systems, nuclear reactors, thermal insulations, drying processes, food processing and storage, fibrous insulation, thermal insulation of buildings, electrochemistry, metallurgy, the design of pebble bed nuclear reactors, underground disposal of nuclear or nonnuclear waste, solar power collectors, nuclear reactors, etc. Early comprehensive reviews of heat transfer in porous media have been first presented by Bejan (1987) and Cheng (1985). Since then, many practical applications, both industrial and environmental, have caused a rapid extension of the research in porous media, and a substantial number of books and papers, which relate the boundary layer flow past surfaces of various configuration, convective flow in cavities filed with a porous medium or by nanofluids, have been published. Bear (2018), Bejan (2013), Bejan and Kraus (2003), Bejan et al. (2004), Ene and Polisevski (1987), Ingham and Pop (1988), Ingham and Pop (2002), Ingham and Pop (2005), Pop and Ingham (2001), Nield and Bejan (2017), Vafai (2005) and Vadasz (2008), as well as in the review papers by Mahdi et al. (2015), Barletta et al. (2015) and Khanafer and Vafai (2019), identified many applications that highlight the directions where further theoretical and experimental developments are required to be investigated. Also, very good references on nanofluids in porous media can be found in the book by Nield and Bejan (2017, pp. 77).

It seems that (Homsy & Sherwood, 1976; Prats, 1966; Wooding, 1963) were among the first who studied the stability of the flow field in a porous medium and established a criteria for this flow. Experimental results were reported only by Combarnous and Bia (1971). Cheng (1977) studied the

Similarity Solutions for the Boundary Layer Flow and Heat Transfer of Viscous Fluids, Nanofluids, Porous Media, and Micropolar Fluids
ISBN 978-0-12-821188-5
https://doi.org/10.1016/B978-0-12-821188-5.00003-5

combined free and forced convection flow about inclined surfaces in porous media. He looked for the possible surface temperature distributions for which the governing equations have a similarity solution. Later on, (Merkin, 1980) considered the analogous mixed convection problem described by Cheng (1977), considering both the cases of constant surface temperature and constant surface heat flux from the plate, and considering also both the cases of assisting and opposing flows. Multiple (dual) solutions were found in the case of opposing flow. Numerical results of mixed convection in vertical and horizontal porous layers with nonuniform heating on the boundary have been further reported by Lai et al. (1988). Then, (Lai & Kulacki, 1990) considered the influence of surface mass flux on mixed convection flow over horizontal plates embedded in saturated porous media. Magyari et al. (2001) presented, in a technical note, an exact solution for the steady mixed convection boundary layer flow over a plane or axisymmetric body of arbitrary shape embedded in a porous medium. The unsteady mixed convection boundary layer flow near the region of the stagnation point on a vertical surface embedded in a porous medium has been studied by Nazar et al. (2004). Merrill et al. (2006) improved further the results obtained by Nazar et al. (2004), and found that dual solutions exist for all values of the negative values of the mixed convection parameter (opposing flow). The steady mixed convection flow over a permeable vertical thin cylinder in a porous medium has been considered by Magyari et al. (2005), who have presented exact analytic solutions for the ordinary (similarity) differential equations. It is worth mentioning to this end that (Jafarimoghaddam et al., 2019) have presented highly accurate perturbation solutions for the problem studied by Merkin (1980) via various techniques, e.g., homotopy analysis method, homotopy perturbation method, variational iteration method, minimum residual methods (see Jafarimoghaddam et al. (2019)), each with certain advantages.

6.2 Mixed convection boundary layer flow on a vertical surface in a saturated porous medium

6.2.1 Problem formulation

Following (Merkin, 1980), we consider the steady boundary layer flow past a vertical impermeable flat surface embedded in a saturated porous medium over which is flowing a uniform stream U_∞, the plate being maintained at a constant temperature T_w, as is shown in Fig. 6.1, where x and y are Cartesian coordinates measured along the plate and normal to it, respectively. It

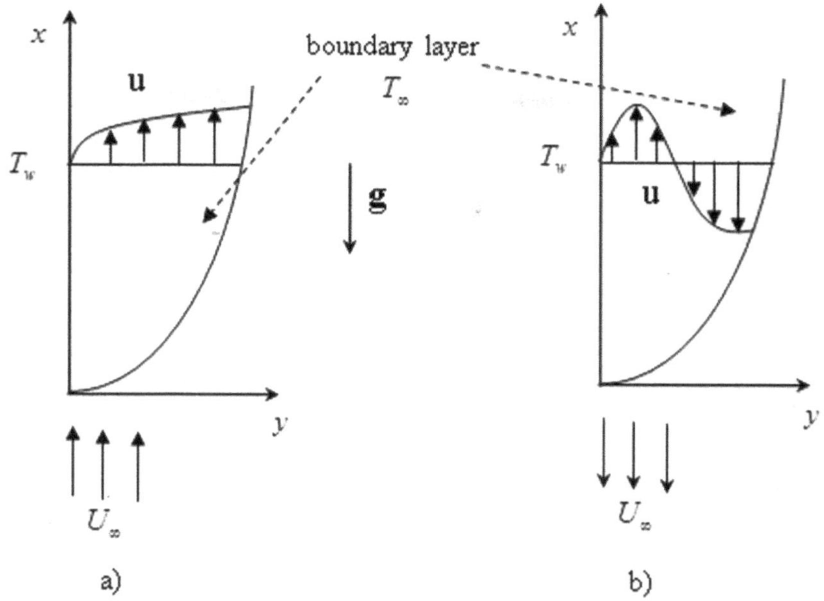

Figure 6.1 Physical model and coordinate system: (A) assisting flow, (B) opposing flow.

is assumed that the Rayleigh number is large, so that the usual boundary layer simplifications can be applied (Wooding, 1963). There is an inviscid flow far from the plate, with the velocity components $v = 0, u = U_\infty$, pressure gradient $\partial p/\partial x = -U_\infty \mu/K$, and a uniform temperature T_∞, and $\partial p/\partial y = 0$ close to the plate.

Using Eqs. (6.1) to (6.3), the following boundary layer equations exist for this problem (see Merkin (1980)):

$$\frac{\partial u}{\partial x} + \frac{\partial v}{\partial y} = 0 \tag{6.1}$$

$$u = U_\infty + \frac{g\beta K}{\nu}(T - T_\infty) \tag{6.2}$$

$$u\frac{\partial T}{\partial x} + v\frac{\partial T}{\partial y} = \alpha\frac{\partial^2 T}{\partial y^2} \tag{6.3}$$

subject to the boundary conditions:

$$\begin{aligned} &v = 0, \ T = T_w \text{ on } \quad y = 0 \\ &u \rightarrow U_\infty, T \rightarrow T_\infty \text{ as } \quad y \rightarrow \infty \end{aligned} \tag{6.4}$$

where T is the fluid temperature, K is the permeability of the porous medium, and the physical meaning of the other quantities is mentioned in the ***Nomenclature***.

Equations (6.1) to (6.3) subject to the boundary conditions (6.4) have a similarity solution of the form

$$\psi = (2\alpha U_\infty)^{1/2} f(\eta),\ \theta(\eta) = (T - T_\infty)/(T_w - T_\infty),\ \eta = (U_\infty/2\alpha x)^{1/2} y \tag{6.5}$$

where ψ is the stream function. Merkin (1980) has shown that using (6.5), Eqs. (6.2) and (6.3) can be reduced to the single equation

$$f''' + ff'' = 0 \tag{6.6}$$

along with the boundary conditions

$$f(0) = 0, \quad f'(0) = 1 + \lambda, \quad f'(\eta) \to 1 \text{ as } \eta \to \infty \tag{6.7}$$

Here primes denote differentiation with respect to the constant nondimensional mixed convection parameter, which is defined as

$$\lambda = \frac{Ra_x}{Pe_x} \tag{6.8}$$

with $Ra_x = gK\beta(T_w - T_\infty)x/_{\alpha\nu}$ being the local Rayleigh number and $Pe_x = U_\infty x/_{\alpha}$ is the local Péclet number. It should be mentioned that $\lambda > 0$ corresponds to assisting flow, $\lambda < 0$ corresponds to opposing flow, and $\lambda = 0$ describes the forced convection flow, respectively.

6.2.2 Comments

Merkin (1980) has shown that for $\lambda > 0$ (assisting flow), solution of the boundary value problem (6.6,6.7) can be obtained for all values of λ and these are given by Cheng (1977). However, for $\lambda < 0$ (opposing flow) this boundary value problem has solutions only in the range $-1.354 \le \lambda \le 0$ and for λ in the range $-1.354 < \lambda < -1$, the solution is not unique, there being multiple (dual), upper branch, $f_u''(0)$ and lower branch $f_l''(0)$, solutions for a given $\lambda(< 0)$, as can be seen from Fig. 6.2, where $f''(0)$ is plotted against $\lambda(< 0)$. However, for $\lambda < -1.354$, the boundary value problem (6.6, 6.7) has no solutions, so that the full Darcy and energy equations have to be solved. Further, when $\lambda = -1$ one gets $f'(0) = 0$ and we have the well-known (Blasius, 1908) solution (with $f''(0) = 0.46960$). For $\lambda < -1$ the solution $f_u(\eta)$ continues from the Blasius solution.

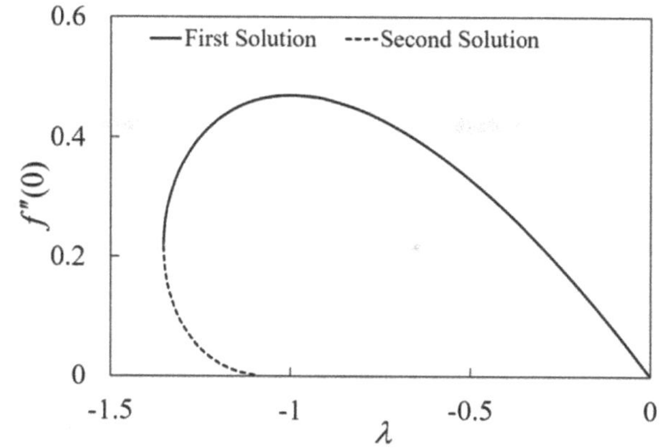

Figure 6.2 Variation of $f''(0)$ with $\lambda<0$ (opposing flow). *(Modified from Aly, E., Elliott, L., & Ingham, D. (2003). Mixed convection boundary-layer flow over a vertical surface embedded in a porous medium.* European Journal of Mechanics - B/Fluids, *22(6), 529–543. https://doi.org/10.1016/S0997-7546(03)00059-1 532.)*

The other set of solutions $f_l(\eta)$ have $f_l''(0) < f_u''(0)$ for a given $\lambda(<0)$ and are such that $f_u''(0) \to 0$ as $\lambda \to -1$. It appears from these results that the solution presented by Cheng (1977) is relevant to the physical problem only for $\lambda \geq -1$.

6.3 Steady mixed convection flow over a permeable vertical thin cylinder in a porous medium

6.3.1 Basic equations

The problem of similarity solutions for the steady mixed convection boundary layer flow along a permeable semiinfinite vertical thin cylinder of radius r_0 embedded in a porous medium was studied by Magyari et al. (2005). The geometry of the problem is shown in Fig. 6.3, where x and r are the axial and radial coordinates. It is assumed that both the mainstream velocity $u_e(x)$ and the surface temperature of the cylinder $T_w(x)$ are defined as

$$u_e(x) = U_\infty\left(x/_L\right), \quad T_w(x) = T_\infty + T_0\left(x/_L\right) \tag{6.9}$$

where $U_\infty > 0$ and $T_\infty > 0$ are the constant velocity and temperature of the ambient fluid, and T_0 and L are the characteristic temperature and length of the cylinder, with $T_0 > 0$ corresponding to a hot cylinder $T_w > T_\infty$ or assisting flow and $T_0 < 0$ to a cold cylinder $(T_w < T_\infty)$ or

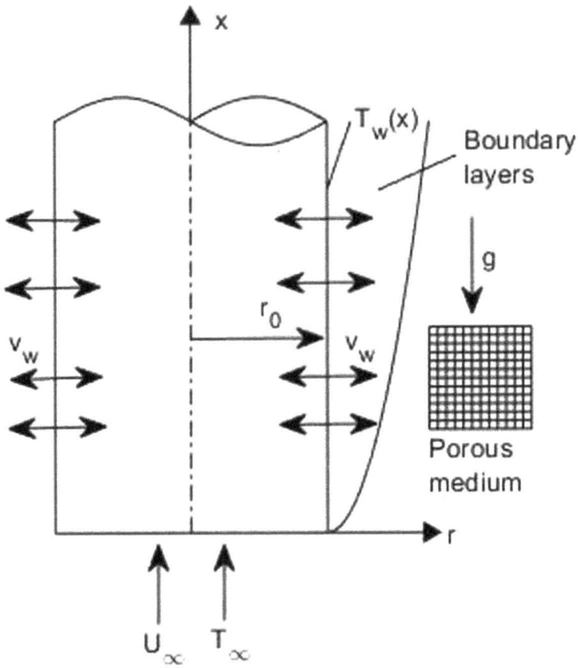

Figure 6.3 Geometry of flow and temperature boundary layer domain. *(From Magyari, E., Pop, I., & Keller, B. (2005). Exact solutions for a longitudinal steady mixed convection flow over a permeable vertical thin cylinder in a porous medium.* International Journal of Heat and Mass Transfer, *48(16), 3435–3442. https://doi.org/10.1016/j.ijheatmasstransfer. 2005.01.043 3441.)*

opposing flow, respectively. We, also assume that the radial mass flux velocity at the cylinder surface is $v(x, r_0) \equiv v_w(x)$ such that $v_w(x) < 0$ corresponds to suction, $v_w(x) > 0$ corresponds to injection, and $v_w(x) = 0$ corresponds to an impermeable surface.

The governing equations, in the usual boundary layer approximations, are of the form (see Magyari et al. (2005))

$$\frac{\partial}{\partial x}(ru) + \frac{\partial}{\partial r}(rv) = 0 \tag{6.10}$$

$$u = U(x) + \frac{g\beta K}{\nu}(T - T_\infty) \tag{6.11}$$

$$u\frac{\partial T}{\partial x} + v\frac{\partial T}{\partial r} = \alpha_m\left(\frac{\partial^2 T}{\partial r^2} + \frac{1}{r}\frac{\partial T}{\partial r}\right) \tag{6.12}$$

subject to the boundary conditions:

$$\begin{aligned} &v = v_w(x), \quad T = T_w(x) \text{ on } \quad r = r_0 \\ &u \to u_e(x), \quad T \to T_\infty \text{ as } \quad r \to \infty \end{aligned} \tag{6.13}$$

where we assume that $u_e(x) = ax$, a being a positive constant. The physical meaning of the quantities in these equations is mentioned in the ***Nomenclature***. It should be mentioned that flow over cylinders is considered to be two dimensional if the body radius r_0 is large compared to the boundary layer thickness. For a thin or slender cylinder, r_0 may be of the same order as the boundary layer thickness, so that the flow may be considered as axisymmetric instead of two dimensional. In this case, the governing Eqs. (6.11) and (6.12) contain the transverse curvature term which influences the flow characteristics. The effect of the transverse curvature is important in some practical applications such as wire or fiber drawing where accurate prediction of flow and heat transfer is required and thick boundary layer can exist on slender or near the slender bodies.

6.3.2 Similarity transformations

Following Magyari et al. (2005), we introduce the similarity variables

$$\psi = 2a\,\alpha_m x f(\eta), \ \theta(\eta) = T_\infty + T_0\left(x/L\right)\theta(\eta), \ \eta = b\frac{r^2 - r_0^2}{r_0^2} \tag{6.14}$$

where ψ is the stream function defined in the usual way as $u = \frac{1}{r}\frac{\partial \psi}{\partial r}$ and $u = -\frac{1}{r}\frac{\partial \psi}{\partial x}$, and a and b are dimensionless scale factors for ψ and η, and $L = U_\infty r_0^2 / 4\alpha_m a\, b$. Thus, the velocity components u and v are obtained in the form

$$u(x,r) = axf'(\eta), \ v(x,r) = -\frac{2a\alpha_m}{r} f(\eta) \tag{6.15}$$

where prime denotes differentiation with respect to η. Therefore, v_w is given by

$$v_w = -\frac{2a\alpha_m}{r_0} S \tag{6.16}$$

and S is the constant mass flux parameter.

Substituting (6.14) into Eqs. (6.11) and (6.12), the following ordinary (similarity) differential equations are obtained

$$f' = 1 + \theta\lambda \tag{6.17}$$

$$(\eta + b)\theta'' + \theta' + a(f\theta' - f'\theta) = 0 \tag{6.18}$$

subject to the boundary conditions

$$f(0) = S, \quad \theta(0) = 1, \quad \theta(\eta) \to 0 \text{ as } \eta \to \infty \tag{6.19}$$

where λ is the mixed convection parameter defined by (6.8).

6.3.3 Exact solution

Eqs. (6.17) and (6.18) subject to the boundary conditions (6.19) admit the exact solution

$$\theta(\eta) = e^{-a\eta}, \; f(\eta) = S + \eta + \frac{\lambda}{a}(1 - \theta) \tag{6.20}$$

where

$$a\eta = ab\frac{r^2 - r_0^2}{r_0^2}, \qquad ab > 0 \tag{6.21}$$

This solution is valid if the scale factors satisfy (in addition to $a\,b > 0$) the equation

$$a(b - f_w) - (\lambda + 2) = 0 \tag{6.22}$$

The corresponding dimensional velocity components and the temperature field are then given by

$$\begin{aligned} u(x, r) &= U_\infty(x/L)(1 + \lambda e^{-a\eta}) \\ v(x, r) &= -\frac{2a\alpha_m}{r}\left[S + \eta + \frac{\lambda}{a}(1 - e^{-a\eta})\right] \\ T(x, r) &= T_\infty + T_0(x/L)e^{-a\eta} \end{aligned} \tag{6.23}$$

Thus, the radial wall heat flux $q_w(x)$ is obtained as

$$q_w(x) = 2abk_m(T_0 / L)(x/r_0) \tag{6.24}$$

6.3.4 Comments

A very detailed discussion on the flow and heat transfer characteristics of this problem can be found in (Magyari et al., 2005). Thus, we will limit presenting results only for the case when $a = a(\gamma)$ and $b = b(\gamma)$ where γ is an arbitrary (real) parameter. Thus, if we put in Eq. (6.22) $a = a(\gamma)$ and $b = b(\gamma)$, we obtain the equation

$$a(\gamma)\,[b(\gamma) - S] - (\lambda + 2) = 0 \tag{6.25}$$

It is always possible to choose the functions $a(\gamma)$ and $b = b(\gamma)$ in such a way that Eq. (6.25) admits arbitrarily many solutions $\gamma = \gamma_n$, which all satisfy the condition $a(\gamma_n)\cdot b(\gamma_n) > 0$ and in some domain of the parameter plane (S, λ) all of them are real and distinct. In this way, Eq. (6.20) yields n (in general distinct) solutions for $\theta(\eta)$,

$$\theta \equiv \theta_n(\eta) = exp\left[- a(\gamma_n)\cdot b(\gamma_n)\frac{r^2 - r_0^2}{r_0^2}\right] \tag{6.26}$$

In order to be more specific, we consider a simple example in which Eq. (6.25) is an algebraic equation of the second degree in such that (at most) dual solutions can arise. To this end we choose

$$a(\gamma) = b(\gamma) \equiv \frac{1}{\gamma} \tag{6.27}$$

In this case, $a(\gamma_n)\cdot b(\gamma_n) = 1/\gamma_n^2 > 0$ satisfied for any real solution γ_n of Eq. (6.25), which now becomes

$$\frac{1}{\gamma^2} - \frac{S}{\gamma} - (\lambda - 2) = 0 \tag{6.28}$$

The solutions of this equation are

$$\frac{1}{\gamma_\pm} = \frac{1}{2}\left(S \pm \sqrt{S^2 + 4\lambda + 8}\,\right) \tag{6.29}$$

In terms of $\gamma_\pm$ the temperature profiles (6.26) become

$$\theta_\pm(\eta) = exp\left(\frac{r^2 - r_0^2}{\gamma_\pm^2 r_0^2}\right) = exp\left[\frac{1}{4}\left(S \pm \sqrt{S^2 + 4\lambda + 8}\,\right)^2\left(1 - \frac{r^2}{r_0^2}\right)\right] \tag{6.30}$$

The existence domain of the solutions corresponding to the parametrization (6.27) of the scale factors a and b is the region of the parameter plane (S,λ) where

$$\lambda \geq \lambda_c(S) = -2 - \frac{S^2}{4} \tag{6.31}$$

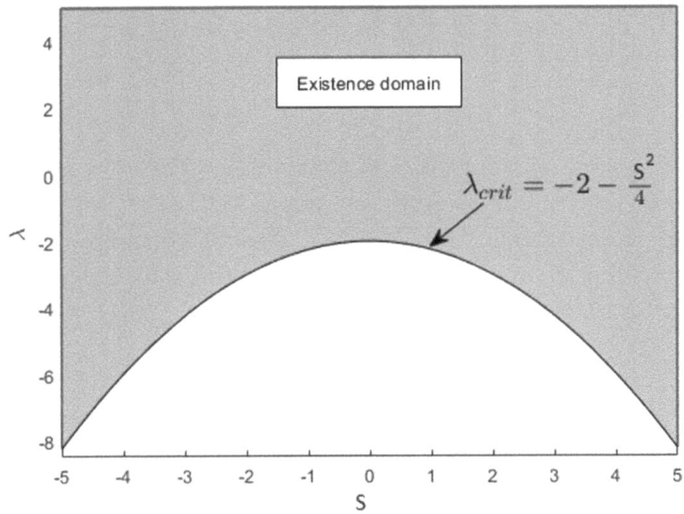

Figure 6.4 The existence domain of the dual solutions are realized by a lateral suction or injection of the liquid, respectively. *(From Magyari, E., Pop, I., & Keller, B. (2005). Exact solutions for a longitudinal steady mixed convection flow over a permeable vertical thin cylinder in a porous medium.* International Journal of Heat and Mass Transfer, 48*(16), 3435–3442. https://doi.org/10.1016/j.ijheatmasstransfer.2005.01.043 3441.)*

holds, without the point $(S, \lambda) = (0, -2)$, which is excluded (see Fig. 6.4), where $\lambda_c < 0$ is the critical value of $\lambda < 0$ beyond which the boundary layer separates from the surface and the solution based upon the boundary layer approximations are not possible. Below the critical parabola (6.30) no solutions of (6.20) and (6.21) exist. Thus, (6.29) can be written as

$$\frac{1}{\gamma_{\pm}} = \frac{S}{2} \pm \sqrt{(\lambda - \lambda_c)} \tag{6.32}$$

The unique, dual, and coincident character of the exact solutions in their existence domain is depicted in Fig. 6.5, while the branching curve (6.32) of the dual solutions as a function of the mixed convection parameter λ is given in Fig. 6.6. Typical dual solutions can be seen in Fig. 6.7.

6.4 Mixed convection boundary layer flow from a vertical flat plate embedded in a porous medium filled with nanofluids

6.4.1 Introduction

Convection heat transfer and fluid flow with porous medium occur in power stations of many engineering applications where cooling or heating

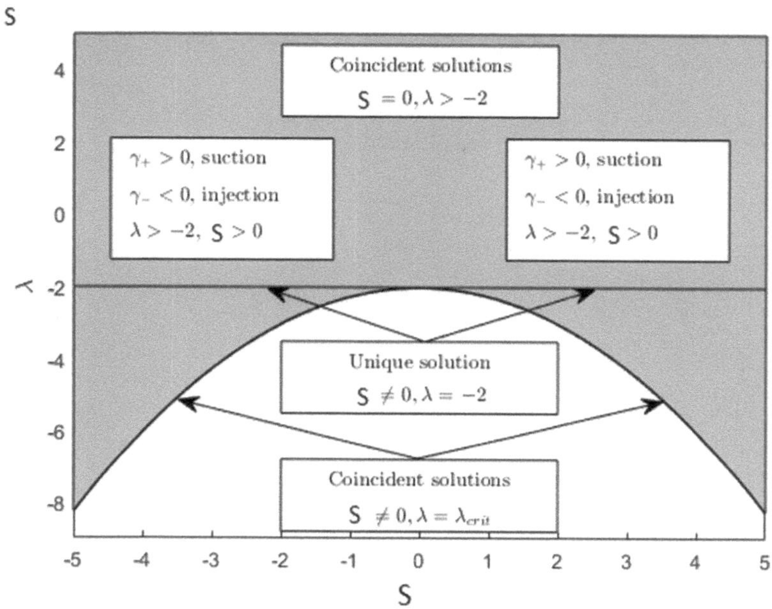

Figure 6.5 The unique, dual, and coincident character of the exact solutions in their existence domain and how these are associated with the corresponding lateral suction/injection velocities of the fluid. *(Modified from Magyari, E., Pop, I., & Keller, B. (2005). Exact solutions for a longitudinal steady mixed convection flow over a permeable vertical thin cylinder in a porous medium.* International Journal of Heat and Mass Transfer, *48(16), 3435–3442. https://doi.org/10.1016/j.ijheatmasstransfer.2005.01.043 3441.)*

is required such as cooling turbine blades, cooling electronic equipment, and combustion systems. The mixing of the low and high energy fluids which occur in these applications significantly affects the performance of these devices. One of the ways to increase heat transfer is to employ porous medium with nanofluid. Many studies on convection heat transfer in porous media with nanofluid have been done during the last several decades. The steady mixed convection boundary layer flow past a vertical flat plate embedded in a porous medium filled with nanofluids has been first considered by Ahmad and Pop (2010) using different types of nanoparticles as Cu (cuprum), Al_2O_3 (aluminum), and TiO_2(titanium).

6.4.2 Basic equations

It is assumed that the free stream velocity and the ambient temperature (far flow from the plate) are U_∞ and T_∞, respectively. It is also assumed that the temperature of the plate is T_w, where $T_w > T_\infty$ corresponds to a heating plate (assisting flow) and $T_w < T_\infty$ corresponds to a cooling plate (opposing

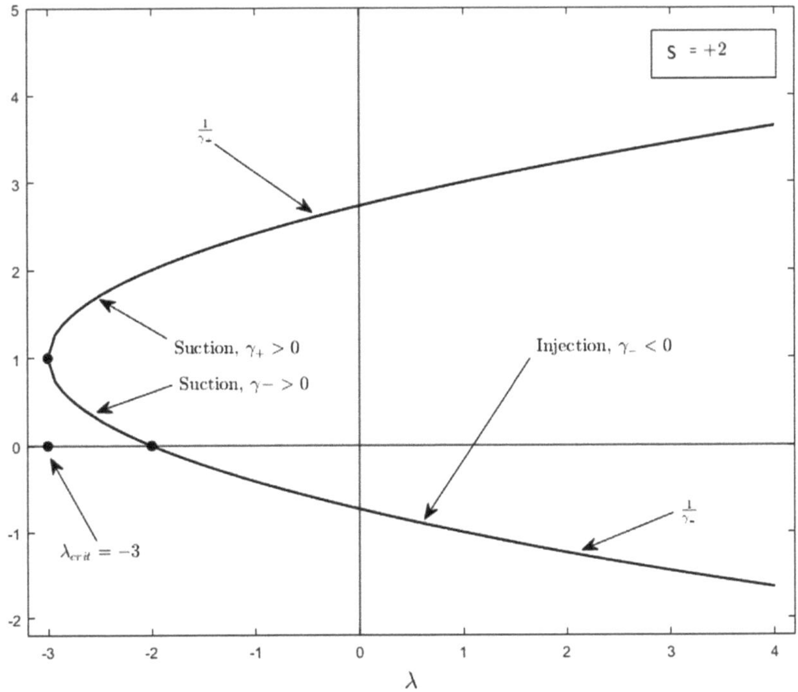

Figure 6.6 The branching curve (6.32) of the dual solutions as a function of the mixed convection parameter λ for $S = +2$. In this case $\lambda_c = 3$. The exact dual solutions can be realized both in the aiding ($\lambda > 0$) and in the opposing ($\lambda_c < \lambda < 0$, $\lambda \neq -2$) flow regimes, respectively. *(From Magyari, E., Pop, I., & Keller, B. (2005). Exact solutions for a longitudinal steady mixed convection flow over a permeable vertical thin cylinder in a porous medium.* International Journal of Heat and Mass Transfer, 48*(16), 3435–3442. https://doi.org/10.1016/j.ijheatmasstransfer.2005.01.043 3441.)*

flows). Fig. 6.1 shows the physical model and coordinate system with assisting and opposing flows.

Following the mathematical nanofluid model proposed by Tiwari and Das (2007) along with the Boussinesq and boundary layer approximations, it is easy to show that the steady boundary layer equations of the present problem are, see also (Nield & Bejan, 2017),

$$\frac{\partial u}{\partial x} + \frac{\partial v}{\partial y} = 0 \tag{6.33}$$

$$\frac{\mu_{nf}}{\mu_f} u = \frac{\mu_{nf}}{\mu_f} U_\infty + \frac{gK\left[\varphi(\rho\beta)_s + (1-\varphi)(\rho\beta)_f\right]}{\mu_f}(T - T_\infty) \tag{6.34}$$

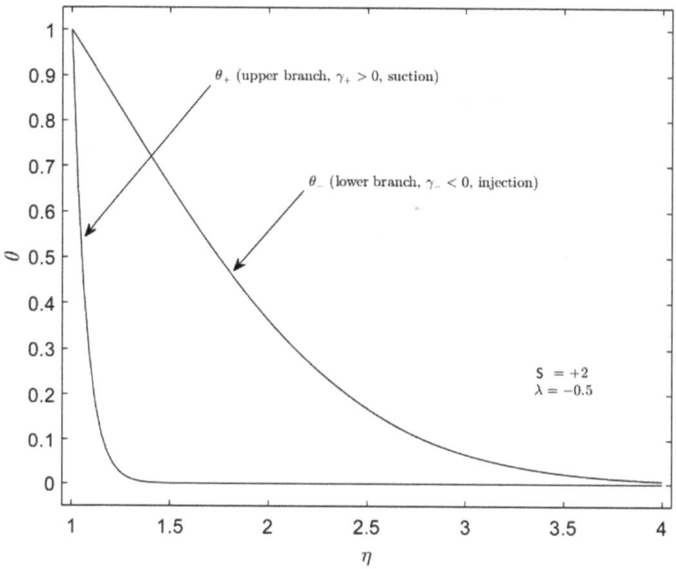

Figure 6.7 The dual temperature profiles (6.31) as functions of the radial distance $r \geq r_0$ are plotted for (S,λ)=(2,−0.5) and $r_0 = 1$. The upper and lower branch solutions correspond to the γ values γ_+ = 0.38742 and γ_− = 1.72076, and can be realized by a lateral suction and injection, respectively. *(From Magyari, E., Pop, I., & Keller, B. (2005). Exact solutions for a longitudinal steady mixed convection flow over a permeable vertical thin cylinder in a porous medium.* International Journal of Heat and Mass Transfer, *48(16), 3435–3442. https://doi.org/10.1016/j.ijheatmasstransfer.2005.01.043 3441.)*

$$u\frac{\partial T}{\partial x} + v\frac{\partial T}{\partial y} = \frac{k_{nf}}{\left(\rho C_p\right)_{nf}}\frac{\partial^2 T}{\partial y^2} \tag{6.35}$$

subject to the boundary conditions

$$\begin{aligned} &v = 0, \quad T = T_w \text{ on } y = 0 \\ &u \to U_\infty, \quad T \to T_\infty \text{ as } y \to \infty \end{aligned} \tag{6.36}$$

The physical meaning of the other quantities is mentioned in the ***Nomenclature***. The thermophysical properties of the nanofluid are given in Table 1.2.

We look for a similarity solution of Eqs. (6.33) to (6.35) of the following form:

$$\psi = \alpha_f(2Pe_x)^{\frac{1}{2}}f(\eta), \quad \theta(\eta) = (T - T_\infty)/(T_w - T_\infty), \quad \eta = Pe_x^{\frac{1}{2}}y/(x\sqrt{2}) \tag{6.37}$$

Substituting variables (6.37) into Eqs. (6.34) and (6.35), it is easy to show that we obtain the following ordinary (similarity) differential equation:

$$\frac{k_{nf}/k_f}{1-\varphi+\varphi\,(\rho C_p)_s/(\rho C_p)_f} f''' + f f'' = 0 \tag{6.38}$$

along with the boundary conditions

$$f(0)=0, \quad \frac{1}{(1-\varphi)^{2.5}} f'(0) = \frac{1}{(1-\varphi)^{2.5}} + \left[1-\varphi+\varphi\frac{\rho_s}{\rho_f}\frac{\beta_s}{\beta_f}\right] \tag{6.39}$$

$$f'(\eta)\to 1 \text{ as } \eta\to\infty$$

It should be noticed that the problem described by Eq. (6.38) subject to the boundary conditions (6.39) reduces to that of a regular fluid–porous medium ($\phi = 0$), first studied by Merkin (1980). Then, we point out that when $\lambda = 0$ (forced convection flow), the boundary value problem (6.38, 6.39) has the solution $f(\eta) = 0$, so that $f''(0) = 0$.

The quantity of practical interest is the skin friction coefficient C_f, which can be easily shown that is given by

$$(2Pe_x)^{\frac{1}{2}} C_f = \frac{1}{(1-\varphi)^{2.5}} f''(0) \tag{6.40}$$

6.4.3 Comments

Numerical solutions to the governing ordinary differential Eq. (6.38) with the boundary conditions (6.39) are obtained using the shooting method. The effects of the solid volume fraction of nanofluid φ and the mixed convection parameter λ are analyzed for a porous medium filled with a regular fluid ($\varphi = 0$) and a porous medium filled with three different nanofluids as Cu–Al_2O_3/water and TiO_2/water as working fluids. The values of φ considered are $\varphi = 0$, 0.1, and 0.2, while λ takes both positive and negative values. The variation with λ of the skin friction coefficient $(2Pe_x)^{\frac{1}{2}} C_f$, given by (6.46), is shown in Figs. 6.8 and 6.9. It is seen that there are regions of unique solution for $\lambda > 0$, dual solutions for $\lambda_c < \lambda \le 0$ and no solution for $\lambda < \lambda_c < 0$. For the regular fluid, the value of λ_c is $\lambda_c = -1.354$ (see Ahmad and Pop (2010) and Merkin (1980)) and the upper and lower branch values of $f''(0)$ denoted by $f_u''(0)$ and $f_l''(0)$ are given in Table 6.1.

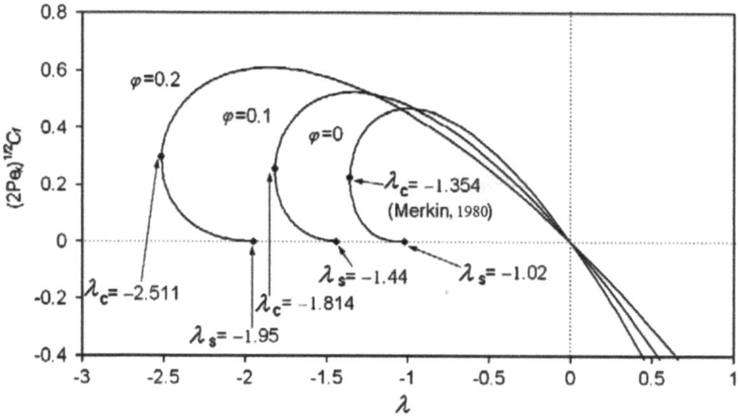

Figure 6.8 Variation of $(2\ \mathrm{Pe}_x)^{1/2}C_f$ with λ for Cu−water when $\varphi = 0$, 0.1, and 0.2. *(Modified from Ahmad, S., & Pop, I. (2010). Mixed convection boundary layer flow from a vertical flat plate embedded in a porous medium filled with nanofluids.* International Communications in Heat and Mass Transfer, *37(8), 987–991. https://doi.org/10.1016/j.icheatmasstransfer.2010.06.004 989.)*

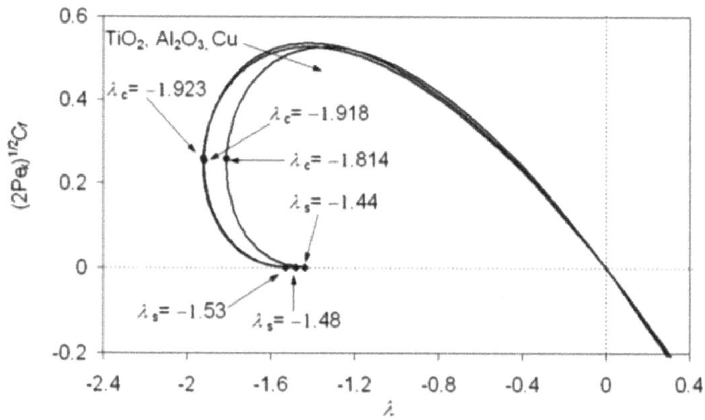

Figure 6.9 Variation of $(2\ \mathrm{Pe}_x)^{1/2}C_f$ with λ for different types of nanoparticles when $\varphi = 0.1$. *(Modified from Ahmad, S., & Pop, I. (2010). Mixed convection boundary layer flow from a vertical flat plate embedded in a porous medium filled with nanofluids.* International Communications in Heat and Mass Transfer, *37(8), 987–991. https://doi.org/10.1016/j.icheatmasstransfer.2010.06.004 989.)*

Table 6.1 Table values of $f_u''(0)$ and $f_l''(0)$ for clear fluid.

	$f_u''(0)$			$f_l''(0)$		
λ_1	**(Merkin, 1980)**	**(Ghalambaz, Grosan, & Pop, 2019a)**	**(Ahmad & Pop, 2010)**	**(Merkin, 1980)**	**(Ghalambaz, Grosan, & Pop, 2019a)**	**(Ahmad & Pop, 2010)**
−1.000	0.46960	0.469600	0.46960			
−1.100	0.46105	0.461049	0.46105	0.00194	0.001936	0.00194
−1.200	0.43015	0.430151	0.43015	0.02219	0.022188	0.022188
−1.300	0.35664	0.356638	0.35664	0.08497	0.084889	0.084889
−1.350	0.25758	0.257581	0.25758	0.17856	0.178597	0.178597
−1.354	0.22428	0.224296	0.22428			

6.5 Mixed convection boundary layer flow along a vertical cylinder embedded in a porous medium filled by a nanofluid

6.5.1 Basic equations

We consider the steady, axisymmetric boundary layer flow along a vertical cylinder of radius which is embedded in a porous medium filled with a nanofluid, where x and r are the cylindrical coordinates measured along the axis of the cylinder and normal to the surface of the cylinder in the radial direction. It is assumed that the velocity of the outer (potential) flow $u_e(x)$ and the surface temperature $T_w(x)$ are of the form $u_e(x) = U_0 x/_a$ and $T_w(x) = T_\infty + T_0 x/_a$, where U_0 and T_0 are constants with $U_0 > 0$ and $T_0 > 0$ for a heated cylinder (assisting flow), $T_0 < 0$ for a cooled cylinder (opposing flow), respectively. Under these assumptions along with the Boussinesq and boundary layer approximations, and using the nanofluid model proposed by Tiwari and Das (2007) (see Eqs. 1.11 to 1.13), the basic equations are

$$\frac{\partial}{\partial x}(ru) + \frac{\partial}{\partial r}(rv) = 0 \tag{6.41}$$

$$\frac{\mu_{nf}}{\mu_f}\frac{\partial u}{\partial r} = \frac{gK\left[(1-\varphi)(\rho\beta)_f + \varphi(\rho\beta)_s\right]}{\mu_f}\frac{\partial T}{\partial r} \tag{6.42}$$

$$u\frac{\partial T}{\partial x} + v\frac{\partial T}{\partial r} = \frac{k_{nf}}{(\rho C_p)_{nf}}\frac{\partial}{\partial r}\left(r\frac{\partial T}{\partial r}\right) \tag{6.43}$$

subject to the boundary conditions

$$\begin{aligned} &v = 0, \quad T = T_\infty + T_0\frac{x}{a} \text{ on } \quad r = a \\ &u \to U_0\frac{x}{a}, \quad T \to T_\infty \text{ as } \quad r \to \infty \end{aligned} \tag{6.44}$$

Eqs. (6.41)-(6.43) with the boundary conditions (6.44) can be transformed into ordinary differential equations by the similarity transformation (see Merkin and Pop (1987))

$$\eta = \frac{r^2 - a^2}{2\,\alpha_f\, a}\left(\frac{U_0\alpha_f}{2\,a}\right)^{1/2}, \quad \psi = \left(2U_0 a\,\alpha_f\right)^{1/2} x f(\eta), \quad \theta(\eta) = \frac{T - T_\infty}{T_w - T_\infty} \tag{6.45}$$

It can be shown that using similarity variables (6.45) Eqs. (6.42) and (6.43) reduce to the ordinary (similarity) differential equation along with the boundary conditions are now

$$A(\varphi)[(1+2\gamma\eta)f'''+2\gamma f'']+2ff''-2f''^2+2f'=0 \tag{6.46}$$

along with the boundary conditions are now

$$f(0)=0,\quad f'(0)=1+B(\varphi)\lambda,\ f'(\eta)\to 1 \text{ as } \eta\to\infty \tag{6.47}$$

where the parameters $A(\varphi)$, $B(\varphi)$ are given by

$$\begin{aligned} A(\varphi) &= \frac{k_{nf}/k_f}{1-\varphi+\varphi(\rho C_p)_s/(\rho C_p)_f} \\ B(\varphi) &= (1-\varphi)^{2.5}\left[1-\varphi+\varphi(\rho\beta)_s/(\rho\beta)_f\right] \end{aligned} \tag{6.48}$$

The constants γ and λ are, respectively, the curvature and mixed convection parameters, which are defined by

$$\gamma=\left(\frac{2\ \alpha_f}{U_0\ a}\right)^{1/2},\ \lambda=\frac{Ra}{Pe} \tag{6.49}$$

where Ra, Pe, and α_f are given in the Nomenclature. Here $\lambda>0$ corresponds to an assisting (or aiding) flow, $\lambda<0$ corresponds to an opposing flow, and $\lambda=0$ corresponds to the forced convection flow past the vertical cylinder. When $\varphi=0$ (regular fluid), $A(\varphi)=B(\varphi)=1$ and, as the nanoparticle volume fraction φ increases, $A(\varphi)$ increases and $B(\varphi)$ decreases, with $B(\varphi)\to 0$ as $\varphi\to 1$. We notice that $\gamma=0$ corresponds to a flat plate. In this case Eqs. (6.46) and (6.47) reduce to those for the problem of steady mixed convection boundary layer flow along a vertical flat plate embedded in a porous medium considered by Aly et al. (2003) when $\lambda=1$. In addition, if $\lambda=1$ and $\gamma=0$, the problem reduces to that studied by Merkin (1980) and Merkin (1986).

The physical quantities of interest are the skin friction coefficient C_f and the local Nusselt number Nu_x, which are defined as

$$C_f=\frac{\mu_{nf}}{\rho_f u_e^2}\left(\frac{\partial u}{\partial r}\right)_{r=a},\quad Nu_x=\frac{k_{nf}}{k_f}\left(-\frac{\partial T}{\partial r}\right)_{r=a} \tag{6.50}$$

Using the similarity variables (6.4) and (6.50), we have

$$2Pe_x^{1/2}\left(\alpha_{nf}/\nu_{nf}\right)C_f=\frac{1}{(1-\varphi)^{2.5}}f''(0),\ \left(2/Pe_x\right)^{1/2}Nu_x=-\frac{k_{nf}}{k_f}\theta''(0) \tag{6.51}$$

6.5.2 Comments

Here we discuss the numerical solution of the boundary value problem (6.46, 6.47) for selected values of the parameters, our main objectives being to assess the effects that curvature and the nanoparticles have on the convective flow and heat transfer. We use the data for Cu nanoparticles in water provided by Oztop and Abu-Nada (2008) with the resulting values for $A(\varphi)$ and $B(\varphi)$, defined in (6.48), for different of φ being given in Table 6.2.

We start by considering the case of a pure fluid ($\varphi = 0$). In Fig. 6.10 the $f''(0)$ against λ for different values of γ has been plotted. It

Table 6.2 Values of the parameters $A(\varphi)$ and $B(\varphi)$ defined in (6.48) using Cu (Oztop & Abu-Nada, 2008).

φ	$A(\varphi)$	$B(\varphi)$
0.05	1.1673	0.8670
0.1	1.3553	0.7463
0.2	1.8089	0.5395

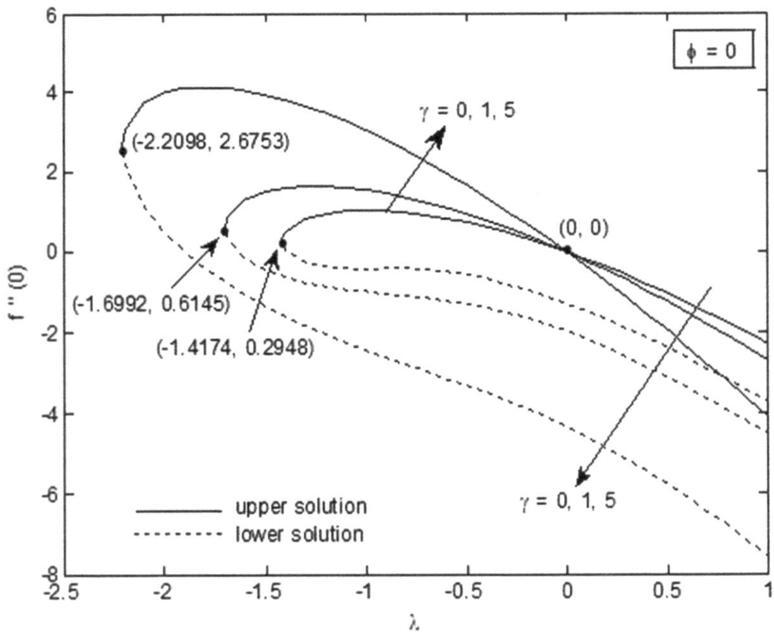

Figure 6.10 Plots of the reduced skin friction $f''(0)$ against the mixed convection parameter λ for different values of curvature parameter γ when $\varphi = 0$. *(From Rohni, A. M., Ahmad, S., Merkin, J. H., & Pop, I. (2013). Mixed convection boundary-layer flow along a vertical cylinder embedded in a porous medium filled by a nanofluid.* Transport in Porous Media, *96(2), 237–253. https://doi.org/10.1007/s11242-012-0085-y 246.)*

is seen that there is a critical value λ_c of λ, with the values of $|\lambda_c|$ increasing with γ. There is a saddle–node bifurcation at $\lambda = \lambda_c$ giving rise to two solution branches for $\lambda > \lambda_c$. Both solution branches continue into the aiding flow, $\lambda > 0$, region with the upper solution branch passing through the forced convection solution $f(\eta) = \eta$ at $\lambda = 0$. On this branch $f''(0) > 0$ for opposing flow, $\lambda_c \le \lambda < 0$ and $f''(0) < 0$ for $\lambda > 0$. The numerical solutions for the lower solution branch (shown by a broken line) also pass smoothly through $\lambda = 0$ without a singularity appearing, as seen, for example, in (Merkin, 1980). Even though a solution to (6.46, 6.47) exists on the lower branch when $\lambda = 0$, it cannot be a physically acceptable solution to the present problem. This can be seen in Fig. 6.11 where $-\theta'(0)$ has been calculated from (6.54) as $\theta'(0) = f''(0)/B\lambda$ clearly showing the discontinuity seen in the figure as $|\lambda| \to 0$. We expect, through the saddle–node bifurcation, that these lower branch solutions to be unstable.

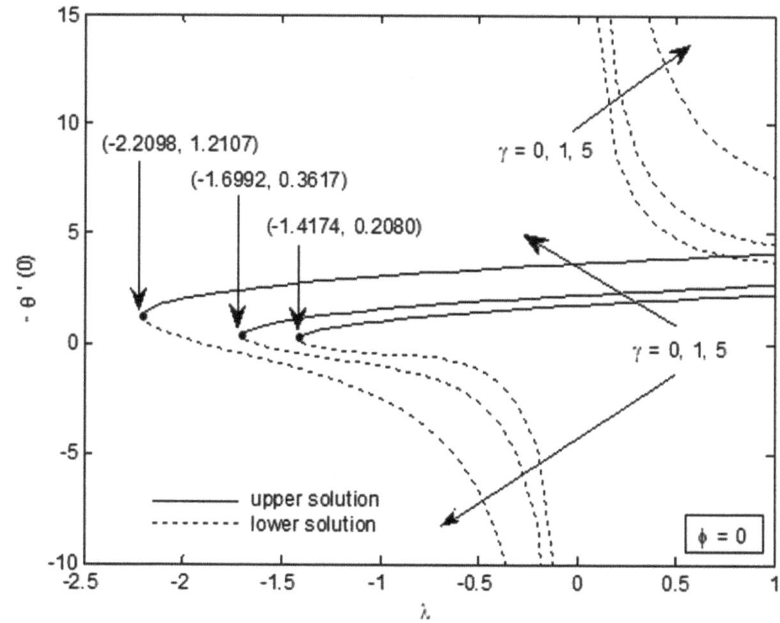

Figure 6.11 Plots of the reduced surface heat transfer $-\theta'(0)$ against the mixed convection parameter λ for different values of curvature parameter γ when $\varphi = 0$. *(From Rohni, A. M., Ahmad, S., Merkin, J. H., & Pop, I. (2013). Mixed convection boundary-layer flow along a vertical cylinder embedded in a porous medium filled by a nanofluid.* Transport in Porous Media, *96(2), 237–253. https://doi.org/10.1007/s11242-012-0085-y 246.)*

In the next figures one evaluates the effects of both curvature and the addition of nanoparticles on the convective flow by plotting $f''(0)$ and $-\theta'(0)$ against λ. In Figs. 6.12 and 6.13, $\gamma = 0$ is taken, giving a flat plate and $\varphi = 0$ (pure fluid), $\varphi = 0.05$, 0.1 and $\varphi = 0.2$ with the corresponding values of the parameters $A(\varphi)$ and $B(\varphi)$. It is see that the effect of adding nanoparticles to the base fluid is to increase the range of existence of solutions in the opposing flow region and to decrease the friction effects on the surface, Fig. 6.12, positive for opposing flow and negative for aiding flow. The addition of nanoparticles also increases the surface heat flux, Fig. 6.13, on the upper solution branch. These features are also seen when the curvature is increased to $\gamma = 1$ in Figs. 6.14 and 6.15 and more noticeably to $\gamma = 5$ in Figs. 6.16 and 6.17. Critical values determining the range of existence of a solution for opposing flow were identified (see Fig. 6.18). It is also seen that $|\lambda_c|$ increases with both increased γ and with the addition of φ.

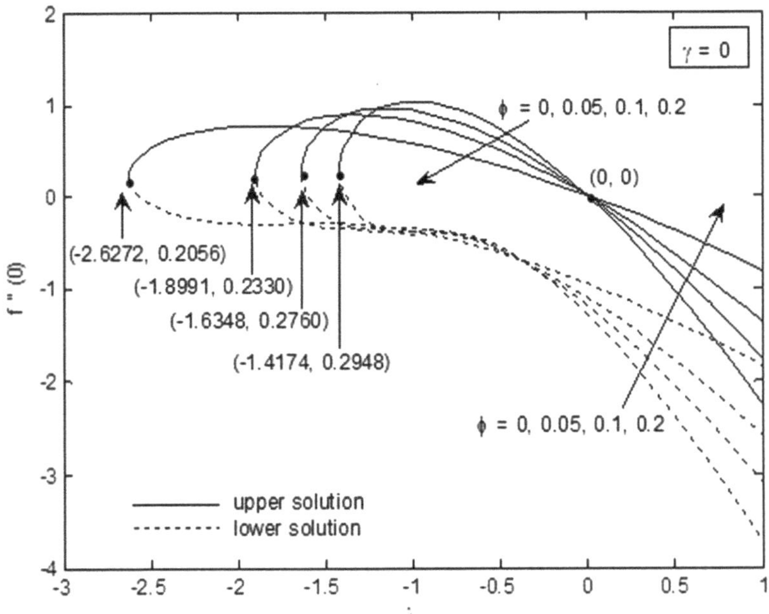

Figure 6.12 Plots of $f''(0)$ against the mixed convection parameter λ for the addition of copper (Cu) nanoparticles with volume fractions $\varphi = 0.0$, 0.5, 0.1, 0.2 and $\gamma = 0$ (flat surface). *(From Rohni, A. M., Ahmad, S., Merkin, J. H., & Pop, I. (2013). Mixed convection boundary-layer flow along a vertical cylinder embedded in a porous medium filled by a nanofluid.* Transport in Porous Media, 96*(2), 237–253. https://doi.org/10.1007/s11242-012-0085-y 246.)*

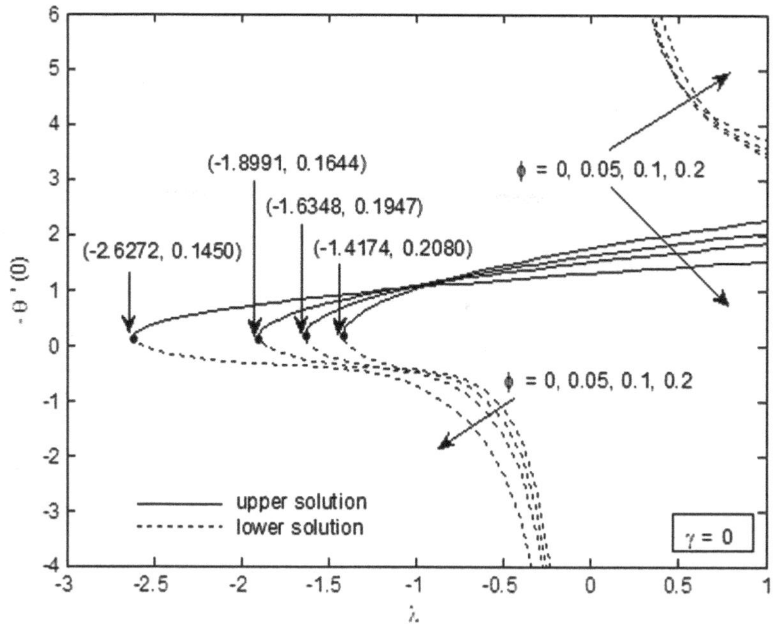

Figure 6.13 Plots of -$\theta'(0)$ against the mixed convection parameter λ for the addition of copper (Cu) nanoparticles with volume fractions $\varphi = 0.0, 0.5, 0.1, 0.2$ and $\gamma = 0$ (flat surface). *(From Rohni, A. M., Ahmad, S., Merkin, J. H., & Pop, I. (2013). Mixed convection boundary-layer flow along a vertical cylinder embedded in a porous medium filled by a nanofluid.* Transport in Porous Media, *96(2), 237–253. https://doi.org/10.1007/s11242-012-0085-y 246.)*

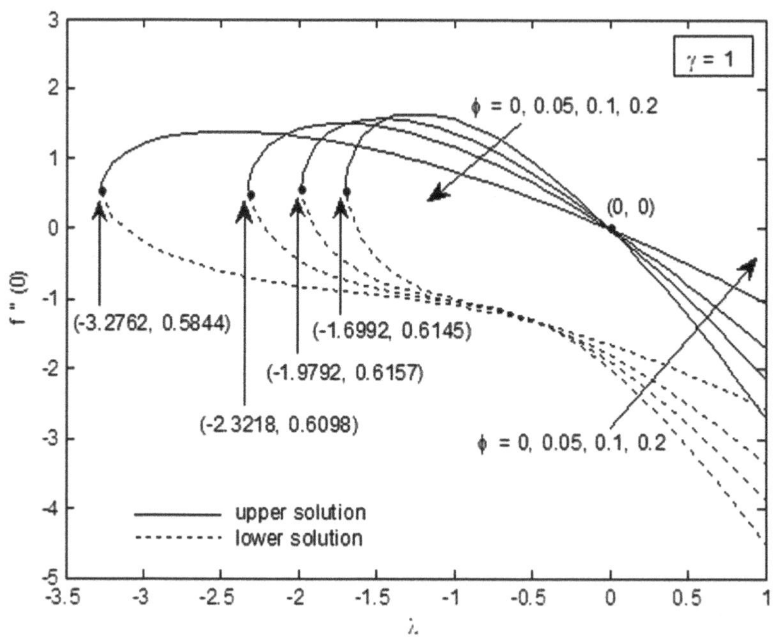

Figure 6.14 Plots of $f''(0)$ against the mixed convection parameter λ for the addition of copper (Cu) nanoparticles with $\varphi = 0.0, 0.05, 0.1, 0.2$ and $\gamma = 1$. *(From Rohni, A. M., Ahmad, S., Merkin, J. H., & Pop, I. (2013). Mixed convection boundary-layer flow along a vertical cylinder embedded in a porous medium filled by a nanofluid.* Transport in Porous Media, *96(2), 237–253. https://doi.org/10.1007/s11242-012-0085-y 246.)*

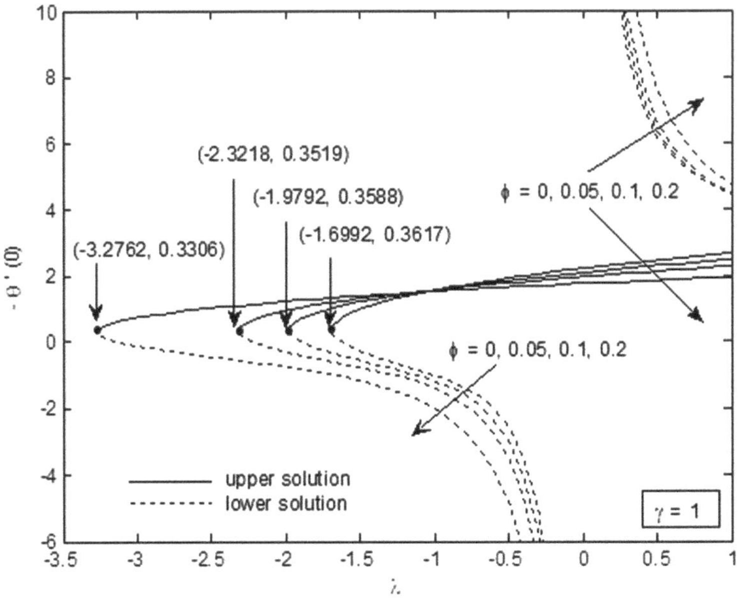

Figure 6.15 Plots of $-\theta'(0)$ against the mixed convection parameter λ for the addition of copper (Cu) nanoparticles with $\varphi = 0.0, 0.05, 0.1, 0.2$ and $\gamma = 1$. *(From Rohni, A. M., Ahmad, S., Merkin, J. H., & Pop, I. (2013). Mixed convection boundary-layer flow along a vertical cylinder embedded in a porous medium filled by a nanofluid.* Transport in Porous Media, *96(2), 237–253. https://doi.org/10.1007/s11242-012-0085-y 246.)*

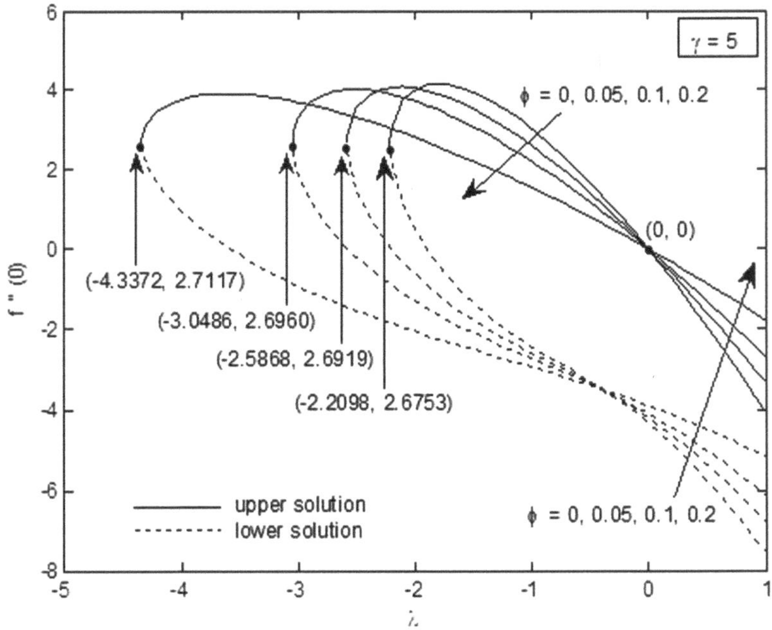

Figure 6.16 Plots of $f''(0)$ against the mixed convection parameter λ for the addition of copper (Cu) nanoparticles with $\varphi = 0.0, 0.05, 0.1, 0.2$ and $\gamma = 5$. *(From Rohni, A. M., Ahmad, S., Merkin, J. H., & Pop, I. (2013). Mixed convection boundary-layer flow along a vertical cylinder embedded in a porous medium filled by a nanofluid.* Transport in Porous Media, *96(2), 237–253. https://doi.org/10.1007/s11242-012-0085-y 246.)*

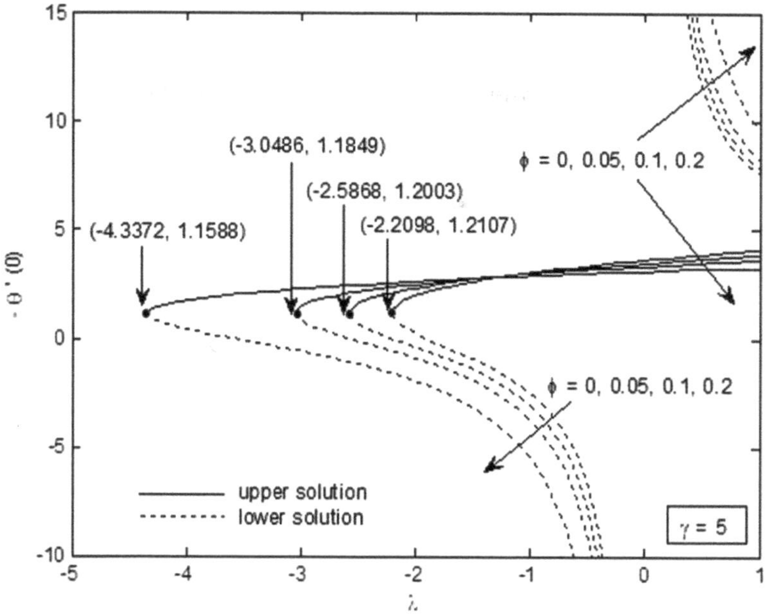

Figure 6.17 Plots of $-\theta'(0)$ against the mixed convection parameter λ for the addition of copper (Cu) nanoparticles with $\varphi = 0.0, 0.05, 0.1, 0.2$ and $\gamma = 5$. *(From Rohni, A. M., Ahmad, S., Merkin, J. H., & Pop, I. (2013). Mixed convection boundary-layer flow along a vertical cylinder embedded in a porous medium filled by a nanofluid.* Transport in Porous Media, *96(2), 237–253. https://doi.org/10.1007/s11242-012-0085-y 246.)*

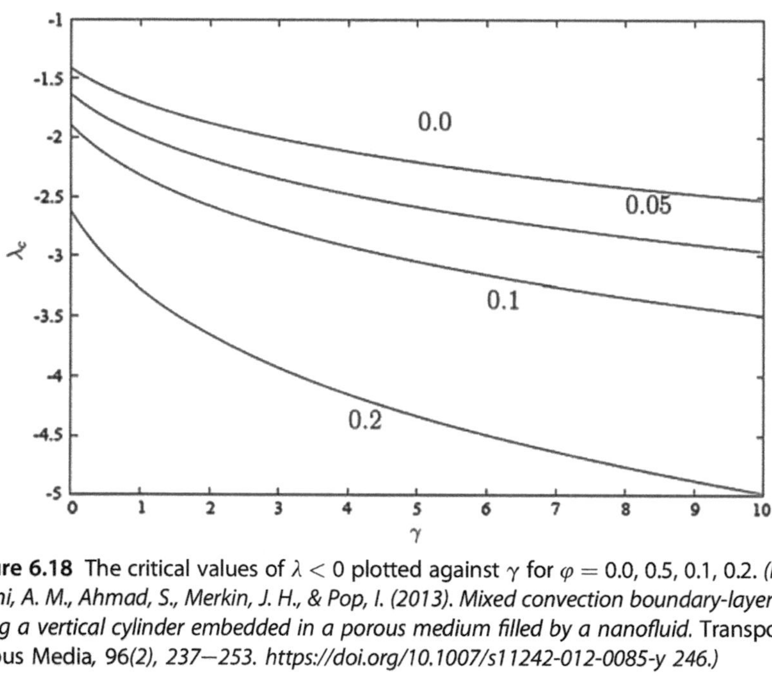

Figure 6.18 The critical values of $\lambda < 0$ plotted against γ for $\varphi = 0.0, 0.5, 0.1, 0.2$. *(From Rohni, A. M., Ahmad, S., Merkin, J. H., & Pop, I. (2013). Mixed convection boundary-layer flow along a vertical cylinder embedded in a porous medium filled by a nanofluid.* Transport in Porous Media, *96(2), 237–253. https://doi.org/10.1007/s11242-012-0085-y 246.)*

6.5.3 A special exact solution

For the corresponding regular fluid model ($\varphi = 0$), that is $A(\varphi) = B(\varphi) = 1$, a special exact solution of Eqs. (6.46) and (6.47) has been presented by Magyari and Keller (2003) when $\gamma = 0$ (flat plate). This exact solution is

$$f(\eta) = \eta + \frac{B\lambda}{c}(1 - e^{-c\eta}), \quad \theta(\eta) = e^{-c\eta} \tag{6.52}$$

where we require $c > 0$ and that

$$c = \sqrt{\frac{2(2 + B\,\lambda)}{A}}, \quad \gamma = \frac{1}{\sqrt{2\,A(2 + B\,\lambda)}} \tag{6.53}$$

From the solution given by (6.63, 6.64) we have

$$f''(0) = -B\lambda\sqrt{\frac{2(2 + B\lambda)}{A}}, \quad \theta'(0) = -\sqrt{\frac{2(2 + B\lambda)}{A}} \tag{6.54}$$

These expressions agree with that for large λ studied in the paper. Expressions (6.54) require that $\gamma > 0$ and $B\lambda > -2$ putting a limit on the range of existence of this solution.

6.6 Mixed convection boundary layer flow over a vertical plate embedded in a porous medium filled with a suspension of nano-encapsulated phase change materials

6.6.1 Mathematical model

Consider the mixed convection flow from a vertical flat plate embedded in a homogenous Darcy porous medium that is filled with a suspension of nano-encapsulated phase change material (NEPCM) nanoparticles. The core of these nanoparticles is a phase change material and it is encapsulated in a shell. This core is in the liquid state when the temperature of the fluid is higher than the fusion temperature ($T > T_f$), and in solid state when the temperature of the fluid is lower than the fusion temperature ($T < T_f$). There is an external flow with the constant velocity U_∞ and temperature T_∞, while the surface of the plate is kept at a constant temperature T_w. It is possible to have an assisting ($T_w > T_\infty$) or an opposing ($T_w < T_\infty$) flow. It is assumed that the fusion temperature fulfilled the condition ($T_w < T_f < T_\infty$), and hence, the nanoparticles will experience phase change in the boundary layer. The physical model is given in Fig. 6.19.

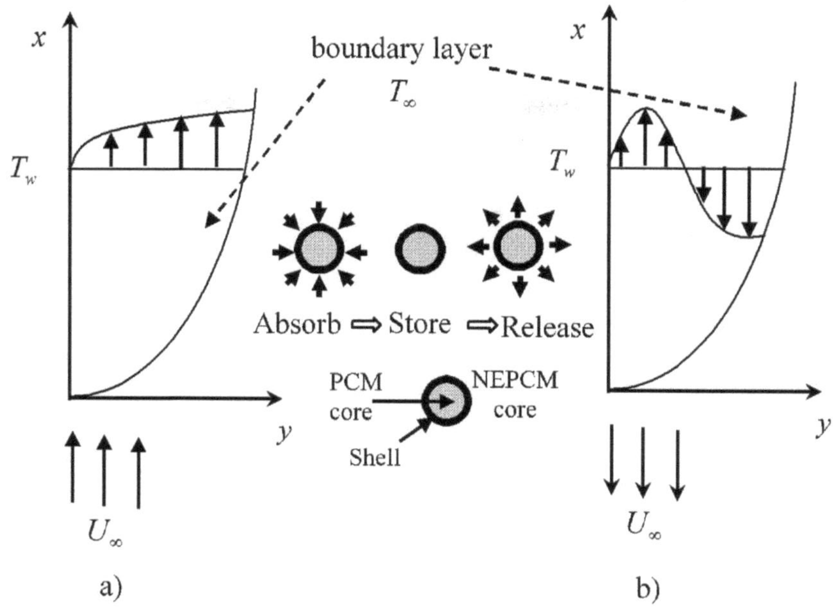

Figure 6.19 Physical model and coordinate system of nano-encapsulated phase change material suspension over a plate embedded in a porous medium: (A) assisting flow; (B) opposing flow. *(From Ghalambaz, M., Grosan, T., & Pop, I. (2019). Mixed convection boundary layer flow and heat transfer over a vertical plate embedded in a porous medium filled with a suspension of nano-encapsulated phase change materials.* Journal of Molecular Liquids, 293, *111432. https://doi.org/10.1016/j.molliq.2019.111432 8.)*

Considering the Oberbeck–Boussinesq approximation and applying the boundary layer approximations, the governing equations for the porous medium saturated by the NEPCM nanofluid are (Ahmad & Pop, 2010; Kuravi et al., 2009; Rehman et al., 2016).

$$\frac{\partial u}{\partial x}+\frac{\partial v}{\partial y}=0 \tag{6.55}$$

$$\frac{\mu_b}{\mu_f}u=\frac{\mu_b}{\mu_f}U_\infty+\frac{gK\left[\varphi(\rho\beta)_p+(1-\varphi)(\rho\beta)_f\right]}{\mu_f}(T-T_\infty) \tag{6.56}$$

$$(\rho c_p)_b\left(u\frac{\partial T}{\partial x}+v\frac{\partial T}{\partial y}\right)=k_{m,b}\frac{\partial^2 T}{\partial y^2} \tag{6.57}$$

along with the boundary conditions

$$v = 0,\ T = T_w \text{ at } y = 0; \quad u \to U_\infty,\ T \to T_\infty \text{ as } y \to \infty \tag{6.58}$$

In these equations the bulk suspension of NEPCM and base fluid is indicated by the subscript b. The subscripts of f and p denote the base fluid and NEPCM particles, respectively. The subscript of m indicates the effective thermophysical properties of the NEPCM suspension and the porous matrix. Here, $k_{m,b}$ is the thermal conductivity of the NEPCM solution and porous matrix. Considering the suspension of a base fluid and NEPCM nanoparticles as a nanofluid, the relations for estimation of thermophysical properties are essential. The density of the NEPCMs' suspension is evaluated as shown in (Chai et al., 2018), while the density of the particle (shell and core) is calculated as shown in (Alisetti & Roy, 2000; Du et al., 2018; Buongiorno et al., 2009):

$$\rho_b = (1-\varphi)\rho_f + \varphi\rho_p, \quad \rho_p = \frac{(1+\iota)\rho_c\rho_s}{\rho_s + \iota\rho_c} \tag{6.59}$$

where φ is the volume fraction of the dispersed NEPCM particles, ι denotes the core–shell weight ratio, and the subscript of c and s represent the core and the shell of NEPCM particles, respectively. The specific heat capacity of a suspension of particles and a host fluid C_{p_b} is computed by Khanafer and Vafai (2011) along with the effective specific heat capacity of the composite NEPCM nanoparticle formed by core and shell (Buongiorno et al., 2009):

$$C_{p_b} = \frac{(1-\varphi)(\rho C_p)_f + \varphi(\rho C_p)_p}{\rho_b}, \quad (\rho C_p)_p = \frac{(C_{p_c} + \iota C_{p_s})\rho_c\rho_s}{(\rho_s + \iota\rho_c} \tag{6.60}$$

where the C_{p_c} denotes the average heat capacity of the phase change core.

The thermal volume expansion of the suspension of a base fluid and nanoparticles (β_b) is given by (Venerus et al., 2010)

$$\beta_b = (1-\varphi)\beta_f + \varphi\beta_p \tag{6.61}$$

A linear dependence between the bulk dynamic viscosity/thermal conductivity and the nanoparticles' volume fraction is considered (see (Venerus et al., 2010), (Zaraki et al., 2015) and (Ghalambaz, Sheremet, et al., 2019b))

$$\mu_b = \mu_f(1 + N\nu\varphi), \quad k_b = k_f(1 + Nc\varphi) \tag{6.62}$$

where $N\nu$ and Nc are the number of dynamic viscosity and the number of thermal conductivity, respectively. The effective thermal conductivity of the NEPCM suspension and porous media can be evaluated using

$$k_{m,b} = \varepsilon k_b + (1-\varepsilon)k_{sp} \tag{6.63}$$

By using this equation, the ratio of $k_{m,b}/k_{m,f}$ can be written as

$$\frac{k_{m,b}}{k_{m,f}} = \frac{\varepsilon\left[k_f(1+Nc\,\varphi)\right] + (1-\varepsilon)k_{sp}}{\varepsilon k_f + (1-\varepsilon)k_{ap}} = 1 + \frac{Nc\,\varphi}{1+\frac{1-\varepsilon}{\varepsilon}\left(\left(k_{sp}/k_f\right)\right)} \tag{6.64}$$

In order to reduce the governing partial differential Eqs. (6.55) to (6.57) to a set of ordinary differential equations, the following similarity variables are introduced:

$$\psi = \alpha_{m,f}\,(2\,Pe_x)^{\frac{1}{2}} f\,(\eta),\ \theta(\eta) = (T - T_\infty)/(T_w - T_\infty),\ \eta = Pe_x^{\frac{1}{2}} y \Big/ \left(x\sqrt{2}\right) \tag{6.65}$$

where ψ is the stream function. Using (6.65), Eqs. (6.55) to (6.57) can be reduced to the following ordinary (similarity) differential equation (see (Ghalambaz, Grosan, & Pop, 2019a)):

$$\frac{k_{m,b}}{k_{m,f}} f''' + \frac{(\rho C_p)_p}{(\rho C_p)_f} f f'' = 0 \tag{6.66}$$

subject to the boundary conditions

$$\begin{gathered} f(0) = 0, \quad f'(0) = 1 + \lambda_1\lambda_2 \\ f'(\eta) \to 1 \text{ as } \eta \to \infty \end{gathered} \tag{6.67}$$

where

$$\lambda_1 = \frac{Ra_x}{Pe_x}, \quad \lambda_2 = \frac{\mu_f}{\mu_b}\left[1 - \varphi + \varphi\frac{(\rho\beta)_s}{(\rho\beta)_p}\right] \tag{6.68}$$

are the constant mixed convection parameters and ξ is the nondimensional specific function related to the specific heat of phase change:

$$\xi = \frac{\pi}{2} sin\left[\frac{\pi}{\delta}\left(\theta - \theta_f + \frac{\delta}{2}\right)\right] \times \begin{cases} 0, & \theta < \theta_f - \frac{\delta}{2} \text{ or } \theta > \theta_f + \frac{\delta}{2} \\ 1, & \theta_f - \frac{\delta}{2} < \theta < \theta_f + \frac{\delta}{2} \end{cases}$$

$$\text{where } \theta_f = \frac{T_f - T_c}{T_w - T_c},\ \delta = \frac{T_{Mr}}{T_w - T_c} \tag{6.69}$$

Here T_f is the fusion temperature, while T_{Mr} is the melting range temperature.

The reduced Nusselt number Nur is given by (see (Ghalambaz, Grosan, & Pop, 2019a))

$$Nur = -\left[1+\frac{Nc\,\varphi}{1+\frac{1-\varepsilon}{\varepsilon}\left(k_{sp}/k_f\right)}\right]\left[\frac{1-\varphi+\varphi\,(\rho\beta)_p/(\rho\beta)_f}{1+Nv\,\varphi}\right]^{-1} f''(0) \tag{6.70}$$

6.6.2 Comments

The third-order nonlinear boundary value problem (6.66, 6.67) is solved numerically by a finite-difference solver as described in (Shampine et al., 2003). The calculations are continued until a relative error less than 10^{-9} is reached. The asymptotic value $\eta_\infty = 12$ is adopted for calculations to ensure that the results are independent of the asymptotic value of the boundary conditions at the edge of the boundary layer. It is found that the present study has a dual solutions, an upper and lower branch solutions. To ensure the performance of the developed code, the results of (Merkin, 1980) is selected as a benchmark study for $\varphi = 0$ (classical porous medium). The results are summarized in Table 6.1. As can be seen the comparison is excellent and it gives us the confidence that the present results are accurate and correct.

Fig. 6.20 shows the dual solutions of the boundary value problem (6.66, 6.67) when $\lambda_1 = -1.4$. Continues lines depict the upper branch (stable) solution, while dashed lines denote the lower branch (unstable) solution. This figure shows that the boundary layer thickness of the stable solution is lower than that of an unstable solution. In an unstable boundary layer, the environmental conditions can easily distribute in the fluid and change the unstable boundary layer to a stable boundary layer. Hence, the unstable boundary layer cannot be existing in regular domains with usual physical assumptions.

Figs. 6.21 and 6.22 show the boundary layer velocity and temperature profiles $f(\eta)$ and $\theta(\eta)$ for various values of φ. As can be seen, the variation of the value of φ would significantly affect these velocity and temperature profiles. When $\eta \cong 1$, $f(\eta)$ all reach a focal point. Fig. 6.21 reveals that $\eta = 1$ corresponds to a $\theta(\eta)$ about $\theta(\eta) = 0.3$, which is the fusion

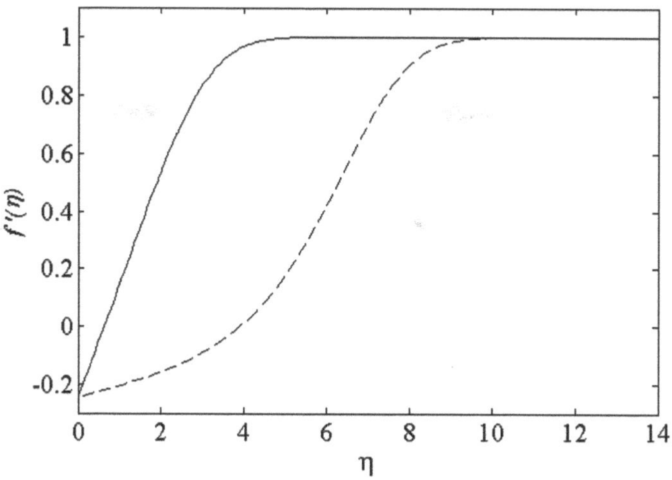

Figure 6.20 Stable (full line) and unstable (broken line) velocity profiles for $\lambda_1 = -1.4$ and $\varphi = 0.01$. *(From Ghalambaz, M., Grosan, T., & Pop, I. (2019). Mixed convection boundary layer flow and heat transfer over a vertical plate embedded in a porous medium filled with a suspension of nano-encapsulated phase change materials.* Journal of Molecular Liquids, 293, *111432. https://doi.org/10.1016/j.molliq.2019.111432 8.)*

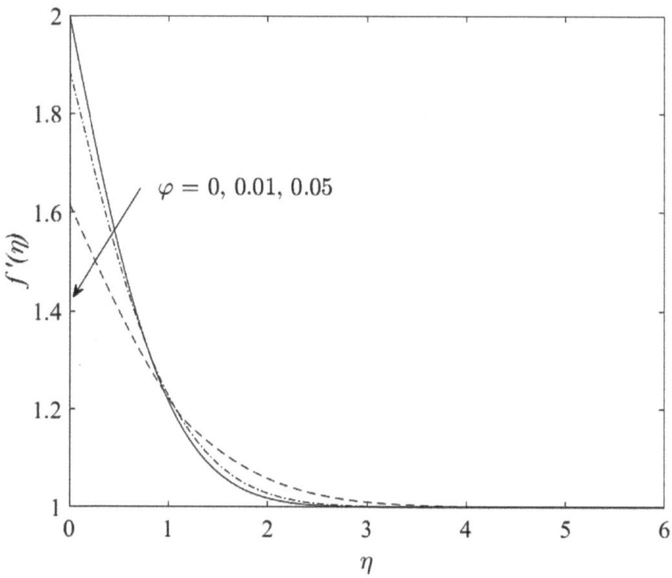

Figure 6.21 Dimensionless velocity profiles $f'(\eta)$ for $\lambda_1 = 1$ and different values of φ. *(From Ghalambaz, M., Grosan, T., & Pop, I. (2019). Mixed convection boundary layer flow and heat transfer over a vertical plate embedded in a porous medium filled with a suspension of nano-encapsulated phase change materials.* Journal of Molecular Liquids, 293, *111432. https://doi.org/10.1016/j.molliq.2019.111432 8.)*

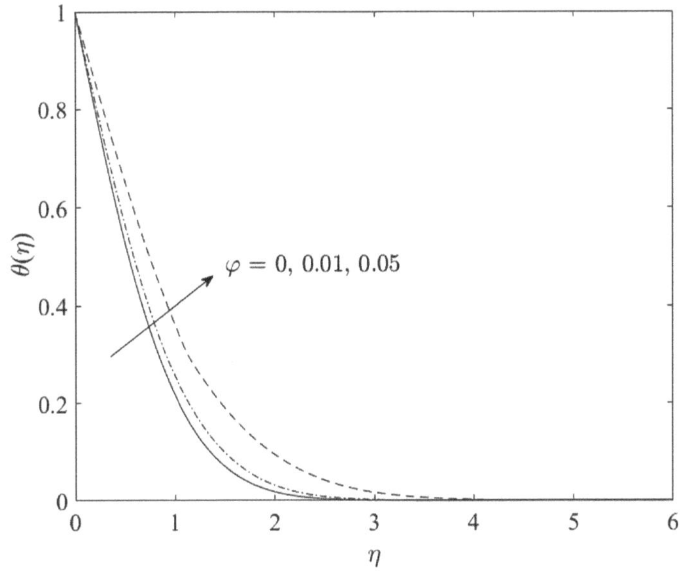

Figure 6.22 Dimensionless temperature profiles $\theta(\eta)$ for $\lambda_1 = 1$ and different values of φ. *(From Ghalambaz, M., Grosan, T., & Pop, I. (2019). Mixed convection boundary layer flow and heat transfer over a vertical plate embedded in a porous medium filled with a suspension of nano-encapsulated phase change materials.* Journal of Molecular Liquids, *293, 111432. https://doi.org/10.1016/j.molliq.2019.111432 8.)*

temperature $\theta_f\ (\eta)\ = 0.3$. Hence, it can be concluded that the behavior of $f'(\eta)$ is dictated by $\theta(\eta)$. The increase of φ reduces the velocity at the wall $f'(0)$. It should be noted that in the Darcy flow, the wall slip velocity at the surface is possible. The reduction of $f'(\eta)$ is mainly due to the increase of dynamic viscosity of nanofluid by the increase of φ. Fig. 6.21 shows that the increase of φ, increases the velocity boundary layer thickens. This is due to the enhancement of the thermal conductivity of the solution. Moreover, in the case of $\eta = 1$, which is about the fusion temperature θ_f a slight change in $\theta(\eta)$ when $\varphi = 0.05$ can be seen. In this case, $\theta(\eta)$ shows that φ tends to increase $\theta'(0)$, see Fig. 6.22.

The values of $f''(0)$ as a function of λ_1 are plotted in Fig. 6.23 for various values of φ. There are two solutions for some negative values of λ_1, an upper and lower solutions, respectively. The critical values λ_{1c} of λ_1 are reported in Table 6.3.

Fig. 6.24 illustrates the variation of *Nur* for several values of λ_1. This figure shows that the increase of λ_1 reduces *Nur*. The increase of λ_1 can be considered as weakening of the forced convection power. When

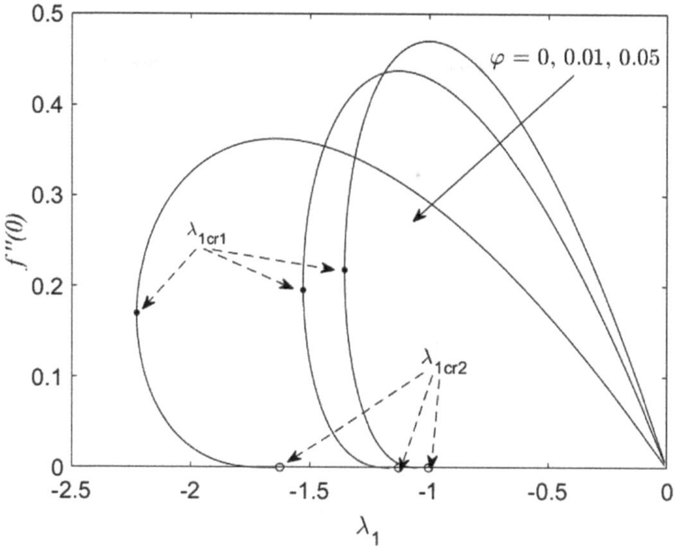

Figure 6.23 Existence domain of the dual solutions for different values of φ. *(From Ghalambaz, M., Grosan, T., & Pop, I. (2019). Mixed convection boundary layer flow and heat transfer over a vertical plate embedded in a porous medium filled with a suspension of nano-encapsulated phase change materials.* Journal of Molecular Liquids, 293, *111432. https://doi.org/10.1016/j.molliq.2019.111432 8.)*

Table 6.3 The critical values of the parameter λ_1.

φ	λ_{1c}	
	Critical 1	Critical 2
0	−1.3541	−1.0000
0.01	−1.5272	−1.1248
0.05	−2.2247	−1.6235

$\lambda_1 = -1.5$, the magnitude of *Nur* in the case of $\varphi = 0$ (classical porous medium) is higher than that of$\varphi = 0.05$. In contrast, when$\lambda_1 = +1$, the magnitude of *Nur* in the case of $\varphi = 0$ is much lower than the case of $\varphi = 0.05$. Fig. 6.25 shows, in addition, that the increase of φ increases *Nur* for $\lambda_1 > \lambda_{1c2}$, while Nur increases for all values of λ_1. It is worth mentioning that for $\lambda_1 < \lambda_{1c2}$, the external flow and the flow due to the buoyancy are opposite (i.e., the temperature of the wall is lower than the ambient temperature) and in this case, the magnitude of Nur increases with the increase of φ. The decrease of the magnitude of *Nur* by the increase of φ for the cases with high values of λ_1 is because the presence

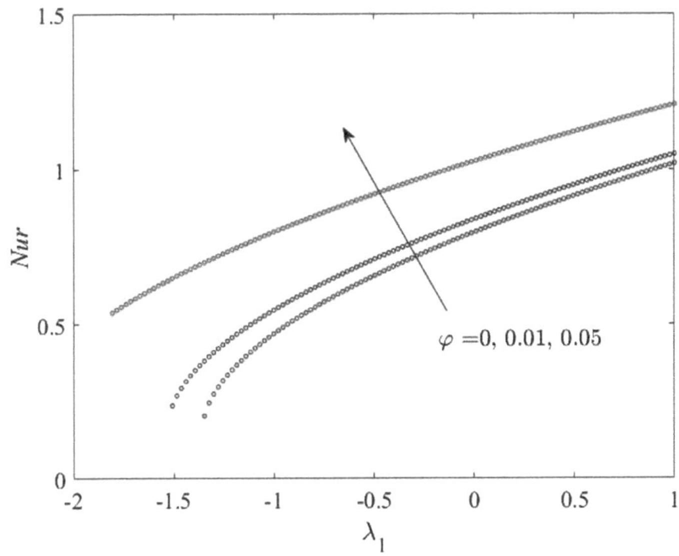

Figure 6.24 Variation of *Nur* with λ_1 when $\theta_f = 0.8$. *(Modified from Ghalambaz, M., Grosan, T., & Pop, I. (2019). Mixed convection boundary layer flow and heat transfer over a vertical plate embedded in a porous medium filled with a suspension of nano-encapsulated phase change materials.* Journal of Molecular Liquids, 293, *111432. https://doi.org/10.1016/j.molliq.2019.111432 8.)*

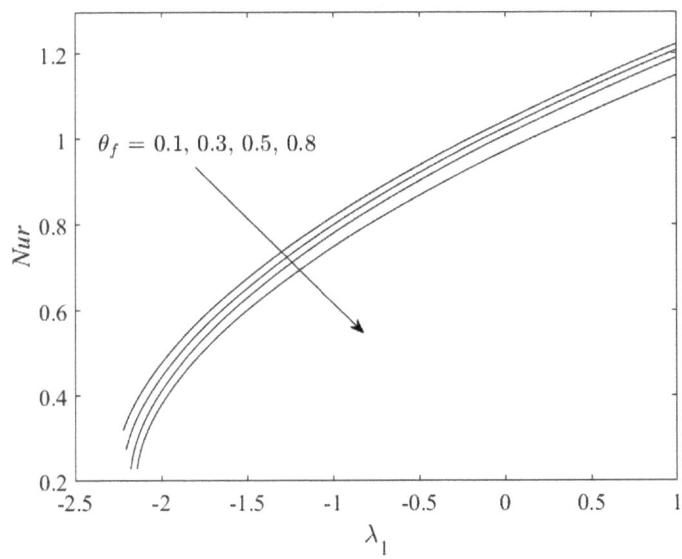

Figure 6.25 Variation of *Nur* with λ_1 when $\varphi = 0.05$. *(Modified from Ghalambaz, M., Grosan, T., & Pop, I. (2019). Mixed convection boundary layer flow and heat transfer over a vertical plate embedded in a porous medium filled with a suspension of nano-encapsulated phase change materials.* Journal of Molecular Liquids, 293, *111432. https://doi.org/10.1016/j.molliq.2019.111432 8.)*

of nanoparticles increases the thermal conductivity of the base fluid. Hence, in a fluid with higher thermal conductivity, *Nur* drops.

It is now commonplace to mention the authors who have published papers on forced, free, and mixed convection boundary layer flow in porous media as well as on nanofluids in porous media: (Ahmad & Pop, 2014; Arifin et al., 2012; Bachok et al., 2010a, 2010b; Chamkha, Al-Mudhaf, & Pop, 2006; Chamkha, Bercea, & Pop, 2006; Chin et al., 2007; Cîmpean et al., 2006; Cortell, 2005; Elbashbeshy & Bazid, 2005; Grosan & Pop, 2001; Hamad & Pop, 2011; Hayat et al., 2011; Ishak et al., 2008a, 2008b, 2008c; Khan & Pop, 2011a, 2011b, 2012a, 2012b, 2012c, 2013a, 2013b; Khan et al., 2014, 2015; Kumaran & Pop, 2006; Ling, Nazar, & Pop, 2007; Ling, Nazar, Pop, & Merkin, 2007; Ling et al., 2009; Lok et al., 2013; Magyari et al., 2006; Merkin & Pop, 2010; Merkin et al., 2012, 2014, 2015, 2019; Nazar et al., 2006; Pătrulescu et al., 2020; Postelnicu et al., 2000, 2009; Ramli et al., 2015; Rohni et al., 2013; Rosali et al., 2011a, 2011b, 2014, 2016; Rosca, Grosan, & Pop, 2012; Rosca, Rosca, et al., 2012; Rosca et al., 2015; Yasin et al., 2013; You et al., 2014), etc.

Nomenclature

a, b, c	constants
A, B	parameters
C_f	skin friction coefficient
$(\rho\, C_p)$	heat capacitance (J/Km^3)
D_B	inertial resistance coefficient (m^2/s)
D_T	thermophoretic diffusion coefficient (m^2/s)
$f'(\eta)$	dimensionless velocity functions
g	gravitational acceleration vector (m/s^2)
k	thermal conductivity (W/mK)
K	permeability of the porous medium (m^2)
L	length scale (m)
Nc	number of thermal viscosity
Nu	Nusselt number
Nv	number of dynamic viscosity
p	pressure (Pa)
Pr	Prandtl number
Pe	Péclet number
q_w	heat flux from the surface (W/m^2)
r_0	body radius
Re	Reynolds number
S	dimensionless mass flux parameter

T	dimensional fluid temperature (K)
T_0	characteristic temperature (K)
T_w	wall (surface) temperature (K)
T_∞	ambient temperature (K)
(u,v)	velocity components (m/s)
$\mathbf{u_e(x)}$	mainstream velocity (m/s)
U_0	scale velocity (m/s)
U_∞	uniform stream velocity (m/s) s
$v_w(x)$	mass flux velocity (m/s)
(x, y)	Cartesian coordinates (m)
(r, x)	radial and axial coordinates (m)

Greek letters

α	thermal diffusivity (m^2/s)
β	coefficient of thermal expansion (1/K)
η	dimensionless similarity variable
γ	parameter (Section 6.2)
γ	curvature (Section 6.4)
λ	mixed convection parameter
λ_c	critical value of λ
μ	dynamic viscosity (Ns/m^2)
ν	kinematic viscosity (m^2/s)
φ	nanoparticle volume fraction
ψ	stream function (m^2/s)
ρ	density of fluid (kg/m^3)
ι	core–shell weight ratio
$\theta(\eta)$	dimensionless temperature function
θ_w	dimensionless wall temperature
τ_w	wall shear stress (N/m^2)
ξ	nondimensional specific function related to the specific heat of phase change

Subscript

b	bulk suspension
c	critical point
c	core (Section 6.5)
e, ∞	condition outside of the boundary layer
f	base fluid
l	lower branch
m	effective thermophysical properties
nf	nanofluid
p	particle
r	in radial direction
s	solid phase

s	shell (Section 6.5)
u	upper branch
w	condition at the wall
x	local numbers
θ	in azimuthal direction

Superscript

′	differentiation with respect to η

References

Ahmad, S., & Pop, I. (2010). Mixed convection boundary layer flow from a vertical flat plate embedded in a porous medium filled with nanofluids. *International Communications in Heat and Mass Transfer, 37*(8), 987–991. https://doi.org/10.1016/j.icheatmasstransfer.2010.06.004

Ahmad, S., & Pop, I. (2014). Melting effect on mixed convection boundary layer flow about a vertical surface embedded in a porous medium: Opposing flows case. *Transport in Porous Media, 102*(3), 317–323. https://doi.org/10.1007/s11242-014-0291-x

Alisetti, L. S., & Roy, S. K. (2000). Forced convection heat transfer to phase change material slurries in circular ducts. *Journal of Thermophysics and Heat Transfer, 14*(1), 115–118. https://doi.org/10.2514/2.6499

Aly, E. H., Elliott, L., & Ingham, D. B. (2003). Mixed convection boundary-layer flow over a vertical surface embedded in a porous medium. *European Journal of Mechanics - B: Fluids, 22*(6), 529–543. https://doi.org/10.1016/S0997-7546(03)00059-1

Arifin, N., Nazar, R., & Pop, I. (2012). Free and mixed convection boundary layer flow past a horizontal surface embedded in a porous medium filled with a nanofluid. *Journal of Thermophysics and Heat Transfer, 26*(2), 375–382. https://doi.org/10.2514/1.T3645

Bachok, N., Ishak, A., & Pop, I. (2010a). Mixed convection boundary layer flow near the stagnation point on a vertical surface embedded in a porous medium with anisotropy effect. *Transport in Porous Media, 82*(2), 363–373. https://doi.org/10.1007/s11242-009-9431-0

Bachok, N., Ishak, A., & Pop, I. (2010b). Mixed convection boundary layer flow over a permeable vertical flat plate embedded in an anisotropic porous medium. *Mathematical Problems in Engineering, 2010*, 659023. https://doi.org/10.1155/2010/659023

Barletta, A., Rossi di Schio, E., & Celli, M. (2015). Convection and instability phenomena in nanofluid-saturated porous media. In V. Bianco, O. Manca, S. Nardini, & K. Vafai (Eds.), *Heat transfer enhancement with nanofluids* (pp. 341–364). CRC Press.

Bear, J. (2018). *Modeling phenomena of flow and transport in porous media*. Springer.

Bejan, A. (1987). Convective heat transfer in porous media. In S. Kakaç, R. K. Shah, & W. Aung (Eds.), *Handbook of single-phase convective heat transfer* (pp. 16.1–16.34). Wiley.

Bejan, A. (2013). *Convective heat transfer* (4th ed.). Wiley.

Bejan, A., Dincer, I., Lorente, S., Miguel, A. F., & Reis, A. H. (2004). *Porous and complex flow structures in modern technologies*. Springer.

Bejan, A., & Kraus, A. D. (Eds.). (2003). *Heat transfer handbook*. Wiley.

Blasius, H. (1908). Grenzschichten in Flüssigkeiten mit kleiner Reibung. *Zeitschrift für Angewandte Mathematik und Physik, 56*, 1–37.

Buongiorno, J., Venerus, D. C., Prabhat, N., McKrell, T., Townsend, J., Christianson, R., Tolmachev, Y. V., Keblinski, P., Hu, L. W., Alvarado, J. L., Bang, I. C., Bishnoi, S. W.,

Bonetti, M., Botz, F., Cecere, A., Chang, Y., Chen, G., Chen, H., Chung, S. J., ... Zhou, S. Q. (2009). A benchmark study on the thermal conductivity of nanofluids. *Journal of Applied Physics, 106*, 094312. https://doi.org/10.1063/1.3245330

Chai, L., Shaukat, R., Wang, L., & Wang, H. S. (2018). A review on heat transfer and hydrodynamic characteristics of nano/microencapsulated phase change slurry (N/MPCS) in mini/microchannel heat sinks. *Applied Thermal Engineering, 135*, 334–349. https://doi.org/10.1016/j.applthermaleng.2018.02.068

Chamkha, A. J., Al-Mudhaf, A. F., & Pop, I. (2006). Effect of heat generation or absorption on thermophoretic free convection boundary layer from a vertical flat plate embedded in a porous medium. *International Communications in Heat and Mass Transfer, 33*(9), 1096–1102. https://doi.org/10.1016/j.icheatmasstransfer.2006.04.009

Chamkha, A. J., Bercea, C., & Pop, I. (2006). Free convection over a truncated cone embedded in a porous medium saturated with pure or saline water at low temperatures. *Mechanics Research Communications, 33*(4), 433–440. https://doi.org/10.1016/j.mechrescom.2005.10.001

Cheng, P. (1977). Combined free and forced convection flow about inclined surfaces in porous media. *International Journal of Heat and Mass Transfer, 20*(8), 807–814. https://doi.org/10.1016/0017-9310(77)90110-7

Cheng, P. (1985). Natural convection in a porous medium: External flows. In S. Kakaç, W. Aung, & W. R. Viskanta (Eds.), *Natural convection: Fundamentals and applications* (pp. 475–513). Hemisphere.

Chin, K. E., Nazar, R., Arifin, N. M., & Pop, I. (2007). Effect of variable viscosity on mixed convection boundary layer flow over a vertical surface embedded in a porous medium. *International Communications in Heat and Mass Transfer, 34*(4), 464–473. https://doi.org/10.1016/j.icheatmasstransfer.2007.01.002

Cîmpean, D., Merkin, J. H., Pop, I., & Ingham, D. B. (2006). On a free convection problem over a vertical flat surface in a porous medium. *Transport in Porous Media, 64*(3), 393–411. https://doi.org/10.1007/s11242-005-5236-y

Combarnous, M. A., & Bia, P. (1971). Combined free and forced convection in porous medium. *Society of Petroleum Engineers Journal, 11*(4), 399–405. https://doi.org/10.2118/3192-PA

Cortell, R. (2005). Flow and heat transfer of a fluid through a porous medium over a stretching surface with internal heat generation/absorption and suction/blowing. *Fluid Dynamics Research, 37*(4), 231–245. https://doi.org/10.1016/j.fluiddyn.2005.05.001

Du, X., Fang, Y., Cheng, X., Du, Z., Zhou, M., & Wang, H. (2018). Fabrication and characterization of flame-retardant nanoencapsulated n-octadecane with melamine–formaldehyde shell for thermal energy storage. *ACS Sustainable Chemistry & Engineering, 6*(11), 15541–15549. https://doi.org/10.1021/acssuschemeng.8b03980

Elbashbeshy, E. M. A., & Bazid, M. A. A. (2005). Heat transfer in a porous medium over a stretching surface with internal heat generation and suction or injection. *Applied Mathematics and Computation, 158*(3), 799–807. https://doi.org/10.1016/j.amc.2003.08.141

Ene, H., & Polisevski, D. (1987). *Thermal flow in porous media*. Reidel.

Ghalambaz, M., Grosan, T., & Pop, I. (2019a). Mixed convection boundary layer flow and heat transfer over a vertical plate embedded in a porous medium filled with a suspension of nano-encapsulated phase change materials. *Journal of Molecular Liquids, 293*, 111432. https://doi.org/10.1016/j.molliq.2019.111432

Ghalambaz, M., Sheremet, M. A., Mehryan, S. A. M., Kashkooli, F. M., & Pop, I. (2019b). Local thermal non-equilibrium analysis of conjugate free convection within a porous enclosure occupied with Ag-MgO hybrid nanofluid. *Journal of Thermal Analysis and Calorimetry, 135*(2), 1381–1398. https://doi.org/10.1007/s10973-018-7472-8

Grosan, T., & Pop, I. (2001). Free convection over a vertical flat plate with a variable wall temperature and internal heat generation in a porous medium saturated with a non-Newtonian fluid. *Technische Mechanik, 21*(4), 313–318.

Hamad, M. A. A., & Pop, I. (2011). Scaling transformations for boundary layer flow near the stagnation-point on a heated permeable stretching surface in a porous medium saturated with a nanofluid and heat generation/absorption effects. *Transport in Porous Media, 87*(1), 25–39. https://doi.org/10.1007/s11242-010-9683-8

Hayat, T., Mustafa, M., Pop, I., & Obaidat, S. (2011). Flow and heat transfer due to stretching surface in an anisotropic porous medium filled with a viscoelastic fluid. *Journal of Porous Media, 14*(6), 523–532. https://doi.org/10.1615/JPorMedia.v14.i6.50

Homsy, G. M., & Sherwood, A. E. (1976). Convective instabilities in porous media with through flow. *AIChE Journal, 22*, 168–174. https://doi.org/10.1002/aic.690220121

Ingham, D. B., & Pop, I. (Eds.). (1988). *Transport phenomena in porous media: Vol. I.* Pergamon.

Ingham, D. B., & Pop, I. (Eds.). (2002). *Transport phenomena in porous media: Vol. II.* Pergamon.

Ingham, D. B., & Pop, I. (Eds.). (2005). *Transport phenomena in porous media: Vol. III.* Elsevier.

Ishak, A., Nazar, R., & Pop, I. (2008a). Dual solutions in mixed convection flow near a stagnation point on a vertical surface in a porous medium. *International Journal of Heat and Mass Transfer, 51*(5–6), 1150–1155. https://doi.org/10.1016/j.ijheatmasstransfer.2007.04.029

Ishak, A., Nazar, R., & Pop, I. (2008b). Mixed convection boundary layer flow adjacent to a vertical surface embedded in a stable stratified medium. *International Journal of Heat and Mass Transfer, 51*(13–14), 3693–3695. https://doi.org/10.1016/j.ijheatmasstransfer.2008.01.028

Ishak, A., Nazar, R., & Pop, I. (2008c). Mixed convection boundary layer flow over a vertical surface embedded in a thermally stratified porous medium. *Physics Letters A, 372*(14), 2355–2358. https://doi.org/10.1016/j.physleta.2007.12.006

Jafarimoghaddam, A., Merkin, J. H., & Pop, I. (2019). Mixed convection boundary layer flow on a vertical surface in a saturated porous medium: New perturbation solutions. *Transport in Porous Media, 128*(2), 741–753. https://doi.org/10.1007/s11242-019-01268-0

Khanafer, K., & Vafai, K. (2011). A critical synthesis of thermophysical characteristics of nanofluids. *International Journal of Heat and Mass Transfer, 54*, 4410–4428. https://doi.org/10.1016/j.ijheatmasstransfer.2011.04.048

Khanafer, K., & Vafai, K. (2019). Applications of nanofluids in porous medium. *Journal of Thermal Analysis and Calorimetry, 135*(2), 1479–1492. https://doi.org/10.1007/s10973-018-7565-4

Khan, Z. H., Culham, J. R., Khan, W. A., & Pop, I. (2015). Triple convective-diffusion boundary layer along a vertical flat plate in a porous medium saturated by a water-based nanofluid. *International Journal of Thermal Sciences, 90*, 53–61. https://doi.org/10.1016/j.ijthermalsci.2014.12.002

Khan, W. A., Culham, J. R., Lhan, Z. H., & Pop, I. (2014). Triple diffusion along a horizontal plate in a porous medium with convective boundary condition. *International Journal of Thermal Sciences, 86*, 60–67. https://doi.org/10.1016/j.ijthermalsci.2014.06.035

Khan, W. A., & Pop, I. (2011a). Flow and heat transfer over a continuously moving flat plate in a porous medium. *Journal of Heat Transfer, 133*(5), 054501. https://doi.org/10.1115/1.4003012

Khan, W. A., & Pop, I. (2011b). Free convection boundary layer flow past a horizontal flat plate embedded in a porous medium filled with a nanofluid. *Journal of Heat Transfer, 133*(9), 094501. https://doi.org/10.1115/1.4003834

Khan, W. A., & Pop, I. (2012a). Boundary layer flow past a stretching surface in a porous medium saturated by a nanofluid: Brinkman-Forchheimer model. *PloS One, 7*(10), e47031. https://doi.org/10.1371/journal.pone.0047031

Khan, W. A., & Pop, I. (2012b). Boundary layer stagnation-point flow towards stretching surface in a porous nanofluid-filled medium. *Journal of Thermophysics and Heat Transfer, 26*(1), 147–153. https://doi.org/10.2514/1.T3680

Khan, W. A., & Pop, I. (2012c). Heat transfer near stretching surface in porous medium using thermal nonequilibrium model. *Journal of Thermophysics and Heat Transfer, 26*(4), 681–685. https://doi.org/10.2514/1.T3841

Khan, W. A., & Pop, I. (2013a). The Cheng-Minkowycz problem for the triple- diffusive natural convection boundary layer flow past a vertical plate in a porous medium. *Journal of Porous Media, 16*(7), 637–646. https://doi.org/10.1615/JPorMedia.v16.i7.50

Khan, W. A., & Pop, I. (2013b). Triple diffusive free convection along a horizontal plate in porous media saturated by a nanofluid with convective boundary condition. *International Journal of Heat and Mass Transfer, 66*, 603–612. https://doi.org/10.1016/j.ijheatmasstransfer.2013.07.074

Kumaran, V., & Pop, I. (2006). Steady free convection boundary layer over a vertical flat plate embedded in a porous medium filled with water at 4°C. *International Journal of Heat and Mass Transfer, 49*(17–18), 3240–3252. https://doi.org/10.1016/j.ijheatmasstransfer.2006.01.026

Kuravi, S., Kota, K. M., Du, J., & Chow, L. C. (2009). Numerical investigation of flow and heat transfer performance of nano-encapsulated phase change material slurry in microchannels. *Journal of Heat Transfer, 131*, 062901. https://doi.org/10.1115/1.3084123

Lai, F. C., & Kulacki, F. A. (1990). The influence of surface mass flux on mixed convection over horizontal plates in saturated porous media. *International Journal of Heat and Mass Transfer, 33*(3), 576–579. https://doi.org/10.1016/0017-9310(90)90191-V

Lai, F. C., Prasad, V., & Kulacki, F. A. (1988). Aiding and opposing mixed convection in a vertical porous layer with a finite wall heat source. *International Journal of Heat and Mass Transfer, 31*(5), 1049–1061. https://doi.org/10.1016/0017-9310(88)90093-2

Ling, S. C., Nazar, R., & Pop, I. (2007a). Steady mixed convection boundary layer flow over a vertical flat plate in a porous medium filled with water at 4°C: Case of variable wall temperature. *Transport in Porous Media, 69*(3), 359–372. https://doi.org/10.1007/s11242-006-9077-0

Ling, S. C., Nazar, R., Pop, I., & Merkin, J. H. (2007b). Steady mixed convection boundary- layer flow over a vertical flat surface in a porous medium filled with water at 4°C: Variable surface heat flux. *Transport in Porous Media, 70*(3), 307–321. https://doi.org/10.1007/s11242-007-9101-z

Ling, S. C., Nazar, R., Pop, I., & Merkin, J. H. (2009). Mixed convection boundary- layer flow in a porous medium filled with water close its maximum density. *Transport in Porous Media, 76*(1), 139–151. https://doi.org/10.1007/s11242-008-9240-x

Lok, Y. Y., Merkin, J. H., & Pop, I. (2013). Mixed convection boundary-layer flow over a vertical surface embedded in a porous material subject to a convective boundary condition. *Transport in Porous Media, 98*(2), 451–463. https://doi.org/10.1007/s11242-013-0153-y

Magyari, E., & Keller, B. (2003). Boundary-layer similarity flows driven by a power-law shear over a permeable plane surface. *Acta Mechanica, 163*, 139–146. https://doi.org/10.1007/s00707-003-0001-1

Magyari, E., Pop, I., & Keller, B. (2001). Exact dual solutions occurring in the Darcy mixed convection flow. *International Journal of Heat and Mass Transfer, 44*(23), 4563–4566. https://doi.org/10.1016/S0017-9310(01)00054-0

Magyari, E., Pop, I., & Keller, B. (2005). Exact solutions for a longitudinal steady mixed convection flow over a permeable vertical thin cylinder in a porous medium. *International Journal of Heat and Mass Transfer, 48*(16), 3435–3442. https://doi.org/10.1016/j.ijheatmasstransfer.2005.01.043

Magyari, E., Pop, I., & Keller, B. (2006). Unsteady free convection along an infinite vertical flat plate embedded in a stably stratified fluid-saturated porous medium. *Transport in Porous Media, 62*(2), 233–249. https://doi.org/10.1007/s11242-005-1292-6

Mahdi, R. A., Mohammed, H. A., Munisamy, K. M., & Saeid, N. H. (2015). Review of convection heat transfer and fluid flow in porous media with nanofluid. *Renewable and Sustainable Energy Reviews, 41*, 715–734. https://doi.org/10.1016/j.rser.2014.08.040

Merkin, J. H. (1980). Mixed convection boundary layer flow on a vertical surface in a saturated porous medium. *Journal of Engineering Mathematics, 14*(4), 301–313. https://doi.org/10.1007/BF00052913

Merkin, J. H. (1986). Free convection from a vertical cylinder embedded in a saturated porous medium. *Acta Mechanica, 62*, 19–28. https://doi.org/10.1007/BF01175850

Merkin, J. H., Bachok, N., Ishak, A., & Pop, I. (2014). A comment on the flow and heat transfer past a permeable stretching/shrinking surface in a porous medium: Brinkman model. In *Transport in porous media*. https://doi.org/10.1007/s11242-013-0249-4

Merkin, J. H., Lok, Y. Y., & Pop, I. (2019). On an equation arising in natural convection boundary layer flow in a porous medium. *Zeitschrift fur Angewandte Mathematik und Physik*. https://doi.org/10.1007/s00033-019-1187-y

Merkin, J. H., & Pop, I. (1987). Mixed convection boundary-layer on a vertical cylinder embedded in a saturated porous medium. *Acta Mechanica, 66*, 251–262. https://doi.org/10.1007/BF01184297

Merkin, J. H., & Pop, I. (2010). Natural convection boundary-layer flow in a porous medium with temperature-dependent boundary conditions. *Transport in Porous Media, 85*(2), 397–414. https://doi.org/10.1007/s11242-010-9569-9

Merkin, J. H., Pop, I., & Ahmad, S. (2015). Note on the melting effect on mixed convection boundary-layer flow over a vertical flat surface embedded in a porous medium. *International Journal of Heat and Mass Transfer*. https://doi.org/10.1016/j.ijheatmasstransfer.2014.12.040

Merkin, J. H., Rohni, A. M., Ahmad, S., & Pop, I. (2012). On the temperature slip boundary condition in a mixed convection boundary-layer flow in a porous medium. *Transport in Porous Media, 94*(1), 133–147. https://doi.org/10.1007/s11242-012-9993-0

Merrill, K., Beauchesne, M., Previte, J., Paullet, J., & Weidman, P. (2006). Final steady flow near a stagnation point on a vertical surface in a porous medium. *International Journal of Heat and Mass Transfer, 49*, 4681–4686. https://doi.org/10.1016/j.ijheatmasstransfer.2006.02.056

Nazar, R., Amin, N., & Pop, I. (2004). Unsteady mixed convection boundary layer flow near the stagnation point on a vertical surface in a porous medium. *International Journal of Heat and Mass Transfer, 47*(12–13), 2681–2688. https://doi.org/10.1016/j.ijheatmasstransfer.2004.01.002

Nazar, R., Arifin, N. M., & Pop, I. (2006). Free convection boundary layer flow over vertical and horizontal flat plates embedded in a porous medium under mixed thermal boundary conditions. *International Communications in Heat and Mass Transfer, 33*(1), 87–93. https://doi.org/10.1016/j.icheatmasstransfer.2005.09.003

Nield, D. A., & Bejan, A. (2017). *Convection in porous media* (5th ed., pp. 629–982). Springer. https://doi.org/10.1007/978-3-319-49562-0

Oztop, H. F., & Abu-Nada, E. (2008). Numerical study of natural convection in partially heated rectangular enclosures filled with nanofluids. *International Journal of Heat and Fluid Flow, 29*(5), 1326–1336. https://doi.org/10.1016/j.ijheatfluidflow.2008.04.009

Pop, I., & Ingham, D. B. (2001). *Convective heat transfer: Mathematical and computational modelling of viscous fluids and porous media*. Pergamon.
Postelnicu, A., Grosan, T., & Pop, I. (2000). Free convection boundary-layer over a vertical permeable flat plate in a porous medium with internal heat generation. *International Communications in Heat and Mass Transfer, 27*(5), 729–738. https://doi.org/10.1016/S0735-1933(00)00153-6
Postelnicu, A., Magyari, E., & Pop, I. (2009). Effect of a uniform horizontal throughflow on the Darcy free convection over a permeable vertical plate with volumetric heat generation. *Transport in Porous Media, 80*(1), 101–115. https://doi.org/10.1007/s11242-009-9346-9
Prats, M. (1966). The effect of horizontal fluid flow on thermally induced convection currents in porous mediums. *Journal of Geophysical Research, 71*(20), 4835–4838. https://doi.org/10.1029/jz071i020p04835
Pătrulescu, F. O., Grosan, T., & Pop, I. (2020). Natural convection from a vertical plate embedded in a non-Darcy bidisperse porous medium. *Journal of Heat Transfer, 142*(1), 012504. https://doi.org/10.1115/1.4045067
Ramli, N., Merkin, J. H., & Pop, I. (2015). Mixed convection boundary-layer flow past a vertical surface embedded in a porous medium: Exponential case. *Transport in Porous Media, 106*(2), 425–438. https://doi.org/10.1007/s11242-014-0408-2
Rehman, M. M. U., Qu, Z. G., & Fu, R. P. (2016). Three-dimensional numerical study of laminar confined slot jet impingement cooling using slurry of nano-encapsulated phase change material. *Journal of Thermal Science, 25*, 431–439. https://doi.org/10.1007/s11630-016-0881-8
Rohni, A. M., Ahmad, S., Merkin, J. H., & Pop, I. (2013). Mixed convection boundary-layer flow along a vertical cylinder embedded in a porous medium filled by a nanofluid. *Transport in Porous Media, 96*(2), 237–253. https://doi.org/10.1007/s11242-012-0085-y
Rosali, H., Ishak, A., Nazar, R., Merkin, J. H., & Pop, I. (2014). The effect of unsteadiness on mixed convection boundary-layer stagnation-point flow over a vertical flat surface embedded in a porous medium. *International Journal of Heat and Mass Transfer, 77*, 147–156. https://doi.org/10.1016/j.ijheatmasstransfer.2014.04.059
Rosali, H., Ishak, A., & Pop, I. (2011a). Mixed convection stagnation-point flow over a vertical plate with prescribed heat flux embedded in a porous medium: Brinkman-extended Darcy formulation. *Transport in Porous Media, 90*(3), 709–719. https://doi.org/10.1007/s11242-011-9809-7
Rosali, H., Ishak, A., & Pop, I. (2011b). Stagnation point flow and heat transfer over a stretching/shrinking sheet in a porous medium. *International Communications in Heat and Mass Transfer, 38*(8), 1029–1032. https://doi.org/10.1016/j.icheatmasstransfer.2011.04.031
Rosali, H., Ishak, A., & Pop, I. (2016). Mixed convection boundary layer flow near the lower stagnation point of a cylinder embedded in a porous medium using a thermal nonequilibrium model. *Journal of Heat Transfer, 138*(8). https://doi.org/10.1115/1.4033164
Rosca, N. C., Grosan, T., & Pop, I. (2012). Stagnation-point flow and mass transfer with chemical reaction past a permeable stretching/shrinking sheet in a nanofluid. *Sains Malaysiana, 41*(10), 1271–1279.
Rosca, A. V., Rosca, N. C., Grosan, T., & Pop, I. (2012). Non-Darcy mixed convection from a horizontal plate embedded in a nanofluid saturated porous media. *International Communications in Heat and Mass Transfer, 39*(8), 1080–1085. https://doi.org/10.1016/j.icheatmasstransfer.2012.06.025
Rosca, A. V., Rosca, N. C., & Pop, I. (2015). Mixed convection heat and mass transfer from a vertical surface embedded in a porous medium. *Transport in Porous Media, 109*(2), 279–295. https://doi.org/10.1007/s11242-015-0517-6

Shampine, L. F., Gladwell, I., & Thompson, S. (2003). Solving ODEs with MATLAB. In *Solving ODEs with MATLAB*. Cambridge University Press. https://doi.org/10.1017/cbo9780511615542

Tiwari, R. K., & Das, M. K. (2007). Heat transfer augmentation in a two-sided lid-driven deferentially heated square cavity utilizing nanofluids. *International Journal of Heat and Mass Transfer, 50*, 2002–2018. https://doi.org/10.1016/j.ijheatmasstransfer.2006.09.034

Vadasz, P. (Ed.). (2008). *Emerging topics in heat and mass transfer in porous media*. Springer.

Vafai, K. (Ed.). (2005). *Handbook of porous media* (2nd ed.). CRC Press.

Venerus, D. C., Buongiorno, J., Christianson, R., Townsend, J., Bang, I. C., Chen, G., Chung, S. J., Chyu, M., Chen, H., Ding, Y., Dubois, F., Dzido, G., Funfschilling, D., Galand, Q., Gao, J., Hong, H., Horton, M., Hu, L., Iorio, C. S., ... Zhou, S. Q. (2010). Viscosity measurements on colloidal dispersions (nanofluids) for heat transfer applications. *Applied Rheology, 20*(4). https://doi.org/10.3933/ApplRheol-20-44582, 44582-1-44582–44587.

Wooding, R. A. (1963). Convection in a saturated porous medium at large Rayleigh number or Péclet number. *Journal of Fluid Mechanics, 15*, 527–544. https://doi.org/10.1017/S0022112063000434

Yasin, M. H. M., Arifin, N. M., Nazar, R., Ismail, F., & Pop, I. (2013). Mixed convection boundary layer flow embedded in a thermally stratified porous medium saturated by a nanofluid. *Advances in Mechanical Engineering, 2013*, Article 121943. https://doi.org/10.1155/2013/121943

You, X., Xu, H., & Pop, I. (2014). Free convection along a convectively heated vertical flat sheet embedded in a saturated porous medium. *International Communications in Heat and Mass Transfer, 55*, 102–108. https://doi.org/10.1016/j.icheatmasstransfer.2014.04.016

Zaraki, A., Ghalambaz, M., Chamkha, A. J., Ghalambaz, M., & De Rossi, D. (2015). Theoretical analysis of natural convection boundary layer heat and mass transfer of nanofluids: Effects of size, shape and type of nanoparticles, type of base fluid and working temperature. *Advanced Powder Technology, 26*(3), 935–946. https://doi.org/10.1016/j.apt.2015.03.012

CHAPTER 7

Convective flows with internal heat generation in porous media

7.1 Introduction

Convective flows in porous media, both natural and mixed, usually arise through the surface being maintained at a prescribed temperature or the surface having a prescribed heat flux. These have been much studied (Gebhart et al., 1988; Ingham & Pop, 2005; Nield & Bejan, 2017; Vafai, 2005), for example, for a variety of geometries and surface conditions. There are flows where the surface heat flux is related to the surface temperature, convective boundary conditions, which again have received considerable attention. An alternative way in which a convective flow can develop is through local heat generation within the porous medium. There are two ways in which this can be modeled, either by prescribing a spatial variation to the heat generation or through the heat generation being dependent on the local fluid temperature. These latter problems have many practical applications, for example when there is radioactive decay within the porous material or through the local heating within organic materials, as in stock piles of coal (Brooks & Glasser, 1986; Brooks et al., 1988), stores of damp wool or other fibrous materials (Gray & Wake, 1990; Gray et al., 2002; Sisson et al., 1992, 1993 or piles of bagasse (the residue after sugar has been extracted from sugar cane (Sexton et al., 2001)). This heating can lead to self-ignition and even to thermal runaway, the possible cause of hazardous situations for storage and transportation (Allali et al., 2011).

7.2 Flows with temperature dependent heat generation

We start by considering when the local heat transfer has a power-law temperature dependence proportional to $(T - T_\infty)^p$, $(p \geq 1)$, where T_∞ is the constant ambient temperature, initially describing the work of

Similarity Solutions for the Boundary Layer Flow and Heat Transfer of Viscous Fluids, Nanofluids, Porous Media, and Micropolar Fluids
ISBN 978-0-12-821188-5
https://doi.org/10.1016/B978-0-12-821188-5.00004-7

(Merkin, 2013) for the unsteady convection near a forward stagnation point. We assume that the flow is modeled by Darcy's law, and there is a constant heat flux q_w on the surface.

$$\frac{\partial u}{\partial x}+\frac{\partial v}{\partial y}=0 \tag{7.1}$$

$$u=\frac{g\beta K}{\nu}(T-T_{\infty})\frac{x}{l} \tag{7.2}$$

$$\frac{\partial T}{\partial t}+u\frac{\partial T}{\partial x}+v\frac{\partial T}{\partial y}=\alpha_m\frac{\partial^2 T}{\partial y^2}+\gamma\,(T-T_{\infty})^p \tag{7.3}$$

subject to the boundary conditions for an impermeable surface

$$v=0,\qquad \frac{\partial T}{\partial y}=-q_w \text{ on } y=0,\qquad u\to 0,\qquad T\to T_{\infty} \text{ as } y\to\infty \tag{7.4}$$

where x measures distance along the body surface and y normal to it, u and v are, respectively, the velocity components in the x and y directions. Also K is the permeability of the porous media, ν the kinematic viscosity of the fluid, β the coefficient of thermal expansion, g the acceleration due to gravity, l the characteristic length, γ the measure of the local heat generation, and α_m the effective thermal diffusivity.

To make Eqs. (7.1)–(7.4) dimensionless we introduce the temperature T_s, velocity U_s, and time scales τ_0

$$T_s=\left(\frac{g\beta K}{l\,\nu\,\gamma}\right)^{1/(p-2)},\qquad U_s=\left(\frac{g\beta K}{\nu}\right)^{(p-1)/(p-2)}\frac{1}{(l\,\gamma)^{1/(p-2)}},\qquad \tau_s=\frac{l}{U_s} \tag{7.5}$$

We then put

$$\overline{x}=\frac{x}{l},\ \overline{y}=Ra^{1/2}\frac{y}{l},\ \overline{t}=\frac{t}{\tau_0},\ u=U_s\overline{u},\ v=U_sRa^{-1/2}\overline{v},\ T-T_{\infty}=T_s\theta \tag{7.6}$$

where $Ra=U_s\,l/\alpha$ is the Rayleigh number for the porous medium. We apply (7.5, 7.6) in Eqs. (7.1)–(7.3) and introduce a stream function ψ to satisfy Eq. (7.1). We then look for a solution near the forward stagnation point by writing

$$\psi=xf(y,t),\qquad \theta=\theta(y,t) \tag{7.7}$$

Eq. (7.2) gives $\theta = \partial f/\partial y$ resulting in the equation for the flow as

$$\frac{\partial^3 f}{\partial y^3} + f\frac{\partial^2 f}{\partial y^2} + \left(\frac{\partial f}{\partial y}\right)^p = \frac{\partial^2 f}{\partial y \partial t} \tag{7.8}$$

subject to

$$f = 0, \quad \frac{\partial^2 f}{\partial y^2} = -h_s \text{ on } y = 0, \frac{\partial f}{\partial y} \to 0 \text{ as } y \to \infty \tag{7.9}$$

where $h_s = (q_w l/T_s)Ra^{-1/2}$.

When the surface is thermally insulated, i.e., $h_s = 0$, there are steady-state solutions for all $p \geq 1$, with $f'(0)$ (a dimensionless measure of the wall temperature) starting with $f'(0) = 1.5$ at $p = 1$ and decreasing slowly to $f'(0) = 1$ as $p \to \infty$ as is shown in Fig. 7.1. For the initial-value problem (7.8, 7.9), the solution approaches the corresponding steady state for $1 \leq p < 2$ as time increases. For $p > 2$, however, the solution becomes zero, i.e., $f(y, t) \to 0$. We can compare these cases in Fig. 7.2, where we plot θ_w, i.e., $(\partial f/\partial y)$ on $y = 0$ against t obtained from the numerical solution of Eqs. (7.8), (7.9) for $p = 1.5$, steady state approached, and for $p = 3.0$, solution decays relatively slowly to zero.

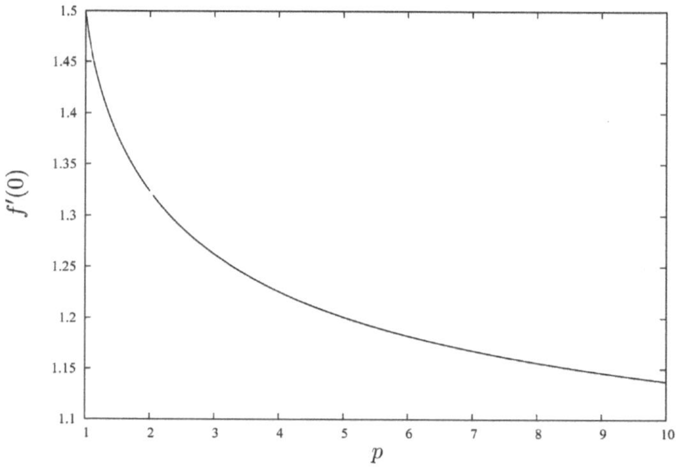

Figure 7.1 Thermally insulated surface, $h_s = 0$, steady states: a plot of $f'(0)$ against p obtained from the steady-state versions of Eqs. (7.8) and (7.9). *(From Merkin, J. H. (2013). Unsteady free convective boundary-layer flow near a stagnation point in a heat-generating porous medium.* Journal of Engineering Mathematics, *79(1), 73–89. https://doi.org/10.1007/s10665-012-9560-2 76)*

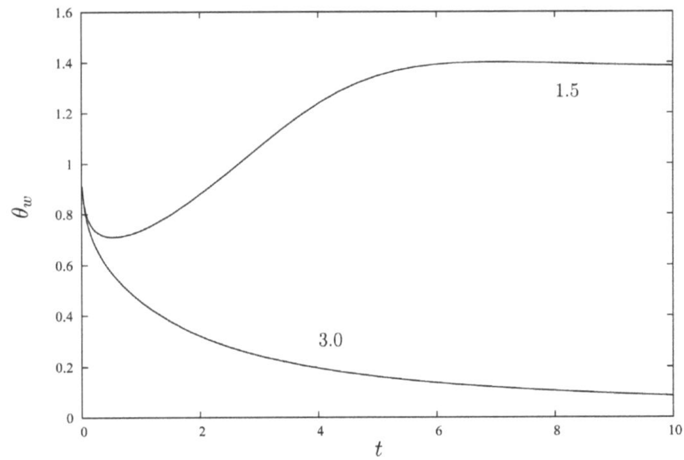

Figure 7.2 Thermally insulated surface, $h_s = 0$, the time evolution of the initial-value problem Eqs. (7.8) and (7.9) for $p = 1.5$ and $p = 3.0$.

When there is thermal input from the boundary, i.e., $h_s > 0$, there are steady states for all h_s when $1 \leq p < 2$. However, for $p > 2$ there is a critical value $h_{s,crit}$ of h_s with steady-state solutions only for $0 \leq h_s \leq h_{s,crit}$ with no steady states for $h_s > h_{s,crit}$. This can seen in Fig. 7.3, where we plot $f'(0)$ against h_s for $p = 1.5$ and $p = 4$. The consequences of this for the initial-value problem are that, when $1 \leq p < 2$, the corresponding steady state is reached at large times. When $p > 2$, a steady state is again reached at large times when $h_s < h_{s,crit}$, now corresponding to the lower branch solutions. For $h_s > h_{s,crit}$ the initial-value problem leads to a finite-time singularity

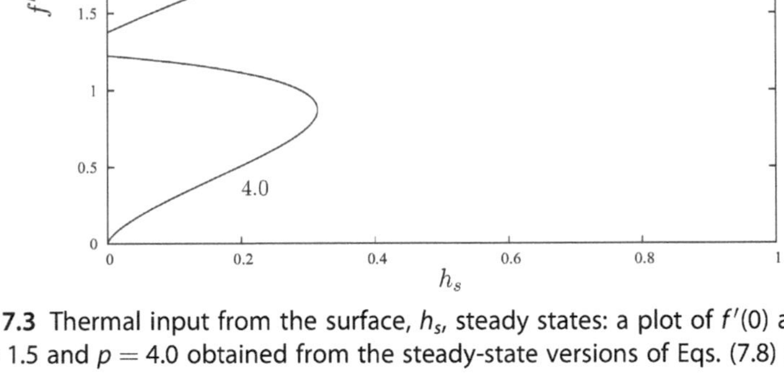

Figure 7.3 Thermal input from the surface, h_s, steady states: a plot of $f'(0)$ against h_s for $p = 1.5$ and $p = 4.0$ obtained from the steady-state versions of Eqs. (7.8) and (7.9).

giving increasingly larger temperatures and to thermal runaway, at least within the context of the present model. This is illustrated in Fig. 7.4, where we plot θ_w against t for $p = 4$ for which $h_{s,crit} \simeq 0.3143$. When $h_s = 0.3$, we see that the corresponding steady state is being approached as t increases; however, for $h_s = 0.4$, the values of θ_w become increasingly larger showing the existence of a finite-time singularity. Further details of this problem, including the asymptotic solutions for large p and large h_s as well as the nature of the finite-time singularity, are presented in (Merkin, 2013). Fig. 7.5 show the variation of $f'(0)$ against p for the thermally insulated surface with mass flux and steady states when $\gamma = 1.0$ and -1.0 obtained from the steady-state versions of Eq. (7.8).

The above discussion illustrates clearly the nature of convective flows with local, temperature-dependent internal heating. The evolution of the flow and heat transfer depends on local heat generation term, the exponent p, and whether the surface is thermally insulated, $h_s = 0$, or not, $h_s > 0$, in (Merkin, 2013). The evolution of the solution can also depend on the initial state, as can be seen in (Merkin & Pop, 2018) where the effects of mass transfer through a thermally insulated wall are considered. This case is also given by Eq. (7.8) though with boundary condition (7.9) having $h_s = 0$ and $f(0) = -\gamma$, where $\gamma = v_w Ra^{1/2}/U_s$ with v_w being the surface mass flux. We again see critical values in the steady states and the existence of dual solutions. The development of the initial-value problem can depend on the initial condition. In the numerical simulations $\partial f/\partial y = a_0 e^{-Y}$ was taken at t = 0. As an illustration θ_w is plotted against t in Fig. 7.6 for

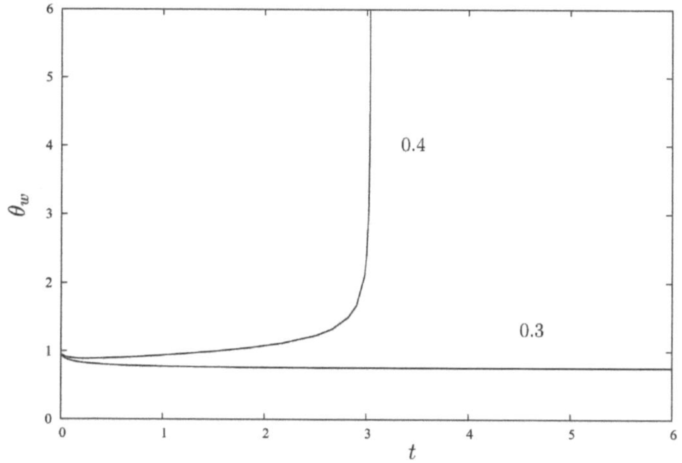

Figure 7.4 Thermal input from the surface, $h_s > 0$, the time evolution of the initial-value problem Eqs. (7.8) and (7.9) for $p = 4.0$ with $h_s = 0.3$ and $h_s = 0.4$.

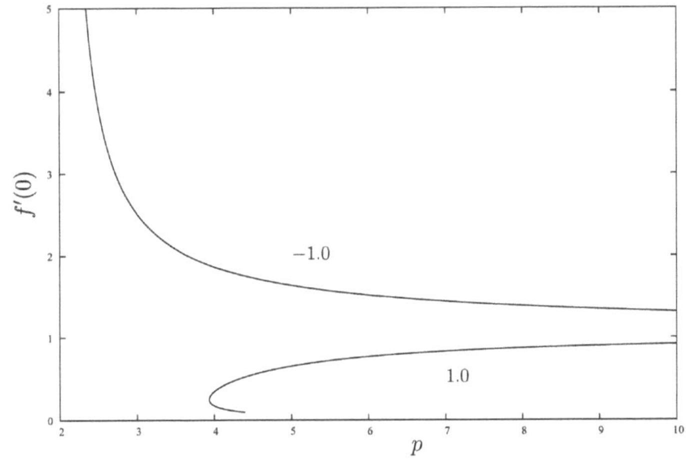

Figure 7.5 Thermally insulated surface with mass flux, steady states: a plot of $f'(0)$ against p for $\gamma = 1.0, -1.0$ obtained from the steady-state versions of Eq. (7.8). *(From Merkin, J. (2018). The effects of wall transpiration on the unsteady free convection boundary-layer flow near a stagnation point in a heat generating porous medium.* Meccanica, 53*(759), 772. https://doi.org/10.1007/s11012-017-0754-6 769)*

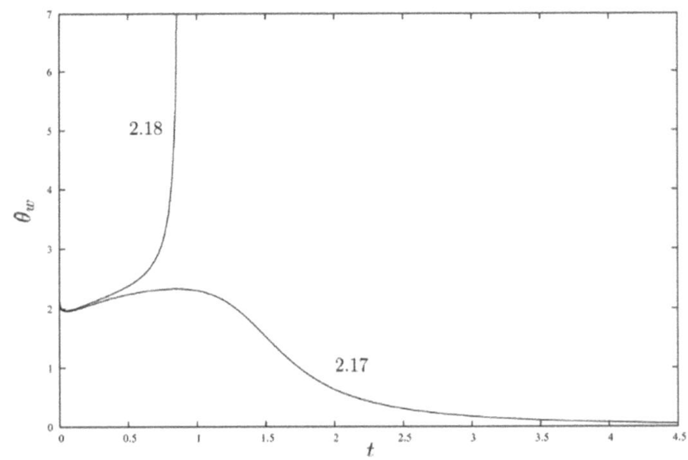

Figure 7.6 Thermally insulated surface with mass flux, plots of θ_w against t for $p = 4.0$, $\gamma = -1.0$, and $a_0 = 2.17$ and $a_0 = 2.18$. *(From Merkin, J. (2018). The effects of wall transpiration on the unsteady free convection boundary-layer flow near a stagnation point in a heat generating porous medium.* Meccanica, 53*(759), 772. https://doi.org/10.1007/s11012-017-0754-6 769)*

$\gamma = 1.0$ and $p = 3$. There is radically different behavior between having $a_0 = 2.17$, where the solution approaches zero, and $a_0 = 2.18$ where a finite-time singularity develops.

The effects of an outer flow U_∞ on the boundary layer flow and heat transfer in a porous medium with local power-law heat generation are discussed in (Merkin et al., 2014) for a thermally insulated surface. The dimensionless equations are now modified to

$$\frac{\partial^3 f}{\partial y^3} + f\frac{\partial^2 f}{\partial y^2} + \left(\frac{\partial f}{\partial y} - \lambda\right)^p = \frac{\partial^2 f}{\partial y \partial t} \tag{7.10}$$

subject to

$$f = 0, \quad \frac{\partial^2 f}{\partial y^2} = 0 \text{ on } y = 0, \quad \frac{\partial f}{\partial y} \rightarrow \lambda \text{ as } y \rightarrow \infty \tag{7.11}$$

where $\lambda = U_\infty / U_s$. The basic nature of this case can be gleaned from the large λ solution. For $1 \leq p < 2$, we find that the steady-state problem has only the trivial solution. For $p > 2$, we put $f = \lambda^{(3-p)/2}\overline{f}$, $\overline{y} = \lambda^{(p-1)/2}y$, $\overline{t} = \lambda^{p-1}t$, and $\overline{\theta} = (\overline{f}_{\overline{y}} - 1)$ then to obtain

$$\frac{\partial^2 \overline{\theta}}{\partial \overline{y}^2} + \overline{\theta}^p = \frac{\partial \overline{\theta}}{\partial \overline{t}} \tag{7.12}$$

This basic problem was treated originally by Fujita (1966) and reviewed fully by Levine (1990). A finite-time singularity develops if $1 < p < 3$, at least for sufficiently small inputs, otherwise the solution dies away to zero. Thus, for the general problem (7.10) and (7.11), we expect the solution to die away to zero when $p > 3$, at least for small inputs. For $p > 3$ and sufficiently large initial inputs, Fujita's problem (7.12) does have a finite-time singularity so we expect different behavior in the general problem (7.10 and 7.11) depending on whether $1 < p < 3$ or $p > 3$. In the former case we might expect a finite-time singularity to develop for all initial inputs, whereas in the latter case we could expect a critical input, below which the solution dies away as t increases and above which a finite-time singularity develops.

This large λ behavior is reflected in the general case in several ways. First, for $1 < p < 2$ there is a critical value λ_c of λ which limits the range of nontrivial solutions to $0 \leq \lambda < \lambda_c$, whereas there is no such limit for $p > 2$. We show this in Fig. 7.7 with plots of $u_w = \theta_w + \lambda$ against t obtained from numerical solutions of Eqs. (7.10) and (7.11) for $p = 1.5$. Here $\lambda_c \simeq 0.2417$ and we see that, for $\lambda < \lambda_c$, the solution approaches the corresponding

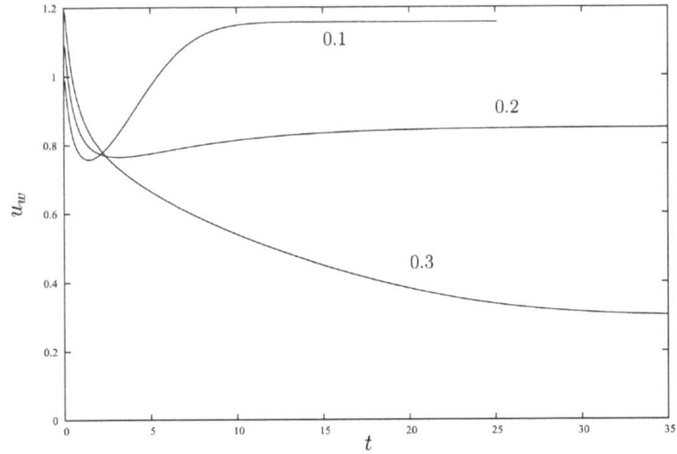

Figure 7.7 Thermally insulated surface with an outer flow, plots of $u_w = \theta_w + \lambda$ against t for $p = 1.5$ and $\lambda = 0.1$, 0.2, 0.3 obtained from numerical integrations of initial-value problem (7.10), (7.11). *(From Merkin, J. H. (2014). The effects of an outer flow on the unsteady free convection boundary layer near a stagnation point in a heat-generating porous medium.* Quarterly Journal of Mechanics and Applied Mathematics, *67(3), 419–444. https://doi.org/10.1093/qjmam/hbu012 428)*

nontrivial steady state and for $\lambda > \lambda_c$, it approaches the trivial solution $f' = \lambda$. There is no such limit for $p > 2$ as we show in Fig. 7.8 with steady-state plots of $f'(0)$ against λ for $p = 2.5$ and 4.0. Here the solution continues to large values of λ. Second the above discussion suggests that, for a sufficiently

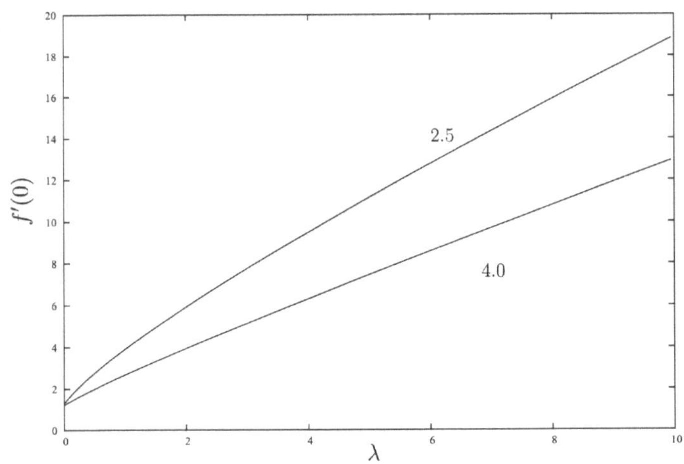

Figure 7.8 Thermally insulated surface with an outer flow—steady states, plots of $f'(0)$ against λ for $p = 2.5$, 4.0. *(From Merkin, J. H. (2014). The effects of an outer flow on the unsteady free convection boundary layer near a stagnation point in a heat-generating porous medium.* Quarterly Journal of Mechanics and Applied Mathematics, *67(3), 419–444. https://doi.org/10.1093/qjmam/hbu012 428)*

large input the solution could develop a finite-time singularity even when $p < 3$. This is illustrated in Fig. 7.9 with plots of u_w against t for $p = 2.5$ where the initial temperature $T_i = a_0 e^{-Y^2}$ for $a_0 = 1.785$, the solution approaches a nontrivial steady state, and $a_0 = 1.79$ where a finite-time singularity develops. The nature of the finite-time singularity is discussed fully by Merkin (2014) where it is seen that region where large temperatures arise in a region close to the wall, as can be seen in Fig. 7.10 showing profile plot at increasing times.

An alternative form for the local heating and one which perhaps more reflects the chemical nature of the heating reaction is to take a modified form of Arrhenius kinetics. Merkin (2016) considered a local heating Q_0 at a rate given by

$$Q_0(T - T_\infty)e^{-E/RT} \tag{7.13}$$

where E is the activation energy, R the universal gas constant, and $Q_0 > 0$ is a constant. The reason for taking this form is that it is zero at the edge of the boundary layer, which is not the case for the original Arrhenius term, as an assumption implicit in the boundary layer approximation is the outer conditions should be unaffected by the flow/heat transfer within the boundary layer. This was also the main reason for taking the power-law form mentioned above, this rate law, at least for p large, has a similar form.

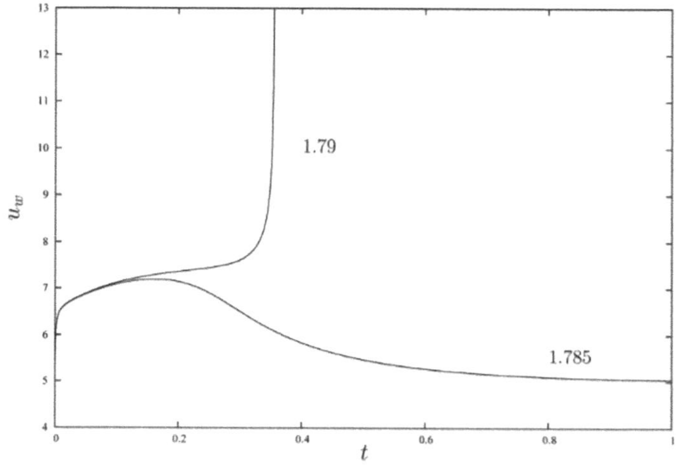

Figure 7.9 Thermally insulated surface with an outer flow, plots of $u_w = \theta_w + \lambda$ against t for $p = 4.0$, $\lambda = 5.0$, and $a_0 = 1.785$, 1.79 obtained from numerical integrations of initial-value problem (7.10, 7.11) with initial temperature $T_i(y) = a_0 \times$ e^{−y^2}. *(From Merkin, J. H. (2014). The effects of an outer flow on the unsteady free convection boundary layer near a stagnation point in a heat-generating porous medium.* Quarterly Journal of Mechanics and Applied Mathematics, *67(3), 419–444. https://doi.org/10.1093/qjmam/hbu012 428)*

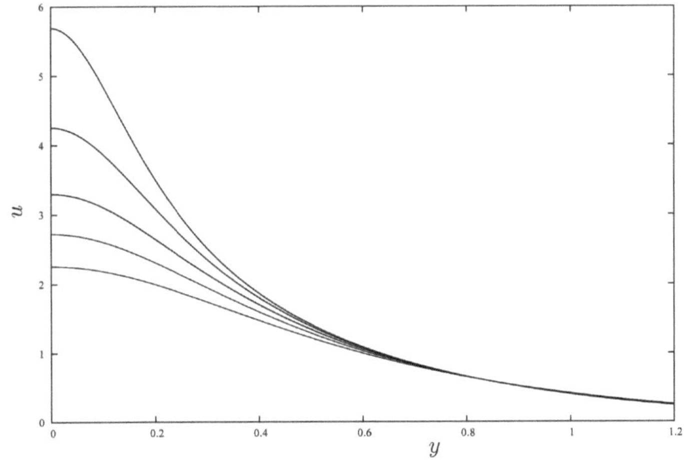

Figure 7.10 Thermally insulated surface with an outer flow, profile plots of u against y for $p = 4.0$, $\lambda = 1.0$ taken at equal time intervals from $t = 0.2431$ to $t = 0.2915$ obtained from numerical integrations of initial-value problem (7.10, 7.11). *(From Merkin, J. H. (2014). The effects of an outer flow on the unsteady free convection boundary layer near a stagnation point in a heat-generating porous medium.* Quarterly Journal of Mechanics and Applied Mathematics, *67(3), 419–444. https://doi.org/10.1093/qjmam/hbu012 428)*

The flow near a forward stagnation point on an impermeable, thermally insulated surface is again considered, resulting in the dimensionless equation

$$\frac{\partial f}{\partial y} = \theta, \quad \frac{\partial^2 \theta}{\partial y^2} + f\frac{\partial \theta}{\partial y} + q_0 \theta e^{\theta/(1+\varepsilon\theta)} = \frac{\partial \theta}{\partial t} \tag{7.14}$$

subject to the boundary conditions

$$f = 0, \quad \frac{\partial \theta}{\partial y} = 0 \quad \text{on} \quad y = 0, \quad \theta \to 0 \quad \text{as} \quad y \to \infty \tag{7.15}$$

where f and θ are, respectively, dimensionless forms of the stream function and temperature and where

$$q_0 = \frac{Q_0 l}{U_s} e^{-E/RT_\infty}, \quad \varepsilon = \frac{RT_\infty}{E} \tag{7.16}$$

In Fig. 7.11, the wall temperature θ_w against t for a range of values of q_0 for $\varepsilon = 0.0$ is illustrated. For the smaller values of q_0 the solution approaches a nontrivial steady state as t is increased. However, for the larger values of q_0 the solution becomes singular at a finite time t_s. Here the wall temperature

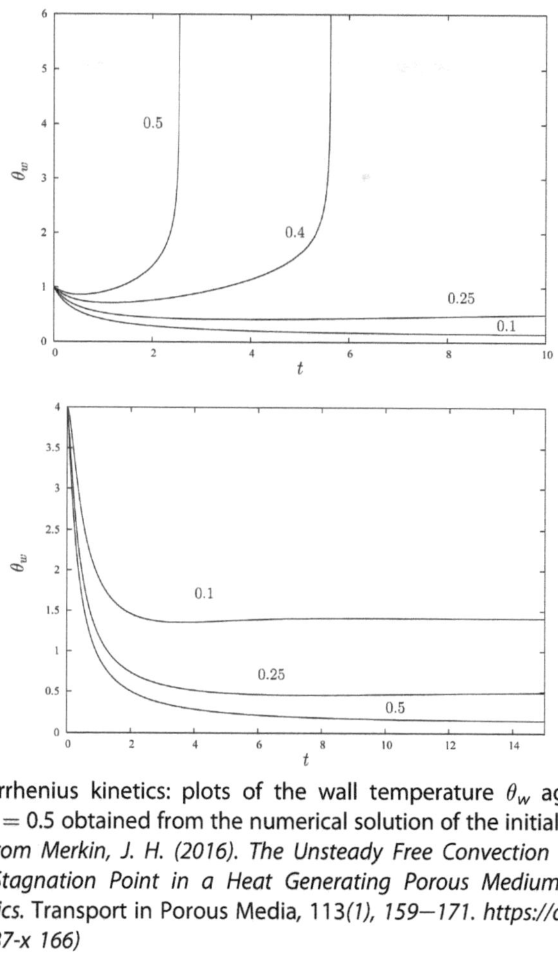

Figure 7.11 Arrhenius kinetics: plots of the wall temperature θ_w against t for (A) $\varepsilon = 0$ and (B) $\varepsilon = 0.5$ obtained from the numerical solution of the initial-value problem (7.14, 7.15). *(From Merkin, J. H. (2016). The Unsteady Free Convection Boundary-Layer Flow Near a Stagnation Point in a Heat Generating Porous Medium with Modified Arrhenius Kinetics.* Transport in Porous Media, 113*(1), 159–171. https://doi.org/10.1007/s11242-016-0687-x 166)*

and, more generally, the temperature and velocity fields become increasingly larger as $t \rightarrow t_s$, leading to a thermal runaway. The dimensionless parameter ε is usually quite small and we see with similar value to that seen in Fig. 7.11 for small values of ε. For much larger values of ε we see different behavior and in Fig. 7.11B we take $\varepsilon = 0.5$, again plotting θ_w against t for representative values of q_0. Here we see that the numerical solution always approaches a steady state, as was also seen for higher values of q_0 tried (not plotted). These steady-state values are somewhat larger than those seen for the corresponding values of q_0 in Fig. 7.11A. When we consider the steady-state versions of Eqs. (7.14 and 7.15) we again see critical values in that there is a critical value $q_{0,crit}$ of q_0, where $q_{0,crit}$ depends on ε, limiting the range of solutions to $q_0 \leq q_{0,crit}$. We illustrate this in Fig. 7.12 with a plot of $\theta(0)$ against q_0 for $\varepsilon = 0$ for which $q_{0,crit} \simeq 0.3395$. The situation is summarized in Fig. 7.13 where $q_{0,crit}$ has been plotted against ε. We see that there is limit on the possibility of having a critical point (and hence dual solutions) at $\varepsilon = \varepsilon_0 \simeq 0.24073$, where $q_{0,crit} \approx 0.5135$. This indicates that for values of $\varepsilon > \varepsilon_0$ there are solutions for all q_0 and in the time-dependent problem, a steady state will be attained at large times without the possibility of a finite-time singularity occurring.

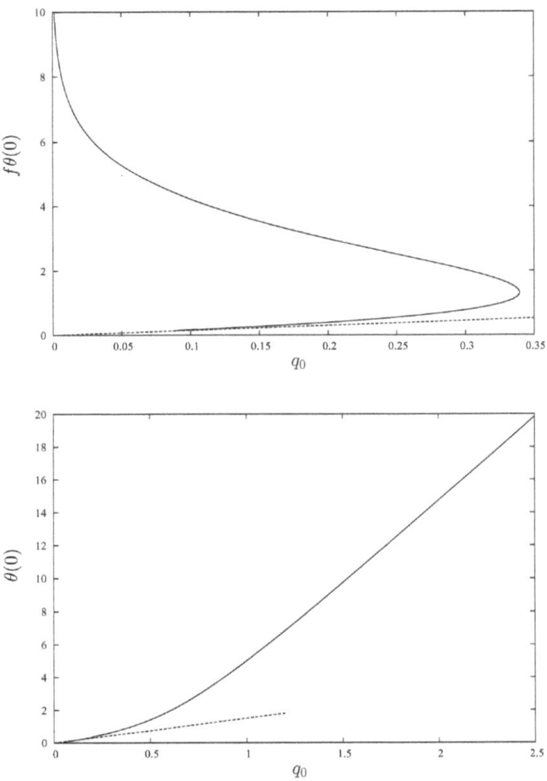

Figure 7.12 Arrhenius kinetics: steady states, plots of the wall temperature $\theta(0)$ against q_0 for (A) $\varepsilon = 0.0$, (B) $\varepsilon = 0.5$. Asymptotic expression for q_0 small is shown by a broken line. *(From Merkin, J. H. (2016). The Unsteady Free Convection Boundary-Layer Flow Near a Stagnation Point in a Heat Generating Porous Medium with Modified Arrhenius Kinetics.* Transport in Porous Media, 113*(1), 159–171. https://doi.org/10.1007/s11242-016-0687-x 166)*

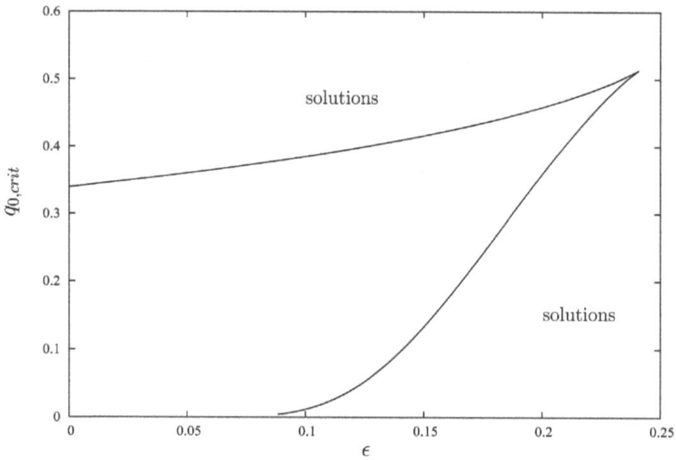

Figure 7.13 Arrhenius kinetics: Steady states a plot of the critical values $q_{0,\text{crit}}$ against ε. *(From Merkin, J. H. (2016). The Unsteady Free Convection Boundary-Layer Flow Near a Stagnation Point in a Heat Generating Porous Medium with Modified Arrhenius Kinetics.* Transport in Porous Media, 113*(1), 159–171. https://doi.org/10.1007/s11242-016-0687-x 166)*

7.3 Flows with spatially dependent heat generation

We start by considering the work of (Postelnicu et al., 2000) and (Postelnicu & Pop, 1999) where a spatially dependent exponential form is taken for the local internal heating. Both vertical and horizontal surfaces are treated and a surface temperature above ambient proportional to x^{μ}, where x measures distance along the surface, is taken to enable the problem to be reduced to similarity form. Here we consider only a vertical surface in which case the similarity problem becomes

$$f''' + \left(\frac{\mu+1}{2}\right) f f'' - \mu f'^2 + e^{-\eta} = 0 \tag{7.17}$$

subject to

$$f(0) = -S, f'(0) = 1, \quad f' \to 0 \text{ as } \eta \to \infty \tag{7.18}$$

where primes denote differentiation with respect to the similarity variable η and where $f(\eta)$ and $f'(\eta)$ are dimensionless versions of the stream function and temperature, respectively. An impermeable, $S = 0$, was treated by Postelnicu and Pop (1999) with wall mass transfer being included in (Postelnicu et al., 2000).

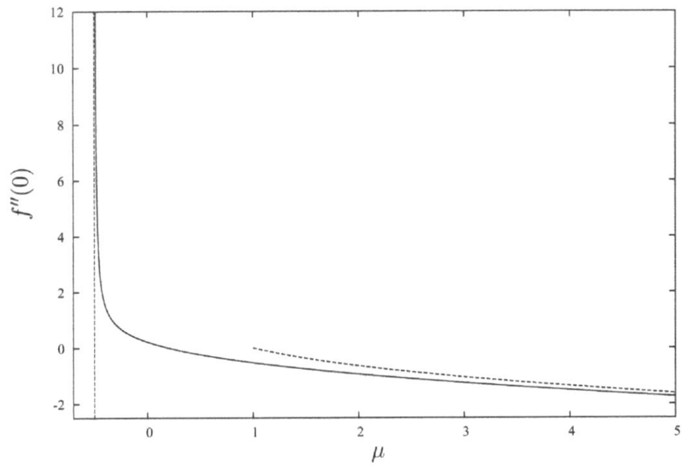

Figure 7.14 Impermeable surface: plot of $f''(0)$ against μ obtained from the numerical solution of Eqs. (7.17 and 7.18). Asymptotic expression (7.21) is shown by a broken line.

The results for an impermeable surface are presented in Fig. 7.14 where plots of $f''(0)$ against μ are shown. The solution continues to large values of μ but does become singular, with $f''(0)$, and consequently the fluid temperature increasing, as $\mu \rightarrow -1/2$, again putting a limit on the range of possible solutions. When $\mu = -1/3$, Eq. (7.16) can be integrated to $f'' + \frac{1}{3} f f' = e^{-\eta}$ on satisfying the outer boundary condition, from which it follows that $f''(0) = 1$. To obtain a solution for large μ we put $f = \mu^{-1/2} F(\zeta)$, where $\zeta = \mu^{1/2}\, \eta$ and expand in powers of μ^{-1}. The leading-order term F_0 satisfies

$$\left.\begin{aligned} & F_0''' + \frac{1}{2} F_0 F_0'' - F_0'^2 = 0 \\ & F_0(0) = 0, \quad F_0'(0) = 1, \quad F_0' \rightarrow 0 \ as\ \zeta \rightarrow \infty \end{aligned}\right\} \tag{7.19}$$

A numerical solution of Eq. (7.19) gives $F_0''(0) = -0.90638$ so that

$$f''(0) \sim -0.90638\, \mu^{1/2} + \cdots \ as\ \mu \rightarrow \infty \tag{7.20}$$

We show asymptotic expression (7.20) in Fig. 7.14 by a broken line showing very good agreement with the numerically determined values even at relatively small values of μ.

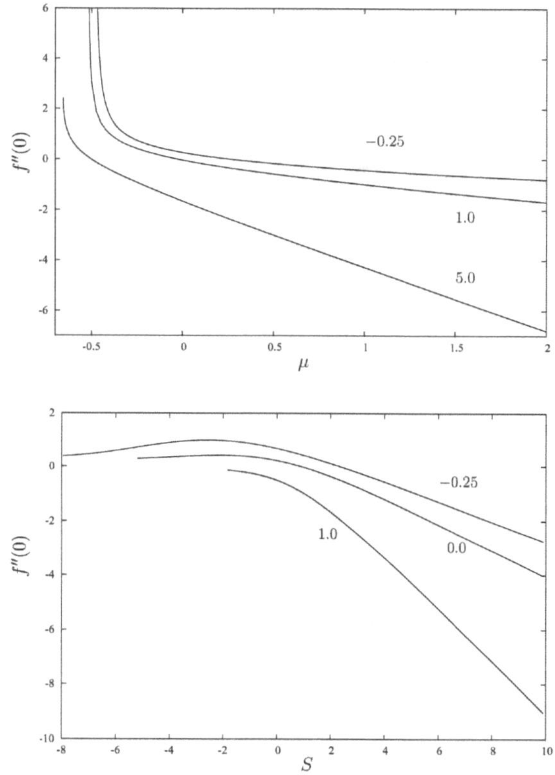

Figure 7.15 Permeable surface: plots of $f''(0)$ (A) against μ for $S = -0.25$, 1.0, 5.0, (B) against S for $\mu = -0.25$, 0.0, 5.0 obtained from the numerical solution of Eqs. (7.16 and 7.17).

The results for a permeable surface, $S \neq 0$, are shown in Fig. 7.15 with plots of $f''(0)$ against μ for representative values of S, Fig. 7.15A, and against S for representative values of μ, Fig. 7.15B. Fig. 7.15A shows that the solution continues to large positive μ, as was seen for an impermeable surface. For negative values of μ the solution becomes singular at a finite value of μ. For negative values of S this is as $\mu \to -1/2$, as for an impermeable surface. However, for positive values of S this singularity is at a larger negative value of μ, appearing to approach $\mu = -1$ as S is increased. In Fig. 7.15B, we see that the solution continues to both large positive and large negative values of S with $f''(0)$ decreasing linearly for larger S. For large S we put $f = S + S^{-1}G, \ Y = S\eta$ and expand in inverse powers of S, the leading-order term G_0 being

$$G_0 = \frac{2}{1+\mu}\left[1 - e^{(1+\mu)Y/2}\right] \tag{7.21}$$

Giving

$$f''(0) \sim -\left(\frac{1+\mu}{2}\right) S + \cdots \text{ as } S \to \infty \tag{7.22}$$

Expression is consistent with the results shown in Fig. 7.15B and indicates that the solution cannot be continued past $\mu = -1$ for large S as seen in Fig. 7.15A.

7.4 Concluding remarks

The effects of internal heating can have profound effects on the flow and heat transfer whether this heating is dependent on the local temperature or takes a specific spatial form. For time-dependent problems a nontrivial steady state can be attained, the temperature reverting back to ambient or a finite-time singularity can develop leading to rapid increases in the temperature indicating that the original model is no longer adequate to describe the flow and heat transfer. These features are dependent on the nature of the internal heating, the original configuration, as well as the input from the bounding surface. The same outcomes apply in steady flows on fixed surfaces. Again how the flow/heat transfer progresses from the leading edge depends critically on the form of the local heating and on the conditions on the bounding surface.

The picture is by no means complete. Situations where the surface heat flux is related to the wall temperature have been considered when there is no internal heating (Merkin & Pop, 2010). The effects of internal heating on these flows have not really been discussed. There can also be a reaction on the surface changing the concentration of the reacting species within the porous material (Mahmood & Merkin, 1998; Merkin & Mahmood, 1998; Merkin & Pop, 2018; Minto et al., 1998). This could be strongly influenced by internal heating. For example, in cases where there is thermal runaway without a reaction, could a reaction within the material inhibit this? Very little work has been reported on the effects of internal heating on enclosed cavities filled with porous material at finite Rayleigh numbers (Mealey & Merkin, 2009). Allali et al. (2011) show that oscillatory behavior is possible. This feature has been noticed for cavities filled with a Newtonian fluid, is it also a more general feature of porous media—filled cavities?

Nomenclature

E	activation energy
f	modified dimensionless stream function
g	acceleration due to gravity
h_s	dimensionless surface heat flux
K	permeability
l	characteristic length
p	exponent
Q_0	measure of the local heat generation
q_w	surface heat flux
R	gas constant
Ra	Rayleigh number
t	time
T	fluid temperature
T_∞	ambient temperature (constant)
U_∞	free stream velocity
u	streamwise velocity component
v	normal velocity component
x	coordinate measured along the surface
y	coordinate measured normal to the surface

Greek symbols

α_m	thermal diffusivity
β	coefficient of thermal expansion
γ	measure of the local heat generation
θ	dimensionless temperature
ν	kinematic viscosity
ψ	stream function
τ_0	time scales

References

Allali, K., Bikany, F., Taik, A., & Volpert, V. (May 23–26, 2011). Interaction between explosion and natural convection in porous media. In *4th international conference on approximation methods and numerical modelling in environment and natural resources*.

Brooks, K., Balakotaiah, V., & Luss, D. (1988). Effect of natural convection on spontaneous combustion of coal stockpiles. *AIChE Journal, 34*(3), 353–365. https://doi.org/10.1002/aic.690340302

Brooks, K., & Glasser, D. (1986). A simplified model of spontaneous combustion in coal stockpiles. *Fuel, 65*(8), 1035–1041. https://doi.org/10.1016/0016-2361(86)90163-8

Fujita, H. (1966). On the blowing up of solutions of the Cauchy problem for u, =\delta u + u^(1+\alpha). *Journal of the Faculty of Science, University of Tokyo, Section I, 13*, 109–124. https://doi.org/10.15083/00039873

Gebhart, B., Jaluria, Y., Mahajan, R. L., & Sammakia, B. (1988). *Buoyancy-induced flows and transport title*. Hemisphere.

Gray, B. F., Sexton, M. J., Halliburton, B., & Macaskill, C. (2002). Wetting-induced ignition in cellulosic materials. *Fire Safety Journal, 37*(5), 465–479. https://doi.org/10.1016/S0379-7112(02)00002-4

Gray, B. F., & Wake, G. C. (1990). The ignition of hygroscopic combustible materials by water. *Combustion and Flame, 79*(1), 2–6. https://doi.org/10.1016/0010-2180(90)90084-5

Ingham, D. B., & Pop, I. (2005). *Transport phenomena in porous media III.* Elsevier. https://doi.org/10.1016/B978-0-08-044490-1.X5003-0

Levine, H. A. (1990). The role of critical exponents in blowup theorems. *SIAM Review, 32*(2), 262–288. https://doi.org/10.1137/1032046

Mahmood, T., & Merkin, J. H. (1998). The convective boundary-layer flow on a reacting surface in a porous medium. *Transport in Porous Media, 32*, 285–298. https://doi.org/10.1023/a:1006576203847

Mealey, L. R., & Merkin, J. H. (2009). Steady finite Rayleigh number convective flows in a porous medium with internal heat generation. *International Journal of Thermal Sciences, 48*(6), 1068–1080. https://doi.org/10.1016/j.ijthermalsci.2008.10.008

Merkin, J. H. (2013). Unsteady free convective boundary-layer flow near a stagnation point in a heat-generating porous medium. *Journal of Engineering Mathematics, 79*(1), 73–89. https://doi.org/10.1007/s10665-012-9560-2

Merkin, J. H. (2014). The effects of an outer flow on the unsteady free convection boundary layer near a stagnation point in a heat-generating porous medium. *Quarterly Journal of Mechanics and Applied Mathematics, 67*(3), 419–444. https://doi.org/10.1093/qjmam/hbu012

Merkin, J. H. (2016). The unsteady free convection boundary-layer flow near a stagnation point in a heat generating porous medium with modified Arrhenius kinetics. *Transport in Porous Media, 113*(1), 159–171. https://doi.org/10.1007/s11242-016-0687-x

Merkin, J. H., Bachok, N., Ishak, A., & Pop, I. (2014). A comment on the flow and heat transfer past a permeable stretching/shrinking surface in a porous medium: Brinkman model. *Transport in Porous Media, 101*(3), 365–371. https://doi.org/10.1007/s11242-013-0249-4

Merkin, J. H., & Mahmood, T. (1998). Convective flows on reactive surfaces in porous media. *Transport in Porous Media, 33*, 279–293. https://doi.org/10.1023/A:1006541819777

Merkin, J. H., & Pop, I. (2010). Natural convection boundary-layer flow in a porous medium with temperature-dependent boundary conditions. *Transport in Porous Media, 85*(2), 397–414. https://doi.org/10.1007/s11242-010-9569-9

Merkin, J. H., & Pop, I. (2018). Stagnation point flow past a stretching/shrinking sheet driven by Arrhenius kinetics. *Applied Mathematics and Computation, 337*, 583–590. https://doi.org/10.1016/j.amc.2018.05.024

Minto, B. J., Ingham, D. B., & Pop, I. (1998). Free convection driven by an exothermic reaction on a vertical surface embedded in porous media. *International Journal of Heat and Mass Transfer, 41*(1), 11–23. https://doi.org/10.1016/S0017-9310(97)00114-2

Nield, D. A., & Bejan, A. (2017). Convection in porous media. In *Convection in porous media* (5th ed.). Springer International Publishing. https://doi.org/10.1007/978-3-319-49562-0

Postelnicu, A., Grosan, T., & Pop, I. (2000). Free convection boundary-layer over a vertical permeable flat plate in a porous medium with internal heat generation. *International Communications in Heat and Mass Transfer, 27*(5), 729–738. https://doi.org/10.1016/S0735-1933(00)00153-6

Postelnicu, A., & Pop, I. (1999). Similarity solutions of free convection boundary layers over vertical and horizontal surfaces in porous media with internal heat generation.

International Communications in Heat and Mass Transfer, 26(8), 1183–1191. https://doi.org/10.1016/S0735-1933(99)00108-6

Sexton, M. J., Macaskill, C., & Gray, B. F. (2001). Self-heating and drying in two-dimensional bagasse piles. *Combustion Theory and Modelling, 5*(4), 517–536. https://doi.org/10.1088/1364-7830/5/4/302

Sisson, R. A., Swift, A., Wake, G. C., & Gray, B. F. (1992). The self-heating of damp cellulosic materials: I. High thermal conductivity and diffusivity. *IMA Journal of Applied Mathematics, 49*(3), 273–291. https://doi.org/10.1093/imamat/49.3.273

Sisson, R. A., Swift, A., Wake, G. C., & Gray, B. F. (1993). The self-heating of damp cellulosic materials: II. On the steady states of the spatially distributed case. *IMA Journal of Applied Mathematics, 50*(3), 285–306. https://doi.org/10.1093/imamat/50.3.285

Vafai, K. (2005). *Handbook of porous media* (2nd ed.). https://doi.org/10.1201/9780824741501

CHAPTER 8

Micropolar fluids over the moving surface

8.1 Introduction

Over the years, fluid flow of micropolar fluids has received considerable attention because of its important applications in engineering. The interest in these fluids, which exhibit the microrotational effects and microrotational inertia, started very soon after the pioneering studies by Eringen (1966, 1972). Under the micropolar fluid theory, the deformation of the fluid microelements is ignored: nevertheless microrotational effects are still present and surface and body couple are permitted ((Agarwal et al., 1990; Ishak et al., 2006b; Kelson & Farrell, 2001); etc.). Examples of industrially relevant flows that can be studied using micropolar theory include the flow of low concentration suspensions, liquid crystals, animal blood, colloidal fluids, polymeric fluids, lubrication, turbulent shear flow, etc. These fluids cannot be explained on the basis of Newtonian fluid flow theory. Extensive reviews of the theory and applications can be found in the review papers by (Ariman et al. (1973) and Ariman et al. (1974)) and in the books by Eringen (2001), Łukaszewicz (1999) and Stokes (1984).

The flow over a moving surface has been investigated by many researchers. The viscous fluid flow over a linearly stretching plate was first considered by Crane (1970) where similarity solution has been obtained in the closed analytical form. Chiam (1994) combined Hiemenz's (Hiemenz, 1911) stagnation flow and Crane's (Crane, 1970) stretching case to study the problem of steady two-dimensional stagnation-point flow of an incompressible viscous fluid toward a stretching plate. In this case, the surface is assumed to be stretched in its own plane. On the other hand, (Miklavčič & Wang, 2006) are the first to investigate the viscous flow problem over a shrinking sheet with imposition of adequate suction. Later, (Wang, 2008) found out that suction is not an essential requirement for the solutions to exist for the stagnation-point flow as such flow has already

Similarity Solutions for the Boundary Layer Flow and Heat Transfer of Viscous Fluids, Nanofluids, Porous Media, and Micropolar Fluids
ISBN 978-0-12-821188-5
https://doi.org/10.1016/B978-0-12-821188-5.00001-1

contained the vorticity field. After the pioneering studies by Crane (1970) and Miklavčič and Wang (2006), many researchers have investigated the problem of stretching or shrinking surface in Newtonian fluids as well as in non-Newtonian fluids with various effects, geometries, and boundary conditions.

Here are the list of some published papers about flow and heat transfer over a stretching or shrinking surface in micropolar fluid: (Ahmad et al., 2012; Aliyu et al., 2017; Aurangzaib et al., 2016; Bhattacharyya et al., 2012; Chun et al., 2014; El-Aziz, 2009; Hassanien & Gorla, 1990; Hayat et al., 2009; Ibrahim, 2017; Ishak et al., 2008, 2010; Kelson & Desseaux, 2001; Khan et al., 2017; Lok et al., 2009, 2018; Mishra et al., 2019; Rosali et al., 2012; Rosca & Pop, 2014; Sharma et al., 2016; Shaw et al., 2013; Turkyilmazoglu, 2014; Yacob & Ishak, 2012a, 2012b; Yahaya et al., 2021; Yasmin et al., 2020; Zheng et al., 2012).

The basic dimensional equations of micropolar model constitutes of continuity, momentum, microrotation, and energy equations, which are given in Cartesian coordinates as below:

$$\frac{1}{\rho}\frac{\partial \rho}{\partial \overline{t}} + \overline{\nabla}\cdot\overline{\mathbf{u}} = 0 \tag{8.1}$$

$$\rho\frac{D\overline{\mathbf{u}}}{D\overline{t}} = -\overline{\nabla}\overline{p} + (\mu+\kappa)\overline{\nabla}^2\overline{\mathbf{u}} + \kappa\left(\overline{\nabla}\times\overline{\mathbf{N}}\right) + \overline{\mathbf{F}} \tag{8.2}$$

$$\rho j\frac{D\overline{\mathbf{N}}}{D\overline{t}} = \gamma\overline{\nabla}^2\overline{\mathbf{N}} + \kappa\left(-2\overline{\mathbf{N}} + \overline{\nabla}\times\overline{\mathbf{u}}\right) \tag{8.3}$$

$$\rho C_p\frac{D\overline{T}}{D\overline{t}} = k\overline{\nabla}^2\overline{T} \tag{8.4}$$

where $\frac{D}{D\overline{t}} = \frac{\partial}{\partial \overline{t}} + \overline{u}\cdot\overline{\nabla}$ with the gradient $\overline{\nabla} = \left(\frac{\partial}{\partial \overline{x}}, \frac{\partial}{\partial \overline{y}}, \frac{\partial}{\partial \overline{z}}\right)$.

In the above equations, $\overline{x}$ is the distance coordinate measured along the wall surface, $\overline{y}$ is measured normal to the $\overline{x}$-axis, while $\overline{z}$ is normal to both $\overline{x}$ and $\overline{y}$ axes. $\overline{\mathbf{u}}$ is the velocity vector, $\overline{\mathbf{N}}$ is the microrotation vector normal to the $\overline{x}-\overline{y}$ plane, $\overline{T}$ is the temperature of fluid, ρ is the density of fluid, $\overline{t}$ is the time, $\overline{p}$ is the pressure, μ is the dynamic viscosity, κ is the vortex viscosity, $\overline{\mathbf{F}}$ is the body force, j is the microinertia density, γ is the spin-gradient viscosity, C_p is the specific heat at constant pressure, and k is the thermal conductivity.

For steady two-dimensional flow with constant fluid properties, Eqs. (8.1)–(8.4) become

$$\frac{\partial \overline{u}}{\partial \overline{x}}+\frac{\partial \overline{v}}{\partial \overline{y}}=0 \tag{8.5}$$

$$\rho\left(\overline{u}\frac{\partial \overline{u}}{\partial \overline{x}}+\overline{v}\frac{\partial \overline{u}}{\partial \overline{y}}\right)=-\frac{\partial \overline{p}}{\partial \overline{x}}+(\mu+\kappa)\left(\frac{\partial^2 \overline{u}}{\partial \overline{x}^2}+\frac{\partial^2 \overline{u}}{\partial \overline{y}^2}\right)+\kappa\frac{\partial \overline{N}}{\partial \overline{y}}+F_x \tag{8.6}$$

$$\rho\left(\overline{u}\frac{\partial \overline{v}}{\partial \overline{x}}+\overline{v}\frac{\partial \overline{v}}{\partial \overline{y}}\right)=-\frac{\partial \overline{p}}{\partial \overline{y}}+(\mu+\kappa)\left(\frac{\partial^2 \overline{v}}{\partial \overline{x}^2}+\frac{\partial^2 \overline{v}}{\partial \overline{y}^2}\right)-\kappa\frac{\partial \overline{N}}{\partial \overline{x}}+F_y \tag{8.7}$$

$$\rho j\left(\overline{u}\frac{\partial \overline{N}}{\partial \overline{x}}+\overline{v}\frac{\partial \overline{N}}{\partial \overline{y}}\right)=\gamma\left(\frac{\partial^2 \overline{N}}{\partial \overline{x}^2}+\frac{\partial^2 \overline{N}}{\partial \overline{y}^2}\right)+\kappa\left(-2\overline{N}+\frac{\partial \overline{v}}{\partial \overline{x}}-\frac{\partial \overline{u}}{\partial \overline{y}}\right) \tag{8.8}$$

$$\rho C_p\left(\overline{u}\frac{\partial \overline{T}}{\partial \overline{x}}+\overline{v}\frac{\partial \overline{T}}{\partial \overline{y}}\right)=k\left(\frac{\partial^2 \overline{T}}{\partial \overline{x}^2}+\frac{\partial^2 \overline{T}}{\partial \overline{y}^2}\right) \tag{8.9}$$

The physical quantities of interest are the wall shear stress and the heat flux from the surface which are defined as

$$\overline{\tau}_w=\left[(\mu+\kappa)\left(\frac{\partial \overline{u}}{\partial \overline{y}}+\frac{\partial \overline{v}}{\partial \overline{x}}\right)+\kappa\overline{N}\right]_{\overline{y}=0},\quad \overline{q}_w=-k\left(\frac{\partial \overline{T}}{\partial \overline{y}}\right)_{\overline{y}=0} \tag{8.10}$$

8.2 Mixed convection flow of a micropolar fluid near a stagnation-point flow over a stretching surface

8.2.1 Problem formulation

The problem of steady two-dimensional mixed convection flow of a micropolar fluid near a stagnation-point flow over a stretching plate had been solved by Lok et al. (2009). The plate is assumed to be stretched with a velocity $\overline{u}_w(x)=c\overline{x}$, where c is the velocity of surface with $c>0$ represents the stretching surface. The temperature of the plate is $\overline{T}_w(\overline{x})=T_\infty+T_0\left(\overline{x}/l\right)$ where l is the characteristic length of the plate, T_0 is the characteristic temperature, while the ambient fluid is at a constant temperature T_∞, where $\overline{T}_w(\overline{x})>T_\infty$ for heated plate or $\overline{T}_w(\overline{x})<T_\infty$ for cooled plate. The governing equations are given by (8.5–8.9) with the body force $F_x=g\rho\beta\left(\overline{T}-T_\infty\right)$ and $F_y=0$ where g is the acceleration due to gravity and β is the coefficient of thermal expansion.

The boundary conditions are

$$\overline{u} = \overline{u}_w(\overline{x}),\ \overline{v} = 0,\ \overline{N} = -m\left(\frac{\partial \overline{u}}{\partial \overline{y}} - \frac{\partial \overline{v}}{\partial \overline{x}}\right),\ \overline{T} = \overline{T}_w(\overline{x}) \text{ on } \overline{y} = 0$$
$$\overline{u} \to \overline{u}_e = a\overline{x},\ \overline{v} \to \overline{v}_e = -a\overline{y},\ \overline{N} \to \overline{N}_e = 0,\ \overline{T} \to T_\infty \text{ as } \overline{y} \to \infty \quad (8.11)$$

where a is positive constant that represents the strength of stagnation flow. Here, m is a constant such that $0 \le m \le 1$, for which the case when $m = 0$ denotes strong concentration fluid, the case corresponding to $m = ½$ represents weak concentration fluid, while the case when $m = 1$ refers to the turbulent boundary layer flow (Peddieson, 1972). A strong concentration fluid will have more micropolar particles and the particles close to the wall are unable to rotate (Guram & Smith, 1980; Jena & Mathur, 1981), while a weak concentration fluid will have less micropolar particles and the particle spin is equal to fluid vorticity at the wall (Ahmadi, 1976). Therefore, when m increases from 0 to 1, the micropolar particles that are not free to rotate become more augmented, hence enhance the velocity of flow.

8.2.2 Similarity transformation

Introducing the nondimensional variables

$$x = \frac{\overline{x}}{l},\ y = \frac{\overline{y}}{l},\ u = \frac{\overline{u}}{cl},\ v = \frac{\overline{v}}{cl},\ N = \frac{\overline{N}}{c},\ T = \frac{\overline{T} - T_\infty}{T_0},\ p = \frac{\overline{p}}{\rho c^2 l^2} \quad (8.12)$$

and take the microinertia density as characteristic length scale $j = \nu/c = l^2$ (Rees & Bassom, 1996) where ν is the kinematic viscosity, and the spin-gradient viscosity is denoted as $\gamma = \left(\mu + \kappa/2\right)j$ (Ahmadi, 1976; Kline, 1977). The nondimensional stream function is defined as

$$u = \frac{\partial \psi}{\partial y},\ v = -\frac{\partial \psi}{\partial x} \quad (8.13)$$

and the similarity variables are

$$\psi = xf(y),\ T = x\theta(y),\ N = xh(y) \quad (8.14)$$

Substituting expression (8.12) into Eqs. (8.5)—(8.9), then write the nondimensional form of equations in terms of stream function Eq. (8.13), and eliminate the pressure p from momentum equations (8.6 and 8.7) before applying the similarity variables (8.14), we obtain, after one integration of momentum equation, the following ordinary differential equations

$$(1+K)f''' + ff'' - f'^2 + \left(\tfrac{a}{c}\right)^2 + Kh' + \lambda\theta = 0 \tag{8.15}$$

$$\left(1+\frac{K}{2}\right)h'' + fh' - f'h - K(f'' + 2h) = 0 \tag{8.16}$$

$$\theta'' + Pr(f\theta' - f'\theta) = 0 \tag{8.17}$$

while the boundary conditions (8.11) become

$$\begin{aligned} &f(0) = 0,\ f'(0) = 1,\ \theta(0) = 1,\ h(0) = -mf''(0) \\ &f'(\infty) = \frac{a}{c},\ \theta(\infty) = 0,\ h(\infty) = 0 \end{aligned} \tag{8.18}$$

where prime denotes the differentiation with respect to y. In the above equations, $K = \kappa/_{\mu}$ is the micropolar parameter, $Pr = \nu/\alpha$ is the Prandtl number, $\lambda = Gr_{\overline{x}}/Re_{\overline{x}}^2$ is the mixed convection parameter with local Grashof number $Gr_{\overline{x}} = g\beta\left(\overline{T}_w - T_\infty\right)\overline{x}^3/\nu^2$, and local Reynolds number $Re_{\overline{x}} = \overline{u}_w\left(\overline{x}\right)\overline{x}/\nu$. If we take $K = 0$, there is no microrotation of fluid elements $(N = 0)$, hence, the above system of equations for micropolar fluid reduced to that of Newtonian fluid which has been considered by Ishak et al. (2006c).

Using the nondimensional variables (8.12) and similarity variables (8.14), the wall shear stress and the heat flux from the stretching surface (8.10) become

$$\tau_w = \frac{\overline{\tau}_w}{\mu c} = [1 + (1-m)K]xf''(0),\ q_w = \frac{\overline{q}_w l}{kT_0} = -x\theta'(0) \tag{8.19}$$

8.2.3 Comments

The mixed convection flow of a micropolar fluid near an orthogonal stagnation—point flow over a stretching surface with no slip condition (8.15—8.18) has been numerically solved by Lok et al. (2009) using the Keller-box method. In their paper, they set the boundary layer thickness 20 with step size 0.005 and the convergence criterion is 5×10^{-7}.

The velocity $f'(y)$ and microrotation $h(y)$ profiles are shown in Figs. 8.1 and 8.2, respectively, for $Pr = 1$, strong concentration $m = 0$, micropolar parameter $K = 0$, 1, 3, and stretching rate $a/c = 0.5$, 1, 1.5 when the flow is assisting ($\lambda = 1$) and opposing ($\lambda = -1$). The temperature profiles are similar to those obtained by Ishak et al. (2006c) for Newtonian viscous

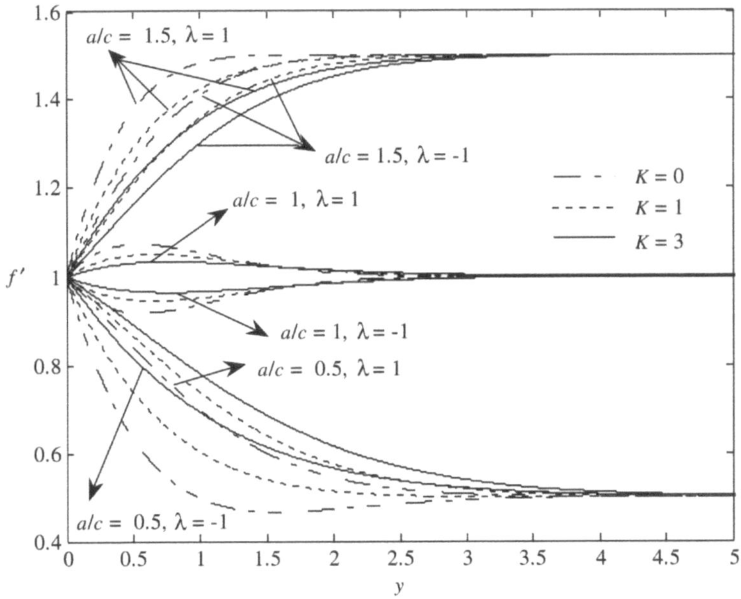

Figure 8.1 Velocity profiles at various values of K and a/c when $Pr = 1$ and $m = 0$ for both assisting and opposing flows. *(From Lok, Y. Y., Pop, I., Ingham, D. B., & Amin, N. (2009). Mixed convection flow of a micropolar fluid near a non-orthogonal stagnation-point on a stretching vertical sheet.* International Journal of Numerical Methods for Heat and Fluid Flow, 19*(3/4), 459–483. https://doi.org/10.1108/09615530910938380 474.)*

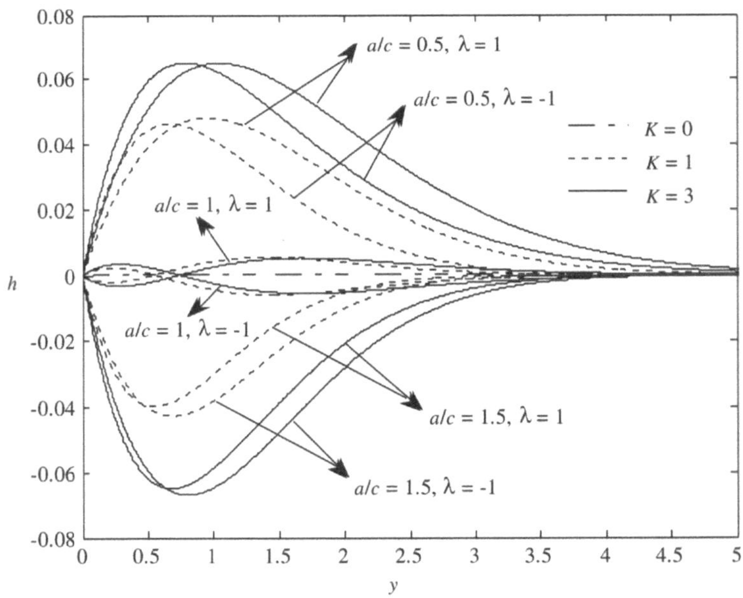

Figure 8.2 Microrotation profiles at various values of K and a/c when $Pr = 1$ and $m = 0$ for both assisting and opposing flows. *(From Lok, Y. Y., Pop, I., Ingham, D. B., & Amin, N. (2009). Mixed convection flow of a micropolar fluid near a non-orthogonal stagnation-point on a stretching vertical sheet.* International Journal of Numerical Methods for Heat and Fluid Flow, 19*(3/4), 459–483. https://doi.org/10.1108/09615530910938380 474.)*

fluid. For the velocity profiles in Fig. 8.1, the curves start at initial value 1, and have positive slopes for $a/c = 1.5$ (both assisting and opposing flow cases, $\lambda = \pm 1$) and $a/c = 1$ (assisting flow case, $\lambda = 1$). The curves reduce with a negative slope for $a/c = 0.5$ (both assisting and opposing flow cases, $\lambda = \pm 1$) and $a/c = 1$ (opposing flow $\lambda = -1$). It is seen from Fig. 8.2 that the microrotation profiles reach a maximum near the wall and then decrease asymptotically for the case $a/c = 0.5$. Opposite trend is observed for the case $a/c = 1.5$. The values of $h(y)$ are all positive for $a/c = 0.5$, and they are all negative for $a/c = 1.5$. When $a/c = 1$, a change in the direction of rotation (reversed flow) occurs at certain locations of y. In general, the velocity decreases while the microrotation profiles increase as K increases.

8.3 Oblique stagnation–slip flow of a micropolar fluid toward a stretching/shrinking surface

8.3.1 Problem formulation

The problem of steady two-dimensional stagnation-point flow impinging obliquely toward a stretching/shrinking surface with slip velocity in a micropolar fluid has been studied by Lok et al. (2018). The governing equations are given by (8.5)–(8.8) without the body force, i.e., $F_x = F_y = 0$.

The boundary conditions are given by

$$\overline{u} = \overline{u}_w + \overline{u}_{slip},\ \overline{v} = 0,\ \overline{N} = -m\left(\frac{\partial \overline{u}}{\partial \overline{y}} - \frac{\partial \overline{v}}{\partial \overline{x}}\right) \quad \text{on } \overline{y} = 0$$
$$\overline{u} \to \overline{u}_e,\ \overline{v} \to \overline{v}_e,\ \overline{N} \to \overline{N}_e \quad \text{as } \overline{y} \to \infty \tag{8.20}$$

The velocity of the surface is given by $\overline{u}_w = c\overline{x}$, where $c > 0$ corresponds to stretching surface and $c < 0$ corresponds to shrinking surface. The slip velocity is of the form $\overline{u}_{slip} = A_t\left(\partial \overline{u}/\partial \overline{y}\right)$, where A_t is defined as the slip length.

It is assumed that the external flow is the combination of an orthogonal stagnation–point flow and a simple shear flow, which is given by Drazin and Riley (2006) and Tooke and Blyth (2008).

$$\overline{\psi} = a\overline{x}\,\overline{y} + \frac{1}{2}b\overline{y}^2 \tag{8.21}$$

where a and b being positive constants which represents the strength of an irrotational flow and the vorticity of a rotational shear flow, respectively. By defining the stream function $\overline{\psi}$ in usual way, $\overline{u} = \partial\overline{\psi}/\partial\overline{y}$ and $\overline{v} = -\partial\overline{\psi}/\partial\overline{x}$, the velocities of the external flow are $\overline{u}_e = a\overline{x} + b\overline{y}$ and $\overline{v}_e = -a\overline{y}$. Besides, from Eq. (8.8) when $\overline{y} \to \infty$, $\overline{N}_e = -b/2\ (b \neq 0)$.

8.3.2 Similarity transformation

The general form of the similarity solutions are given by (Borrelli et al. (2012, 2015), Drazin and Riley (2006), Lok et al. (2015), Tooke and Blyth (2008))

$$\overline{\psi} = (a\nu)^{1/2}\overline{x}f(\eta) + \frac{b\nu}{a}\int_0^\eta g(\eta)d\eta, \quad \overline{N} = \left(\frac{a^3}{\nu}\right)^{1/2}\overline{x}h(\eta) + bq(\eta),$$
$$\eta = \left(\frac{a}{\nu}\right)^{1/2}\overline{y} \tag{8.22}$$

and take $j = \upsilon/a$ as the characteristic length scale. On eliminating the pressure p from Eqs. (8.6) and (8.7) by cross differentiation, then substituting the similarity variables (8.22) into the governing equations, followed by equating the terms in $\bar{x}$ and those of independent of $\bar{x}$, consequently integrate those with the fourth-order equation with respect to η, we get

$$(1+K)f''' + ff'' - f'^2 + Kh' + 1 = 0 \tag{8.23}$$

$$(1+K)g'' + fg' - f'g + Kq' + \alpha - \beta = 0 \tag{8.24}$$

$$\left(1 + \frac{K}{2}\right)h'' + fh' - f'h - K(f'' + 2h) = 0 \tag{8.25}$$

$$\left(1 + \frac{K}{2}\right)q'' + fq' - gh - K(g' + 2q) = 0 \tag{8.26}$$

The boundary conditions (8.20) now become

$$f(0) = 0, \ f'(0) = \varepsilon + Af''(0), \ g(0) = Ag'(0), \ h(0) = -mf''(0),$$
$$q(0) = -mg'(0)$$
$$f' \to 1, \ g' \to 1, \ h \to 0, \ q \to -\frac{1}{2} \ \text{as } \eta \to \infty \tag{8.27}$$

where the prime denotes the differentiation with respect to η, $K = \kappa/\mu$ is the micropolar parameter, $\varepsilon = c/a$ is the stretching ($\varepsilon > 0$) or shrinking ($\varepsilon < 0$) parameter, and $A = A_t\sqrt{a/\nu}$ is the slip parameter. It should be pointed out that the solutions for f and g require that $f \sim \eta - \alpha$ and $g \sim \eta - \beta$ so that they match with the outer flow (8.21). The values of α can be determined when solving orthogonal stagnation—point problem (8.23), (8.25) and boundary conditions (8.27). β is a free parameter and can take any constant value (Drazin & Riley, 2006; Tooke & Blyth, 2008).

The dimensionless stream function is

$$\psi = \frac{\overline{\psi}}{\nu} = \xi f(\eta) + \lambda \int_0^{\eta} g(\eta)d\eta \tag{8.28}$$

where $\xi = (a/\nu)^{1/2}\overline{x}$ and $\lambda = b/a$ is the shear flow parameter. The point where $\psi = 0$ meet the surface is denoted as ξ_0. The dimensionless wall shear stress is

$$\tau_w = \frac{\overline{\tau}_w}{a\mu} = [1 + (1-m)K][\xi f''(0) + \lambda g'(0)] \tag{8.29}$$

Compare to the wall shear stress in Eq. (8.19) for orthogonal stagnation-point flow, there is an extra term g′(0) in Eq. (8.29) for the oblique stagnation-point flow. The point of zero wall shear stress $\tau_w = 0$ is given by

$$\xi_s = -\frac{\lambda g'(0)}{f''(0)} \tag{8.30}$$

8.3.3 Stability analysis

We consider the unsteady case of (8.6) to (8.8), and define a new nondimensional time variable $\tau = at$ (Merkin, 1986; Weidman et al., 2006), hence the assumption of the form of the solutions are similar to (8.22), but f, g, h, and q now are functions in terms of η and τ. The partial differential equations in terms of two independent variables, η and τ, are given by

$$(1+K)\frac{\partial^3 f}{\partial \eta^3} + f\frac{\partial^2 f}{\partial \eta^2} - \left(\frac{\partial f}{\partial \eta}\right)^2 + K\frac{\partial h}{\partial \eta} + 1 - \frac{\partial^2 f}{\partial \tau \partial \eta} = 0 \tag{8.31}$$

$$(1+K)\frac{\partial^2 g}{\partial \eta^2} + f\frac{\partial g}{\partial \eta} - \frac{\partial f}{\partial \eta}g + K\frac{\partial q}{\partial \eta} + (\alpha - \beta) - \frac{\partial g}{\partial \tau} = 0 \tag{8.32}$$

$$\left(1+\frac{K}{2}\right)\frac{\partial^2 h}{\partial\eta^2}+f\frac{\partial h}{\partial\eta}-\frac{\partial f}{\partial\eta}h-K\left(\frac{\partial^2 f}{\partial\eta^2}+2h\right)-\frac{\partial h}{\partial\tau}=0 \quad (8.33)$$

$$\left(1+\frac{K}{2}\right)\frac{\partial^2 q}{\partial\eta^2}+f\frac{\partial q}{\partial\eta}-gh-K\left(\frac{\partial g}{\partial\eta}+2q\right)-\frac{\partial q}{\partial\tau}=0 \quad (8.34)$$

subject to the boundary conditions

$$f(0,\tau)=0,\ \frac{\partial f}{\partial\eta}(0,\tau)=\varepsilon+A\frac{\partial^2 f}{\partial\eta^2}(0,\tau),\ g(0,\tau)=A\frac{\partial g}{\partial\eta}(0,\tau),$$

$$h(0,\tau)=-m\frac{\partial^2 f}{\partial\eta^2}(0,\tau),\ q(0,\tau)=-m\frac{\partial g}{\partial\eta}(0,\tau)$$

$$\frac{\partial f}{\partial\eta}(\eta,\tau)\to 1,\ \frac{\partial g}{\partial\eta}(\eta,\tau)\to 1,\ h(\eta,\tau)\to 0,\ q(\eta,\tau)\to -\frac{1}{2}\ \text{as}\ \eta\to\infty \quad (8.35)$$

Adopting the stability analysis that proposed by Merkin (1986), let

$$\begin{aligned} f(\eta,\tau)&=f_0(\eta)+e^{\delta\tau}F(\eta),\ g(\eta,\tau)=g_0(\eta)+e^{\delta\tau}G(\eta),\\ h(\eta,\tau)&=h_0(\eta)+e^{\delta\tau}H(\eta),\ q(\eta,\tau)=q_0(\eta)+e^{\delta\tau}Q(\eta), \end{aligned} \quad (8.36)$$

where $f_0(\eta)$, $g_0(\eta)$, $h_0(\eta)$, $q_0(\eta)$ are the steady flow solutions of Eqs. (8.23)–(8.26), while $F(\eta)$, $G(\eta)$, $H(\eta)$, $Q(\eta)$ are small relative to $f_0(\eta)$, $g_0(\eta)$, $h_0(\eta)$, $q_0(\eta)$. The exponential terms that appear in (8.36) is considered as disturbance where δ is the eigenvalue parameter. Substituting (8.36) into (8.31)–(8.34), the following linearized equations are obtained:

$$(1+K)F'''+f_0F''-f_0''F-2f_0'F'+KH'+\delta F'=0 \quad (8.37)$$

$$(1+K)G''+f_0G'+g_0'F-f_0'G-g_0F'+KQ'+\delta G=0 \quad (8.38)$$

$$\left(1+\frac{K}{2}\right)H''+f_0H'+h_0'F-f_0'H-h_0F'-K(F''+2H)+\delta H=0 \quad (8.39)$$

$$\left(1+\frac{K}{2}\right)Q''+f_0Q'+q_0'F-g_0H-h_0G-K(G'+2Q)+\delta Q=0 \quad (8.40)$$

and the boundary conditions (8.35) now become

$$\begin{aligned} &F(0)=0,\ F'(0)=AF''(0),\ F''(0)=1,\ G(0)=AG'(0),\\ &H(0)=-mF''(0),\ Q(0)=-mG'(0)\\ &G'\to 0,\ H\to 0,\ Q\to 0\ \ as\ \eta\to\infty \end{aligned} \quad (8.41)$$

Eqs. (8.37)—(8.41) are solved numerically and the range of eigenvalues δ can be identified by relaxing a boundary condition at the far field condition (Harris et al., 2009). Here, the condition $F'(\infty) = 0$ is relaxed, and without loss of generality $F''(0) = 1$ is used in Eq. (8.41) to determine the eigenvalues δ. For a fixed values of K, A, and m, the stability of the solutions $f_0(\eta)$, $g_0(\eta)$, $h_0(\eta)$, $q_0(\eta)$ are identified by examining the smallest eigenvalue that is obtained.

8.3.4 Comments (orthogonal stagnation—point flow)

The orthogonal stagnation—point flow problem (8.23) and (8.25) in micropolar fluid over a shrinking surface with no slip condition $A = 0$ and subject to boundary conditions (8.27) for f and h, has been solved numerically using the Keller-box method by Ishak et al. (2010). Similar to the problem in viscous fluid that was studied by Wang (2008), two different set of solutions are obtained, though both fulfill the far field boundary conditions asymptotically when considering two different values of boundary thickness for some values of governing parameters.

The variation of velocity and angular velocity profiles for different values of micropolar parameter K when $m = 0$ are given in Figs. 8.3 and 8.4, respectively. These figures are obtained for the shrinking case $\varepsilon = -1.15$. It is obvious that the boundary layer thicknesses for the upper

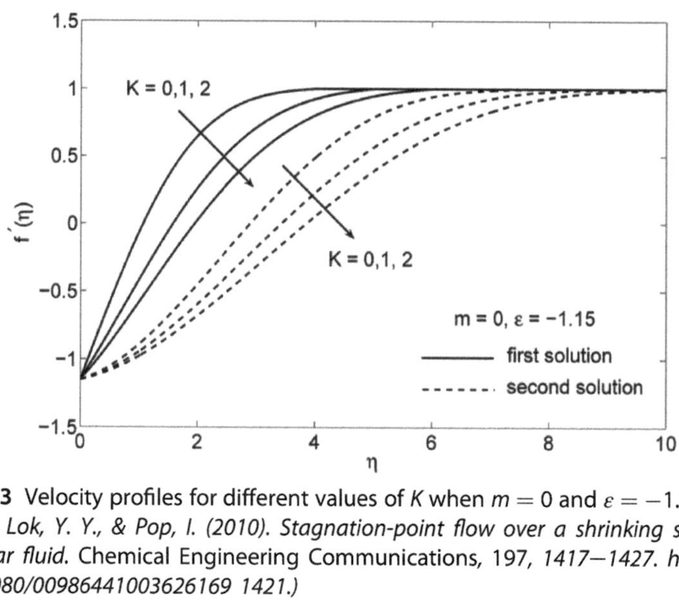

Figure 8.3 Velocity profiles for different values of K when $m = 0$ and $\varepsilon = -1.15$. *(From Ishak, A., Lok, Y. Y., & Pop, I. (2010). Stagnation-point flow over a shrinking sheet in a micropolar fluid.* Chemical Engineering Communications, 197, *1417—1427. https://doi.org/10.1080/00986441003626169 1421.)*

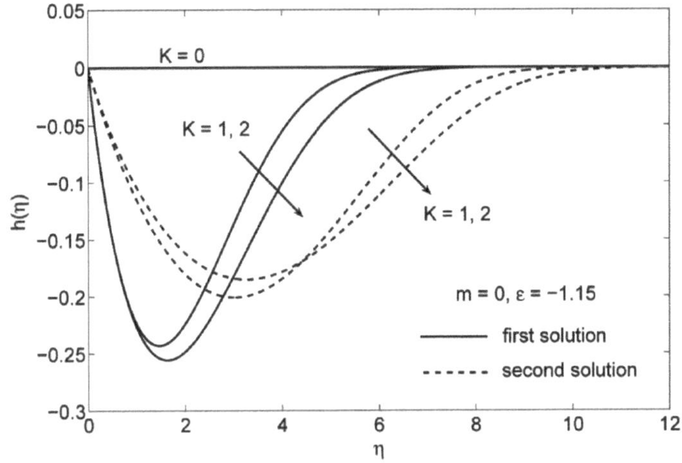

Figure 8.4 Angular velocity profiles for different values of K when $m = 0$ and $\varepsilon = -1.15$. *(From Ishak, A., Lok, Y. Y., & Pop, I. (2010). Stagnation-point flow over a shrinking sheet in a micropolar fluid.* Chemical Engineering Communications, 197, *1417–1427. https://doi.org/10.1080/00986441003626169 1421.)*

solution branch are smaller than those of the lower solution branch. Applying the stability analysis in Section 8.3.3 for the case of orthogonal stagnation–point flow, (8.37), (8.39), and (8.41) are solved and the smallest eigenvalues are identified. The upper solution branch gives negative eigenvalue ($\delta < 0$), while positive eigenvalue ($\delta > 0$) is obtained for the lower solution branch. As $\tau \rightarrow \infty$, if δ is negative, there is an initial decay of disturbance, therefore the upper solution branch is physically stable. If δ is positive, there is an initial growth of disturbance, hence the lower solution branch is unstable.

The effect of shrinking parameter ε on the angular velocity profile $h(\eta)$ has been given in Fig. 8.5 when $K = 1$ and $m = 0.5$ (weak concentration). Further investigation on the effect of shrinking parameter reveals that solution is unique for $\varepsilon > -1$, dual solution are found to exist for $\varepsilon_c < \varepsilon \leq -1$, and solution does not exist for $\varepsilon < \varepsilon_c$, where the critical values ε_c are approximately -1.2465, -1.2084, and -1.1957 for $K = 0$, 1, and 2, respectively (Ishak et al., 2010).

8.3.5 Comments (oblique stagnation–point flow)

The oblique stagnation–point flow (8.23)–(8.27) in micropolar fluid over a stretching or shrinking surface with slip condition had been numerically solved by Lok et al. (2018) using the bvp4c function in Matlab software.

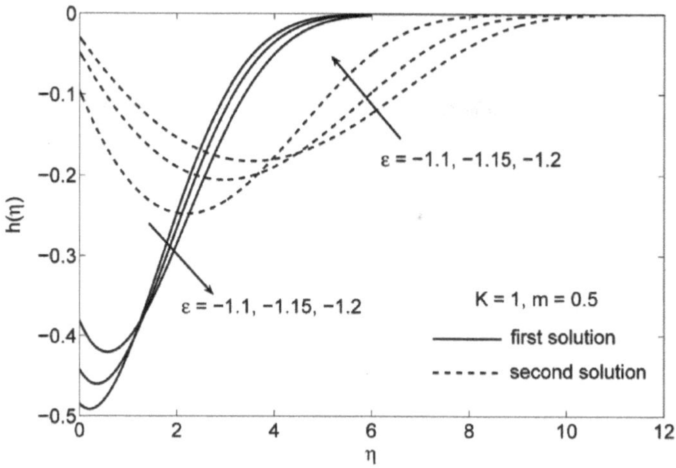

Figure 8.5 Angular velocity profiles for different values of ε when $K = 1$ and $m = 0.5$. *(From Ishak, A., Lok, Y. Y., & Pop, I. (2010). Stagnation-point flow over a shrinking sheet in a micropolar fluid.* Chemical Engineering Communications, 197, *1417–1427. https://doi.org/10.1080/00986441003626169 1421.)*

Figs. 8.6–8.9 show the initial values of $f''(0)$, $g'(0)$, $h'(0)$, and $q'(0)$ which are plotted against the stretching/shrinking parameter ε, respectively, for micropolar parameter $K = 1, 5$; slip parameter $A = 0.1, 0.5$, strong concentration of micropolar fluid $m = 0$ when $\alpha = \beta$. The initial values $f''(0)$ and $g'(0)$ decrease, while the absolute values of $h'(0)$ and $q'(0)$ increase when the value of K increases. For the shrinking case ($\varepsilon < 0$), dual solutions exist, where the critical values for $K = 1$ and $K = 5$ are -1.24784 and -1.20595, respectively, when $A = 0.1$. The critical values increase slightly when $A = 0.5$, which are -1.46652 and -1.28789, respectively. The bifurcation curve terminates at *0* as $\varepsilon \to -1$. Solving the stability problem (8.37)–(8.41) gives the negative smallest eigenvalues for the upper solution branch and positive smallest eigenvalues for the lower solution branch. These led to the conclusion that the upper solution branch is stable, while the lower solution branch is unstable, whereas the stability changes at the saddle-node bifurcation (the critical point). Observation on the effect of slip parameter shows that for stable solution, an increase in slip parameter increases the normalized velocity f' and tangential velocity g'; however, it caused the decrease in the boundary layer thickness.

Since β is a free parameter, the effect of this parameter on $g(\eta)$ is examined by considering some constant values of β. For the upper solution branch, plots of g against η, for shrinking case ($\varepsilon < 0$) and strong

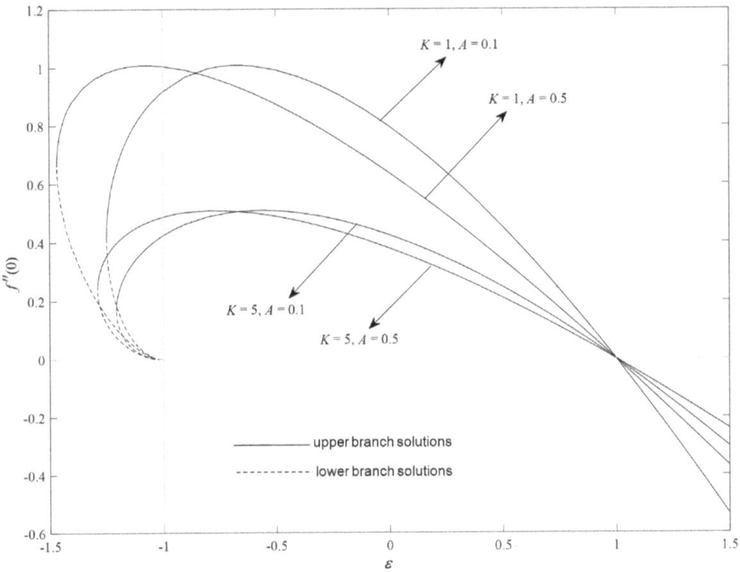

Figure 8.6 Plots of initial values $f''(0)$ against ε for some values of K and A, $\alpha = \beta$, $m = 0$. *(From Lok, Y. Y., Ishak, A., & Pop, I. (2018). Oblique stagnation slip flow of a micropolar fluid towards a stretching/shrinking surface: a stability analysis.* Chinese Journal of Physics, 56, *3062–3072. https://doi.org/10.1016/j.cjph.2018.10.016 3064.)*

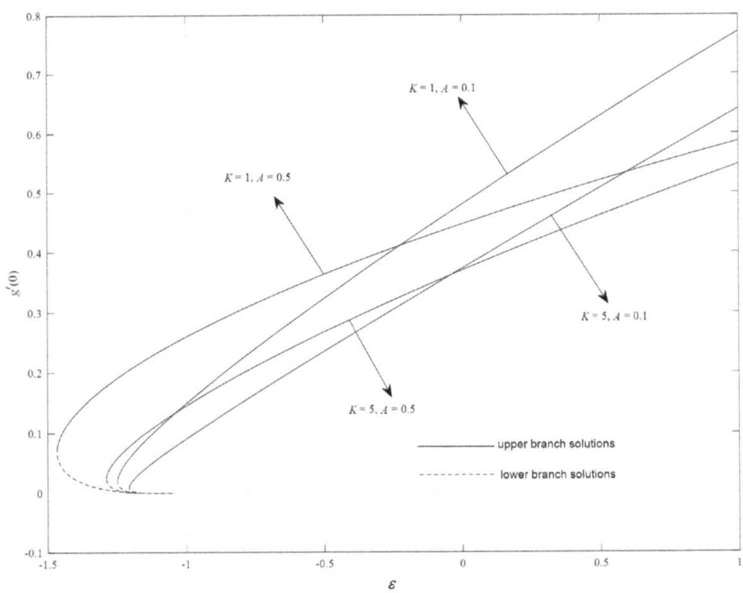

Figure 8.7 Plots of initial values $g'(0)$ against ε for some values of K and A, $\alpha = \beta$, $m = 0$. *(From Lok, Y. Y., Ishak, A., & Pop, I. (2018). Oblique stagnation slip flow of a micropolar fluid towards a stretching/shrinking surface: A stability analysis.* Chinese Journal of Physics, 56, *3062–3072. https://doi.org/10.1016/j.cjph.2018.10.016 3064.)*

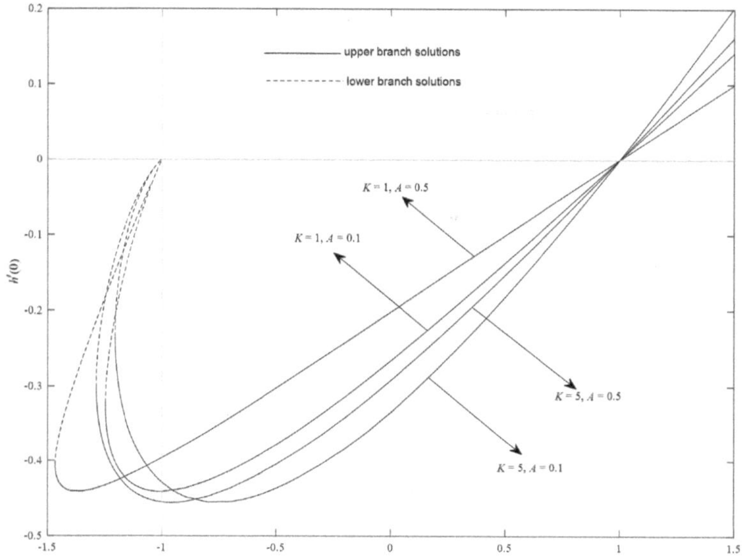

Figure 8.8 Plots of initial values $h'(0)$ against ε for some values of K and A, $\alpha = \beta$, $m = 0$. *(From Lok, Y. Y., Ishak, A., & Pop, I. (2018). Oblique stagnation slip flow of a micropolar fluid towards a stretching/shrinking surface: A stability analysis.* Chinese Journal of Physics, *56, 3062–3072. https://doi.org/10.1016/j.cjph.2018.10.016 3064.)*

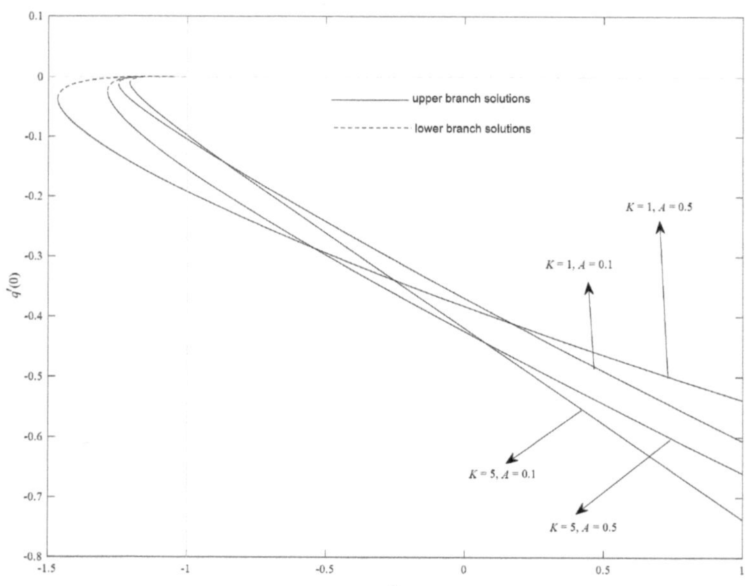

Figure 8.9 Plots of initial values $q'(0)$ against ε for some values of K and A, $\alpha = \beta$, $m = 0$. *(From Lok, Y. Y., Ishak, A., & Pop, I. (2018). Oblique stagnation slip flow of a micropolar fluid towards a stretching/shrinking surface: A stability analysis.* Chinese Journal of Physics, *56, 3062–3072. https://doi.org/10.1016/j.cjph.2018.10.016 3064.)*

concentration ($m = 0$) when $K = 1$ and $A = 0.1$ show that an inverted boundary layer structure is seen for $\beta > \alpha$. The profile for lower solution branch is similar, but the overshoot of the plot is greater for the lower solution branch, consequently the boundary layer thickness is larger for the lower solution branch.

Streamlines pattern for oblique stagnation flow over the shrinking surface is given in Fig. 8.10 for the case of $A = 0.5$ and Fig. 8.11 for the case of $A = 5$. These are the streamlines for upper solution branch with strong concentration micropolar fluid $m = 0$ and taking shrinking parameter is the same as the shear flow parameter, $\varepsilon = \lambda = -1.2$. Similar streamlines pattern is observed for the lower solution branch but its horizontal dividing streamline is further from the wall surface. As expected, no symmetric streamlines for the oblique stagnation flow. Reversed flow is observed for small slip parameter, and this rotating flow near to the wall surface reduces and gradually eliminates as slip parameter increases. Both the points ξ_0 and ξ_s do not fall on the same location due to the moving surface.

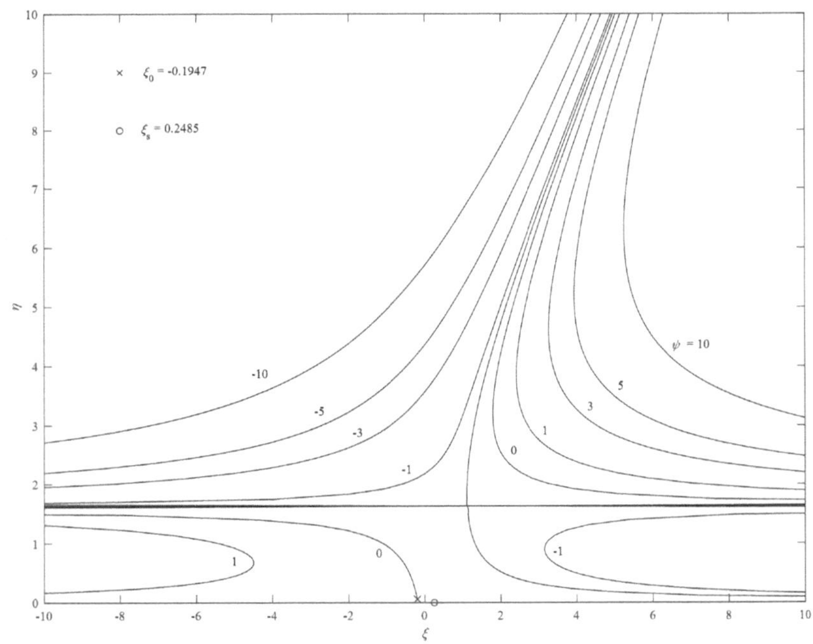

Figure 8.10 Dimensionless streamline plots η against ζ for upper solution branch when $A = 0.5$, $K = 1$, $\varepsilon = \lambda = -1.2$, $m = 0$. *(From Lok, Y. Y., Ishak, A., & Pop, I. (2018). Oblique stagnation slip flow of a micropolar fluid towards a stretching/shrinking surface: A stability analysis.* Chinese Journal of Physics, 56, *3062–3072. https://doi.org/10.1016/j.cjph.2018.10.016 3068.)*

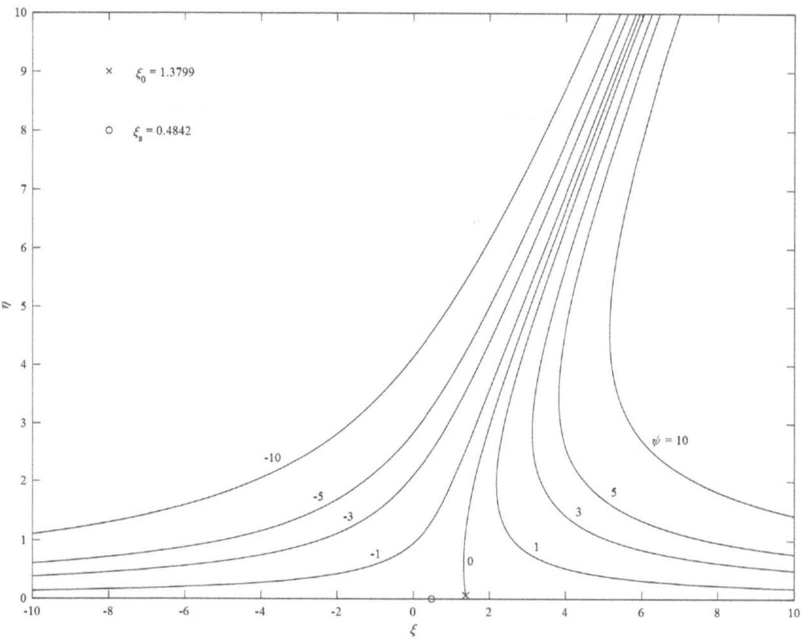

Figure 8.11 Dimensionless streamline plots η against ζ for upper solution branch when $A = 5$, $K = 1$, $\varepsilon = \lambda = -1.2$ and $m = 0$. *(From Lok, Y. Y., Ishak, A., & Pop, I. (2018). Oblique stagnation slip flow of a micropolar fluid towards a stretching/shrinking surface: A stability analysis.* Chinese Journal of Physics, 56, *3062–3072. https://doi.org/10.1016/j.cjph.2018.10.016 3068.)*

8.4 Moving wedge and flat plate in a micropolar fluid

8.4.1 Introduction

Historically, the steady laminar flow passing a fixed wedge was first analyzed in the early 1930s by Falkner and Skan (1931) to illustrate the application of Prandtl's boundary layer theory. With a similarity transformation, the boundary layer equation is reduced to an ordinary differential equation, which is well known as the Falkner and Skan equation. This equation includes nonuniform flow, i.e., outer flows which, when evaluated at the wall, take the form ax^m, where x is the coordinate measured along the wedge wall and a (> 0), and m are constants.

8.4.2 Basic equations

Consider the problem of forced convection boundary layer flow over a wedge which moves with a velocity $u_w(x)$ in a micropolar fluid as it was formulated by Ishak et al. (2006a). The positive x-coordinate is measured along the surface of the wedge with the apex as origin, and the positive y-coordinate is measured normal to the x-axis in the outward direction toward the fluid, see Fig. 8.12.

The governing equations for steady forced convection flow over a continuously moving wedge, after applying the boundary layer approximation can be written as (see Ishak et al., 2006a):

$$\frac{\partial u}{\partial x}+\frac{\partial v}{\partial y}=0 \tag{8.42}$$

$$u\frac{\partial u}{\partial x}+v\frac{\partial u}{\partial y}=u_e\frac{du_e}{dx}+\left(\frac{\mu+\kappa}{\rho}\right)\frac{\partial^2 u}{\partial y^2}+\frac{\kappa}{\rho}\frac{\partial N}{\partial y} \tag{8.43}$$

$$\rho j\left(u\frac{\partial N}{\partial x}+v\frac{\partial N}{\partial y}\right)=\frac{\partial}{\partial y}\left(\gamma\frac{\partial N}{\partial y}\right)-\kappa\left(2N+\frac{\partial u}{\partial y}\right) \tag{8.44}$$

$$u\frac{\partial j}{\partial x}+v\frac{\partial j}{\partial y}=0 \tag{8.45}$$

where the definitions of the variables are given in the Nomenclature. The boundary conditions are of the following form:

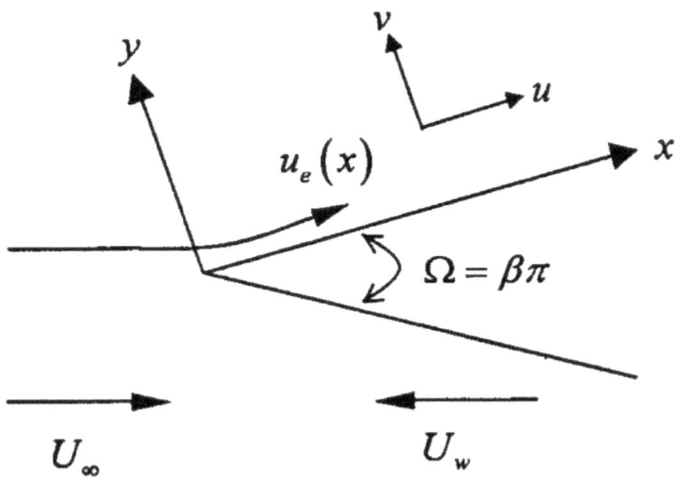

Figure 8.12 Physical model and the coordinate system. *(From Ishak, A., Nazar, R., & Pop, I. (2006). Moving wedge and flat plate in a micropolar fluid.* International Journal of Engineering Science, *44(18), 1225–1236. https://doi.org/10.1016/j.ijengsci.2006.08.005 1227.)*

$$v = 0, \quad u = u_w(x) = -U_w x^m, \quad j = 0, \quad N = -\frac{1}{2}\frac{\partial u}{\partial y} \text{ at } y = 0$$
$$u \to u_e(x) = U_\infty x^m, \quad N \to 0 \text{ as } y \to \infty \tag{8.46}$$

where $m = \beta/(2 - \beta)$, and β is the Hartree pressure gradient parameter which corresponds to $\beta = \Omega/\pi$ for a total wedge angle Ω, and U_w and U_∞ are constants. We notice that $0 \leq m \leq 1$, with $m = 0$ for the flow over a moving flat plate and $m = 1$ for the flow near the stagnation point on a moving wall. As it was shown by Ahmadi (1976), Kline (1977) and Kline and Allen (1970), the spin-gradient viscosity can be defined as

$$\gamma(x,y) = \left(\mu + \frac{\kappa}{2}\right) j(x,y) = \mu\left(1 + \frac{K}{2}\right) j(x,y) \tag{8.47}$$

where $K = \kappa/\mu$ denotes the material parameter. The case $K = 0$ describes the classical Navier–Stokes equations for a viscous and incompressible fluid. Eq. (8.47) has been very successfully used by Gorla (1980), Ishak et al. (2006b) and Yücel (1989)) to study different problems of convective flow of micropolar fluids. It has also been stated by (Ahmadi, 1976) that for non-constant microinertia, it is possible using Eq. (8.47) to find similar and self-similar solutions for a large number of problems of micropolar fluids.

8.4.3 Similarity transformation

Introducing the following similarity variables

$$\psi(x,y) = \left[\frac{2\nu x u_e(x)}{(m+1)}\right]^{1/2} f(\eta), \quad N(x,y) = u_e(x)\left[\frac{(m+1)u_e(x)}{2\nu x}\right]^{1/2} h(\eta),$$
$$j(x,y) = \frac{2\nu x}{(m+1)u_e(x)} i(\eta), \quad \eta = \left[\frac{(m+1)u_e(x)}{2\nu x}\right] y \tag{8.48}$$

where ψ is the stream function which satisfied Eq. (8.42) automatically. Substituting (8.48) into Eqs. (8.43)–(8.45), the following ordinary differential equations are obtained:

$$(1+K)f''' + ff'' + \left(\frac{2m}{m+1}\right)(1 - f'^2) + Kh' = 0 \tag{8.49}$$

$$\left(1 + \frac{K}{2}\right)(i\,h')' + i\left[fh' - \left(\frac{3m-1}{m+1}\right)f'h\right] - K(2h + f'') = 0 \tag{8.50}$$

$$(1-m)f'i - \frac{m+1}{2} f\, i' = 0 \tag{8.51}$$

subject to the boundary conditions

$$\begin{gathered} f(0) = i(0) = 0,\ f'(0) = -\lambda,\ h(0) = -\frac{1}{2} f''(0) \\ f' \to 1,\ h \to 0 \ \ as\ \ \eta \to \infty \end{gathered} \tag{8.52}$$

where $\lambda = U_w / U_\infty$ is the ratio of the wall velocity to the free stream fluid velocity, and corresponds to the situation when the wall moves in the opposite direction to the free stream (and vice versa), while $\lambda = 0$ is for a fixed flat wall. We get, after integrating Eq. (8.51),

$$i = Af^{2(1-m)/(1+m)} \tag{8.53}$$

where A is the nondimensional constant of integration. It is noticed that Eqs. (8.49)−(8.51) were also derived by Kim (1999) and Kim and Kim (2003). However, Eq. (8.49) was wrongly derived in (Kim, 1999) because in his paper Eq. (8.49) contains the extra term $[m/(m+1)]\eta f' h'$.

The physical quantity of interest is the skin friction coefficient which is defined as

$$C_f = \frac{1}{\rho u_w^2} \left[(\mu + \kappa) \frac{\partial u}{\partial y} + \kappa N \right]_{y=0} \tag{8.54}$$

Using the similarity variables (8.48), the skin friction coefficient (8.54) now has the form

$$C_f Re_x^{1/2} = \left(1 + \frac{K}{2} \right) f''(0) \tag{8.55}$$

8.4.4 Flat plate problem

In this case $m = 0$, and thus Eqs. (8.49) and (8.50) reduce to

$$(1+K)f''' + f f'' + Kh' = 0 \tag{8.56}$$

$$\left(1 + \frac{K}{2} \right) (i\, h')' + i(f\, h)' - K(2h + f'') = 0 \tag{8.57}$$

subject to the boundary conditions (8.52), while (8.53) becomes

$$i = Af^2 \tag{8.58}$$

The solution of Eqs. (8.56)−(8.58) subject to boundary conditions (8.52) for the case of a fixed plate can be found in (Ahmadi, 1976). If $K = 0$ (Newtonian fluid), Eq. (8.57) reduces to the problem studied by Klemp and Acrivos (1972). If K is not zero, but $A = 0$, i.e., $i = 0$, then from, we find $h = -\frac{1}{2}f''$, i.e., the gyration is identical to the angular velocity. Then Eq. (8.56) becomes

$$(1 + K/2)f''' + ff'' = 0 \tag{8.59}$$

Following (Rees & Bassom, 1996), take

$$\widehat{f}\left(\widehat{\eta}\right) = (1 + K/2)^{-1/2} f(\eta), \; \widehat{\eta} = (1 + K/2)^{-1/2}\eta \tag{8.60}$$

hence, Eq. (8.59) reduces to the problem of boundary layer flow over a moving and impermeable semi-infinite flat plate in a viscous (Newtonian) fluid as studied by Fang (2003), i.e.,

$$\widehat{f}''' + \widehat{f}\widehat{f}'' = 0 \tag{8.61}$$

with the boundary conditions

$$\widehat{f}(0) = 0, \; \widehat{f}'(0) = -\lambda, \; \widehat{f}'(\infty) \rightarrow 1 \tag{8.62}$$

where now primes denote differentiation with respect to $\widehat{\eta}$. The skin friction coefficient given by Eq. (8.55) becomes

$$C_f Re_x^{1/2} = (1 + K/2)^{-1/2} \widehat{f}''(0) \tag{8.63}$$

8.4.5 Wedge problem

If $A = 0$, i.e., $i = 0$, in Eq. (8.58), it can be easily shown, on using (8.52) and (8.60), Eqs. (8.49) and (8.50) reduce to

$$\widehat{f}''' + \widehat{f}\widehat{f}'' + \beta\left(1 - \widehat{f}'^2\right) = 0 \tag{8.64}$$

where $\beta = 2m/(m+1)$, subject to the boundary conditions (8.62). The solution to this problem can be found in (Riley & Weidman, 1989).

8.4.6 Comments

The boundary value problem (8.49−8.52) has been solved numerically using an implicit finite-difference scheme known as Keller-box method for several values of m, λ, and K by Ishak et al. (2006a)). Following Ahmadi (1976) and Kim (1999), the value of A is considered being unity.

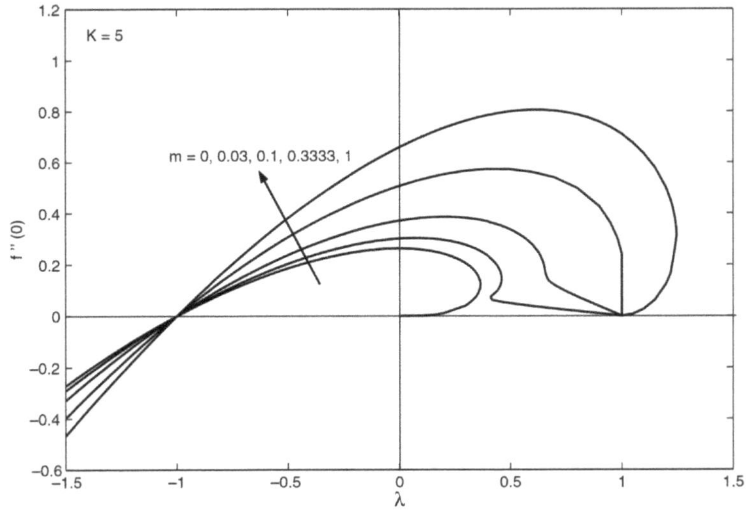

Figure 8.13 Reduced skin friction coefficient as a function of λ for various values of m when $K = 5$. *(From Ishak, A., Nazar, R., & Pop, I. (2006). Moving wedge and flat plate in a micropolar fluid.* International Journal of Engineering Science, *44(18), 1225–1236. https://doi.org/10.1016/j.ijengsci.2006.08.005 1229.)*

Variations of the reduced skin friction coefficient $f''(0)$ with λ are shown in Fig. 8.13 for some values of m when $K = 5$. It shows that multiple solutions are possible for some values of m when $\lambda > 0$, i.e., when the wall is moving in opposite direction to the free stream, while the solution is unique when $\lambda \leq 0$ for all m. For example, there are three solutions when $m = 0.03$ for the values of λ within the range $0.41 \leq \lambda \leq 0.46$, and dual solutions are obtained for $m > 0.3333$, when $\lambda > 1$. The velocity and microrotation profiles that show the existence of three solutions when $m = 0.03$ and $\lambda = 0.45$ are presented in Figs. 8.14 and 8.15, respectively. These figures show that the boundary conditions (8.52) are satisfied. It is shown in Fig. 8.13 that for $m = 1$ ($\Omega = 180$ degrees), there is no solution when $\lambda > \lambda_c$ where $\lambda_c = 1.2467$, while for $m = 0$ (flat plate), there is no solution when $\lambda > \lambda_c$ where $\lambda_c = 0.3645$. At these critical values of λ (λ_c), both solution branches are connected and thus a unique solution is obtained. Beyond these critical values, the boundary layer separates from the surface and thus the solutions based upon the boundary layer approximations are no longer valid. Further, it is seen that all curves in Fig. 8.13 intersect at the point $(-1, 0)$ since there is no shear stress at the surface when the wall and the fluid move with the same velocity. Moreover, all the solution curves for $m > 0$ (wedge) have $(1, 0)$ as the limit point, whereas for

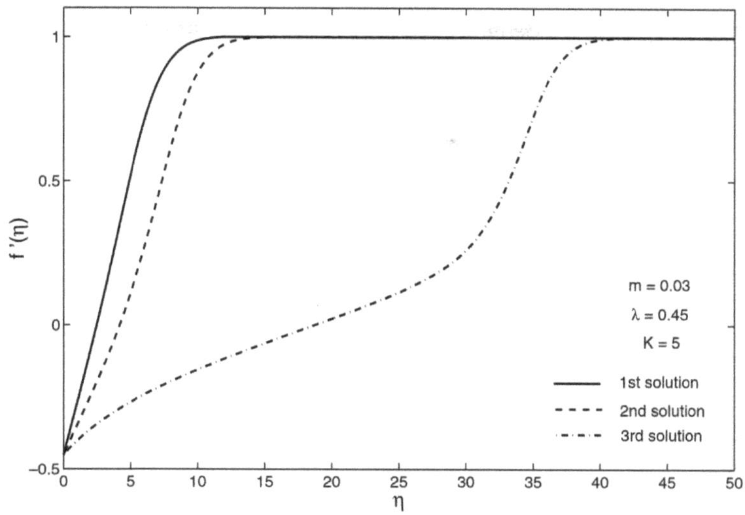

Figure 8.14 Velocity profiles for $m = 0.03$, $\lambda = 0.45$ and $K = 5$. *(From Ishak, A., Nazar, R., & Pop, I. (2006). Moving wedge and flat plate in a micropolar fluid.* International Journal of Engineering Science, *44(18), 1225–1236. https://doi.org/10.1016/j.ijengsci.2006.08.005 1229.)*

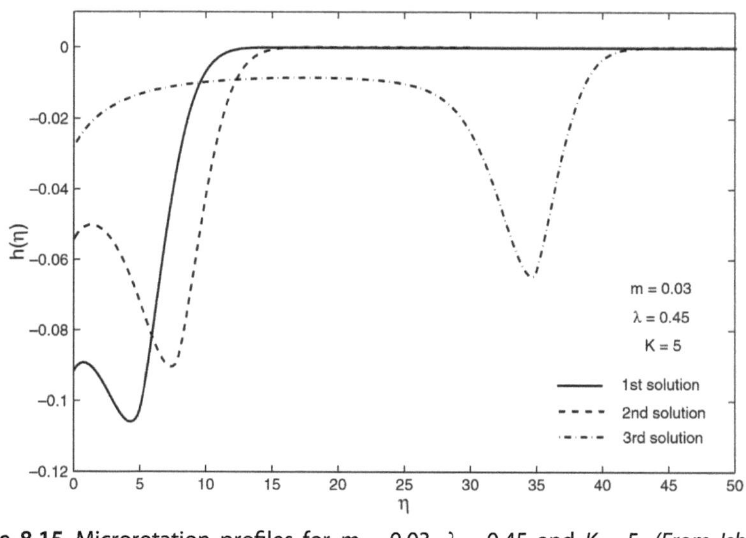

Figure 8.15 Microrotation profiles for $m = 0.03$, $\lambda = 0.45$ and $K = 5$. *(From Ishak, A., Nazar, R., & Pop, I. (2006). Moving wedge and flat plate in a micropolar fluid.* International Journal of Engineering Science, *44(18), 1225–1236. https://doi.org/10.1016/j.ijengsci.2006.08.005 1229.)*

$m = 0$ (flat plate), the curve terminates at the origin. The detail results when $m = 0$ and $K = 0$ can be found in Klemp and Acrivos (1972), while for the case of the permeable plate, they are given in Fang (2003). More important results on this problem can be found in Ishak et al. (2006a).

Nomenclature

Roman letter

a,b,c	positive constant (s^{-1})
A	slip parameter (Section 8.3)
A	dimensionless constant of integration (Section 8.4)
A_t	slip length (m)
C_f	skin friction coefficient
C_p	specific heat at constant pressure (J $kg^{-1}K^{-1}$)
$\mathbf{F}$	body force vector
F_x, F_y	body force along the x and y axes ($kgms^{-2}$)
g	acceleration due to gravity (ms^{-2})
Gr_x	local Grashof number
j	microinertia density (m^2)
k	thermal conductivity $KWm^{-1}K^{-1}$
K	micropolar parameter
l	characteristic length (m)
m	constant (Section 8.4)
m	constant denotes the concentration of fluid (Section 8.2 and Section 8.3)
N	Microrotation component normal to the $x-y$ plane (s^{-1})
$\mathbf{N}$	microrotation vector
p	pressure (Nm^{-2})
Pr	Prandtl number
q_w	heat flux from the surface
Re_x	local Reynolds number
t	dimensional time (s)
T	temperature of fluid (K)
T_0	characteristic temperature
(u, v)	velocity components along x and y axes (ms^{-1})
$\mathbf{u}$	velocity vector
U_w, U_∞	velocity reference
(x, y)	Cartesian coordinates (m)

Greek symbols

ε	stretching or shrinking parameter
α,β	constant parameter (Section 8.3)
β	coefficient of thermal expansion (K^{-1}) (Section 8.2)
β	Hartree pressure gradient parameter (Section 8.4)

δ	eigenvalue
η	dimensionless similarity variable
γ	spin-gradient viscosity (kg ms^{-1})
κ	vortex viscosity (N sm^{-2})
λ	ratio of the wall velocity to the free stream fluid velocity (N sm^{-2}) (Section 8.4)
λ	shear flow parameter (Section 8.3)
μ	dynamic viscosity (N sm^{-2})
ν	kinematic viscosity (m^2s^{-1})
Ω	total wedge angle
ρ	density of fluid (kgm^{-3})
ψ	stream function (m^2s^{-1})
τ	dimensionless time variable
τ_w	wall shear stress (Nm^{-2})
ξ_s	point of zero wall shear stress

Subscripts

e, ∞	condition outside of the boundary layer
w	condition at the wall

References

Agarwal, R. S., Bhargava, R., & Balaji, A. V. S. (1990). Finite element solution of nonsteady three-dimensional micropolar fluid flow at a stagnation-point. *International Journal of Engineering Science, 28*(8), 851–857. https://doi.org/10.1016/0020-7225(90)90030-M

Ahmadi, G. (1976). Self-similar solution of incompressible micropolar boundary layer flow over a semi-infinite plate. *International Journal of Engineering Science, 14*(7), 639–646. https://doi.org/10.1016/0020-7225(76)90006-9

Ahmad, K., Nazar, R., Ishak, A., & Pop, I. (2012). Unsteady three-dimensional boundary layer flow due to a stretching surface in a micropolar fluid. *International Journal for Numerical Methods in Fluids, 68*(12), 1561–1573. https://doi.org/10.1002/fld.2543

Aliyu, M. D., Saleh, S. H. M., Arifin, N. M., Nazar, R., & Pop, I. (2017). Unsteady micropolar fluid over a permeable curved stretching shrinking surface. *Mathematical Problems in Engineering, 2017*, 3085249. https://doi.org/10.1155/2017/3085249

Ariman, T., Turk, M. A., & Sylvester, N. D. (1973). Microcontinuum fluid mechanics—a review. *International Journal of Engineering Science, 11*(8), 905–930. https://doi.org/10.1016/0020-7225(73)90038-4

Ariman, T., Turk, M. A., & Sylvester, N. D. (1974). Applications of microcontinuum fluid mechanics. *International Journal of Engineering Science, 12*(4), 273–293. https://doi.org/10.1016/0020-7225(74)90059-7

Aurangzaib, Sharif Uddin, Md, Bhattacharyya, K., & Shafie, S. (2016). Micropolar fluid flow and heat transfer over an exponentially permeable shrinking sheet. *Propulsion and Power Research, 5*(4), 310–317. https://doi.org/10.1016/j.jppr.2016.11.005

Bhattacharyya, K., Mukhopadhyay, S., Layek, G. C., & Pop, I. (2012). Effects of thermal radiation on micropolar fluid flow and heat transfer over a porous shrinking sheet. *International Journal of Heat and Mass Transfer, 55*(11), 2945–2952. https://doi.org/10.1016/j.ijheatmasstransfer.2012.01.051

Borrelli, A., Giantesio, G., & Patria, M. C. (2012). MHD oblique stagnation-point flow of a Newtonian fluid. *Zeitschrift für Angewandte Mathematik und Physik, 63*(2), 271–294. https://doi.org/10.1007/s00033-011-0174-8

Borrelli, A., Giantesio, G., & Patria, M. C. (2015). Influence of a non-uniform external magnetic field on the oblique stagnation-point flow of a micropolar fluid. *IMA Journal of Applied Mathematics, 80*(3), 747–765. https://doi.org/10.1093/imamat/hxu019
Chiam, T. C. (1994). Stagnation-point flow towards a stretching plate. *Journal of the Physical Society of Japan, 63*(6), 2443–2444. https://doi.org/10.1143/JPSJ.63.2443
Chun, C., Zaimi, K., & Ishak, A. (2014). Stagnation-point flow and heat transfer over a nonlinearly stretching/shrinking sheet in a micropolar fluid. *Abstract and Applied Analysis, 2014*, 261630. https://doi.org/10.1155/2014/261630
Crane, L. J. (1970). Flow past a stretching plate. *Zeitschrift für Angewandte Mathematik und Physik ZAMP, 21*(4), 645–647. https://doi.org/10.1007/BF01587695
Drazin, P. G., & Riley, N. (2006). The Navier-Stokes equations: A classification of flows and exact solutions. In *London mathematical society lecture note series*. Cambridge University Press. https://doi.org/10.1017/CBO9780511526459
El-Aziz, M. A. (2009). Viscous dissipation effect on mixed convection flow of a micropolar fluid over an exponentially stretching sheet. *Canadian Journal of Physics, 87*(4), 359–368. https://doi.org/10.1139/P09-047
Eringen, A. C. (1966). Theory of micropolar fluids. *Journal of Mathematics and Mechanics, 16*(1), 1–18. http://www.jstor.org/stable/24901466.
Eringen, A. C. (1972). Theory of thermomicrofluids. *Journal of Mathematical Analysis and Applications, 38*(2), 480–496. https://doi.org/10.1016/0022-247X(72)90106-0
Eringen, A. C. (2001). *Microcontinuum field theories II: Fluent media.* Springer-Verlag.
Falkner, V. M., & Skan, S. W. (1931). LXXXV. Solutions of the boundary-layer equations. *The London, Edinburgh, and Dublin Philosophical Magazine and Journal of Science, 12*(80), 865–896. https://doi.org/10.1080/14786443109461870
Fang, T. (2003). Further study on a moving-wall boundary-layer problem with mass transfer. *Acta Mechanica, 163*(3), 183–188. https://doi.org/10.1007/s00707-002-0979-9
Gorla, R. S. R. (1980). Thermal boundary layer of a micropolar fluid at a stagnation point. *International Journal of Engineering Science, 18*(4), 611–617. https://doi.org/10.1016/0020-7225(80)90133-0
Guram, G. S., & Smith, A. C. (1980). Stagnation flows of micropolar fluids with strong and weak interactions. *Computers & Mathematics with Applications, 6*(2), 213–233. https://doi.org/10.1016/0898-1221(80)90030-9
Harris, S. D., Ingham, D. B., & Pop, I. (2009). Mixed convection boundary-layer flow near the stagnation point on a vertical surface in a porous medium: Brinkman model with slip. *Transport in Porous Media, 77*(2), 267–285. https://doi.org/10.1007/s11242-008-9309-6
Hassanien, I. A., & Gorla, R. S. R. (1990). Heat transfer to a micropolar fluid from a non-isothermal stretching sheet with suction and blowing. *Acta Mechanica, 84*(1), 191–199. https://doi.org/10.1007/BF01176097
Hayat, T., Javed, T., & Abbas, Z. (2009). MHD flow of a micropolar fluid near a stagnation-point towards a non-linear stretching surface. *Nonlinear Analysis: Real World Applications, 10*(3), 1514–1526. https://doi.org/10.1016/j.nonrwa.2008.01.019
Hiemenz, K. (1911). Die Grenzschicht an einem in den gleichförmigen Flüssigkeitsstrom eingetauchten geraden Kreiszylinder. *Dingler's Polytechnisches Journal, 326*, 321–324.
Ibrahim, W. (2017). MHD boundary layer flow and heat transfer of micropolar fluid past a stretching sheet with second order slip. *Journal of the Brazilian Society of Mechanical Sciences and Engineering, 39*, 791–799. https://doi.org/10.1007/s40430-016-0621-8
Ishak, A., Lok, Y. Y., & Pop, I. (2010). Stagnation-point flow over a shrinking sheet in a micropolar fluid. *Chemical Engineering Communications, 197*, 1417–1427. https://doi.org/10.1080/00986441003626169

Ishak, A., Nazar, R., & Pop, I. (2006a). Moving wedge and flat plate in a micropolar fluid. *International Journal of Engineering Science, 44*(18), 1225–1236. https://doi.org/10.1016/j.ijengsci.2006.08.005

Ishak, A., Nazar, R., & Pop, I. (2006b). The Schneider problem for a micropolar fluid. *Fluid Dynamics Research, 38*(7), 489–502. https://doi.org/10.1016/j.fluiddyn.2006.03.004

Ishak, A., Nazar, R., & Pop, I. (2006c). Mixed convection boundary layers in the stagnation-point flow toward a stretching vertical sheet. *Meccanica, 41*(5), 509–518. https://doi.org/10.1007/s11012-006-0009-4

Ishak, A., Nazar, R., & Pop, I. (2008). Heat transfer over a stretching surface with variable heat flux in micropolar fluids. *Physics Letters A, 372*(5), 559–561. https://doi.org/10.1016/j.physleta.2007.08.003

Jena, S. K., & Mathur, M. N. (1981). Similarity solutions for laminar free convection flow of a thermomicropolar fluid past a nonisothermal flat plate. *International Journal of Engineering Science, 19*(11), 1431–1439. https://doi.org/10.1016/0020-7225(81)90040-9

Kelson, N. A., & Desseaux, A. (2001). Effect of surface conditions on flow of a micropolar fluid driven by a porous stretching sheet. *International Journal of Engineering Science, 39*(16), 1881–1897. https://doi.org/10.1016/S0020-7225(01)00026-X

Kelson, N., & Farrell, T. (2001). Micropolar fluid flow over a porous stretching sheet with strong suction or injection. *International Communications in Heat and Mass Transfer, 28*, 479–488. https://doi.org/10.1016/S0735-1933(01)00252-4

Khan, Z. H., Qasim, M., Ishfaq, N., & Khan, W. A. (2017). Dual solutions of MHD boundary layer flow of a micropolar fluid with weak concentration over a stretching/shrinking sheet. *Communications in Theoretical Physics, 67*, 449. https://doi.org/10.1088/0253-6102/67/4/449

Kim, Y. J. (1999). Thermal boundary layer flow of a micropolar fluid past a wedge with constant wall temperature. *Acta Mechanica, 138*(1), 113–121. https://doi.org/10.1007/BF01179545

Kim, Y. J., & Kim, T. A. (2003). Convective micropolar boundary layer flows over a wedge with constant surface heat flux. *International Journal of Applied Mechanics and Engineering, 8*, 147–153.

Klemp, J. B., & Acrivos, A. (1972). A method for integrating the boundary-layer equations through a region of reverse flow. *Journal of Fluid Mechanics, 53*(1), 177–191. https://doi.org/10.1017/S0022112072000096

Kline, K. A. (1977). A spin-vorticity relation for unidirectional plane flows of micropolar fluids. *International Journal of Engineering Science, 15*(2), 131–134. https://doi.org/10.1016/0020-7225(77)90028-3

Kline, K. A., & Allen, S. J. (1970). Nonsteady flows of fluids with microstructure. *The Physics of Fluids, 13*(2), 263–270. https://doi.org/10.1063/1.1692918

Lok, Y. Y., Ishak, A., & Pop, I. (2018). Oblique stagnation slip flow of a micropolar fluid towards a strecthing/shrinking surface: A stability analysis. *Chinese Journal of Physics, 56*, 3062–3072. https://doi.org/10.1016/j.cjph.2018.10.016

Lok, Y. Y., Merkin, J. H., & Pop, I. (2015). MHD oblique stagnation-point flow towards a stretching/shrinking surface. *Meccanica, 50*(12), 2949–2961. https://doi.org/10.1007/s11012-015-0188-y

Lok, Y. Y., Pop, I., Ingham, D. B., & Amin, N. (2009). Mixed convection flow of a micropolar fluid near a non-orthogonal stagnation-point on a stretching vertical sheet. *International Journal of Numerical Methods for Heat and Fluid Flow, 19*(3/4), 459–483. https://doi.org/10.1108/09615530910938380

Łukaszewicz, G. (1999). *Micropolar fluids: Theory and applications*. Birkhäuser.

Merkin, J. H. (1986). On dual solutions occurring in mixed convection in a porous medium. *Journal of Engineering Mathematics, 20*(2), 171–179. https://doi.org/10.1007/BF00042775

Miklavčič, M., & Wang, C. Y. (2006). Viscous flow due to a shrinking sheet. *Quarterly of Applied Mathematics, 64*(2), 283–290. http://www.jstor.org/stable/43638726.
Mishra, S. R., Hoque, M. M., Mohanty, B., & Anika, N. N. (2019). Heat transfer effect on MHD flow of a micropolar fluid through porous medium with uniform heat source and radiation. *Nonlinear Engineering, 8*(1), 65–73. https://doi.org/10.1515/nleng-2017-0126
Peddieson, J. (1972). An application of the micropolar fluid model to the calculation of turbulent shear flow. *International Journal of Engineering Science, 10*(1), 23–32. https://doi.org/10.1016/0020-7225(72)90072-9
Rees, D. A. S., & Bassom, A. P. (1996). The Blasius boundary-layer flow of a micropolar fluid. *International Journal of Engineering Science, 34*(1), 113–124. https://doi.org/10.1016/0020-7225(95)00058-5
Riley, N., & Weidman, P. D. (1989). Multiple solutions of the Falkner–Skan equation for flow past a stretching boundary. *SIAM Journal on Applied Mathematics, 49*(5), 1350–1358. https://doi.org/10.1137/0149081
Rosali, H., Ishak, A., & Pop, I. (2012). Micropolar fluid flow towards a stretching/shrinking sheet in a porous medium with suction. *International Communications in Heat and Mass Transfer, 39*(6), 826–829. https://doi.org/10.1016/j.icheatmasstransfer.2012.04.008
Rosca, N. C., & Pop, I. (2014). Boundary layer flow past a permeable shrinking sheet in a micropolar fluid with a second order slip flow model. *European Journal of Mechanics - B: Fluids, 48*, 115–122. https://doi.org/10.1016/j.euromechflu.2014.05.004
Sharma, R., Ishak, A., & Pop, I. (2016). Stagnation point flow of a micropolar fluid over a stretching/shrinking sheet with second-order velocity slip. *Journal of Aerospace Engineering, 29*(5), 04016025. https://doi.org/10.1061/(ASCE)AS.1943-5525.0000616
Shaw, S., Kameswaran, P. K., & Sibanda, P. (2013). Homogeneous-heterogeneous reactions in micropolar fluid flow from a permeable stretching or shrinking sheet in a porous medium. *Boundary Value Problems, 2013*(1), 77. https://doi.org/10.1186/1687-2770-2013-77
Stokes, V. K. (1984). *Theories of fluids with microstructure: An introduction*. Springer.
Tooke, R. M., & Blyth, M. G. (2008). A note on oblique stagnation-point flow. *Physics of Fluids, 20*(3), 033101. https://doi.org/10.1063/1.2876070
Turkyilmazoglu, M. (2014). A note on micropolar fluid flow and heat transfer over a porous shrinking sheet. *International Journal of Heat and Mass Transfer, 72*, 388–391. https://doi.org/10.1016/j.ijheatmasstransfer.2014.01.039
Wang, C. Y. (2008). Stagnation flow towards a shrinking sheet. *International Journal of Non-linear Mechanics, 43*(5), 377–382. https://doi.org/10.1016/j.ijnonlinmec.2007.12.021
Weidman, P. D., Kubitschek, D. G., & Davis, A. M. J. (2006). The effect of transpiration on self-similar boundary layer flow over moving surfaces. *International Journal of Engineering Science, 44*(11), 730–737. https://doi.org/10.1016/j.ijengsci.2006.04.005
Yacob, N. A., & Ishak, A. (2012a). Micropolar fluid flow over a shrinking sheet. *Meccanica, 47*(2), 293–299. https://doi.org/10.1007/s11012-011-9439-8
Yacob, N. A., & Ishak, A. (2012b). Stagnation point flow towards a stretching/shrinking sheet in a micropolar fluid with a convective surface boundary condition. *The Canadian Journal of Chemical Engineering, 90*(3), 621–626. https://doi.org/10.1002/cjce.20517
Yahaya, R. I., Md Arifin, N., Mohamed Isa, S. S. P., Rashidi, M. M., et al. (2021). Magnetohydrodynamics boundary layer flow of micropolar fluid over an exponentially shrinking sheet with thermal radiation: triple solutions and stability analysis. *Mathematical Methods in the Applied Sciences, 44*, 10578–10608. https://doi.org/10.1002/mma.7432
Yasmin, A., Ali, K., & Ashraf, M. (2020). Study of heat and mass transfer in MHD flow of micropolar fluid over a curved stretching sheet. *Scientific Reports, 10*(1), 4581. https://doi.org/10.1038/s41598-020-61439-8

Yücel, A. (1989). Mixed convection in micropolar fluid flow over a horizontal plate with surface mass transfer. *International Journal of Engineering Science, 27*(12), 1593–1602. https://doi.org/10.1016/0020-7225(89)90153-5

Zheng, L., Niu, J., Zhang, X., & Ma, L. (2012). Dual solutions for flow and radiative heat transfer of a micropolar fluid over stretching/shrinking sheet. *International Journal of Heat and Mass Transfer, 55*(25), 7577–7586. https://doi.org/10.1016/j.ijheatmasstransfer.2012.07.067

CHAPTER 9

Jets

9.1 Introduction

In the industrial application of jets such as annealing of metal and plastic sheets, drying of textiles, cooling of microelectronic equipment, and so forth, the temperature of the incoming jet should be different from that of the quiescent fluid. Therefore, the solution of the thermal boundary equation is required.

The first reported solution for the classical laminar wall jet is due to (Tetervin, 1948). Independently, (Akatnov, 1953; Glauert, 1956) rediscovered the wall jet solution of (Tetervin, 1948). For this reason, we will use here the abbreviation TAG for the Tetervin–Akatnov–Glauert laminar wall jet solution (see (Magyari & Weidman, 2005; Weidman, 2017)). Glauert (1956) obtained a well-known nontrivial analytical solution, that exhibits exponential decay at infinity (e-jet). Riley (1958) has studied the thermal characteristics of the preheated radial jet. Heat transfer characteristics associated with the Glauert wall jet flow were analyzed initially by Schwarz and Caswell (1961), where constant temperature and constant wall heat flux were examined in closed form analytic manner and in the absence of heat dissipation. Gorla (1973) gave approximate solutions to the unsteady energy equation in the context of Glauert's wall jet flow. Approximate solutions for heat transfer and momentum characteristics of wall jet flows in the presence of heat dissipation were also analyzed by Bansal and Tak (1978) subject to isothermal and adiabatic conditions.

Merkin and Needham (1986) extended Glauert's work to the case when the suction/injection is imposed at the wall and it is simultaneously stretched. They found that Glauert's solution would not be possible if one of the two parameters suction/injection and stretching are zero. Then, (Needham & Merkin, 1987) showed that the dual solutions were possible if reverse flow was allowed. The both solutions discussed by Merkin and Needham (1986) and Needham and Merkin (1987) could be considered as e-jet types. The classical Merkin and Needham's wall-jet problem was very

Similarity Solutions for the Boundary Layer Flow and Heat Transfer of Viscous Fluids, Nanofluids, Porous Media, and Micropolar Fluids
ISBN 978-0-12-821188-5
https://doi.org/10.1016/B978-0-12-821188-5.00010-2

recently studied by Aly and Pop (2020) for copper–titanium dioxide/water hybrid nanofluid with the effects of suction/injection and thermal energy. The similarity equations were solved both analytically and numerically using hypergeometric function. Expressions for both the reduced skin friction coefficient and reduced Nusselt number were obtained. Multiple (dual) upper and lower branch solutions were determined, that are possible only for suction. Moreover, it was shown that copper nanoparticle volume fraction plays a very important role in variation of the velocity. Magyari and Keller (2001a, 2001b) showed that the free laminar jets of the classical hydrodynamics may be identified with certain boundary layer flows induced by continuous surfaces immersed in quiescent incompressible fluid and stretched with well-defined velocities. Later on, (Magyari & Keller, 2004) noticed that it is suffice to retain suction alone in the Glauert's integral condition if the outer flow is assumed to be algebraically decaying (a-jets). By introducing an asymptotic boundary condition at far field, (Magyari & Keller, 2004) found that an infinity number of a-jet solutions can exist when the suction is applied to a stationary wall. Launder and Rodi (1979, 1983) considered in these papers the turbulent wall jets. Astin and Wilks (1996, 1997) and Barker and Wilks (2002) have presented new solutions of the Falkner–Skan equation, which display jet-like behavior. Magyari and Weidman (2005, 2006) have investigated the Airy wall jet, a jet which belongs to the similarity class of the free jet, but its far field behavior is an algebraically decaying rotational flow. Kalteh and Abbassi (2006) have obtained similarity solutions of the laminar axisymmetric jets with effect of viscous dissipation.

Cohen et al. (1992) discovered a new family of solutions describing the mean flow of an incompressible two-dimensional wall jet subjected to boundary suction or blowing, in which the vertical velocity at the wall varies in a power law with downstream distance. Linear stability calculations indicate that two unstable modes, an inviscid inflectional point instability and a viscous instability, may coexist in the flow. It is demonstrated theoretically that the relative importance of each mode can be controlled by subjecting the wall jet to small amounts of blowing or suction. Xu and Wu (2011) extended the problem considered by Cohen et al. (1992) by assuming that the wall is allowed to stretch. They showed that solutions to both e-jet and a-jet are possible. In particular, analytical solutions are given for some special cases, while solutions in general cases are obtained by the homotopy analysis method (HAM), founded by Liao (2003). In fact,

(Xu et al., 2008) found that such a kind of algebraically decaying solution could also be possible for nonzero suction parameter.

Other contributions in the field of wall jet flows can be found in (Schlichting, 1933) and who obtained a closed form analytic solution form of this plane jet. Turkyilmazoglu (2014) presented in a note, the exact temperatures of free two-dimensional and circular jets. Jafarimoghaddam et al. (2019) discussed the propagation of the nonsimilar wall jet flows with suction/injection. In practice, such a wall jet is of course turbulent and the study is mainly of theoretical interest but the understanding of the laminar jet is a necessary preliminary to understanding the turbulent jet, and in any case, there are theories of the turbulent jets in good agreement with experiment, which make extensive use of the arguments used in laminar flow.

It seems that (Rahimi-Esbo et al., 2012) are the first who have numerically studied the two-dimensional incompressible water—nanofluid flow in a confined jet in the laminar flow regime. A finite volume technique on a collocated grid is employed for discretizing the governing equations by applying the SIMPLE algorithm to link the pressure and velocity fields. Turkyilmazoglu (2016, 2019) provided both analytical and numerical solutions for heat transfer characteristics of several nanofluids jets discharged over a plane surface where the effect of heat dissipation was also included. In that study, the thermal boundary conditions were isothermal and adiabatic, and nanofluids were modeled through single-phase framework. In a series of papers, (Jafarimoghaddam, 2017, 2018a) presented closed form analytic solutions to heat and mass transfer characteristics of wall jet flow of nanofluids; a numerical approach is used in (Jafarimoghaddam & Shafizadeh, 2019). Jafarimoghaddam (2018a, 2018b) discussed the two-phase modeling of magnetic nanofluids jets over a stretching/shrinking wall, and the heat transfer characteristics and thermal instabilities of the Glauert type jet in analytic closed form. Jafarimoghaddam and Pop (2019) discussed numerical modeling of Glauert type exponentially decaying wall jet flows of nanofluids using (Tiwari & Das, 2007) mathematical nanofluid model. Very good references on jets can be found in the review paper by Xu and Wu (2011) and in the paper by Rehman et al. (2017), who have stated that the free-surface jet impingement technique operating with two-phase advanced coolants has recently drawn much favorable attention in cooling applications.

Finally at this end, we mention that (Weidman, 2017) studied three problems on plate in uniform motion transverse to the wall jet, a Blasius boundary layer, and the flow generated by uniform shear over a semi-infinite flat plate. Streamwise wall jet velocity profile and cross-flow profile for uniform transverse motion of a wall beneath a TAG wall jet is illustrated in a figure.

9.2 Wall jet

9.2.1 Problem formulation

Consider the steady two-dimensional boundary layer flow and heat transfer of a viscous and incompressible fluid past a permeable flat surface as is shown in Fig. 9.1, where x and y are Cartesian coordinates measured along the wall and normal to it, respectively. It is assumed that the surface is stretched with the velocity $u_w(x)$, the mass velocity is $v_w(x)$, and surface temperature is $T_w(x)$.

The basic equations in dimensionless form are (see(Vulis & Kashkarov, 1965))

$$\frac{\partial u}{\partial x}+\frac{\partial v}{\partial y}=0 \tag{9.1}$$

$$u\frac{\partial u}{\partial x}+v\frac{\partial u}{\partial y}=\frac{\partial^2 u}{\partial y^2} \tag{9.2}$$

$$u\frac{\partial T}{\partial x}+v\frac{\partial T}{\partial y}=\frac{1}{Pr}\frac{\partial^2 T}{\partial y^2} \tag{9.3}$$

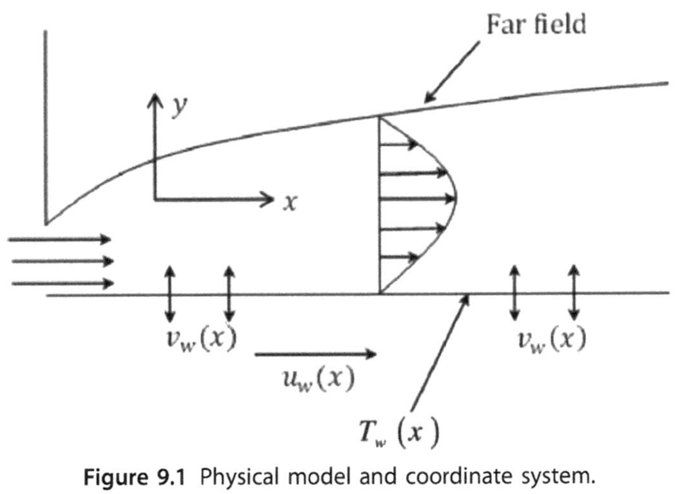

Figure 9.1 Physical model and coordinate system.

where u and v are the velocity components along the axes x and y, T is the fluid temperature, and Pr is the Prandtl number.

Glauert (1956) assumed that Eqs. (9.1) and (9.2) subject to the boundary conditions

$$v(x,0) = 0, \quad u(x,0) = 0, \quad u(x,y) \to 0 \ \text{as} \ y \to \infty \tag{9.4}$$

possess the similarity solution

$$u(x,y) = 4x^{-1/2} f'(\eta), \quad v(x,y) = -x^{-3/4} \left[f(\eta) - 3\eta f'(\eta) \right] \\ \eta = y\, x^{-3/4} \tag{9.5}$$

Substituting (9.5) into Eq. (9.2), we obtain the following ordinary (similarity) differential equation:

$$f''' + ff'' + 2f'^2 = 0 \tag{9.6}$$

along with the boundary conditions

$$f(0) = 0, \ f'(0) = 0, \ f'(\eta) \to 0 \ \text{as} \ \eta \to \infty \tag{9.7}$$

where prime denotes the differentiation with respect to η. A well-known nontrivial analytical solution of Eqs. (9.6) and (9.7) has been obtained by Glauert (1956)

$$\eta = 3^{1/2} arctang \left[\frac{(3f)^{1/2}}{2 + f^{1/2}} \right] + ln \left[\frac{\left(1 + f + f^{1/2}\right)^{1/2}}{1 - f^{1/2}} \right] \tag{9.8}$$

The solution exhibits exponential decay at infinity (e-jet).

9.2.2 Permeable wall jet

Merkin and Needham (1986) have considered the case of the permeable wall jet with suction or injection (blowing) velocity

$$v(x,0) = v_w(x), \quad u(x,0) = 0, \quad u(x,y) \to 0 \ \text{as} \ y \to \infty \tag{9.9}$$

These authors have shown that Eqs. (9.1) and (9.2) possess similarity solutions only when $v_w(x) < 0$ suction through the fixed wall is imposed and assumed that it is of the form $v_w(x) = -(1/4)\, S\, x^{-3/4}$. Merkin and Needham (1986) have looked for a similarity solution of Eqs. (9.1) and (9.2) of the form $\psi = x^{1/4}f(\eta)$ *and* $\eta = y/x^{3/4}$ and obtained the following ordinary (similarity) differential equation:

$$4f''' + ff'' + 2f'^2 = 0 \tag{9.10}$$

subject to the boundary conditions

$$f(0) = S,\ f'(0) = 0,\ f'(\eta) \to 0 \text{ as } \eta \to \infty. \tag{9.11}$$

Merkin and Needham (1986) have then considered the case of the wall jet along a permeable moving wall. Thus, Eqs. (9.1) and (9.2) have to be solved along with the boundary conditions

$$v = v_w(x),\quad u = u_w(x)\varepsilon \text{ on } y = 0,\quad u \to 0 \text{ as } y \to \infty \tag{9.12}$$

On integrating Eq. (9.2) and using (9.1), one gets the condition

$$\frac{dM}{dx} = \frac{1}{2}u_w^2 + v_w \int_0^\infty u^2 dy \tag{9.13}$$

where

$$M = \int_0^\infty u \left(\int_y^\infty u^2 ds \right) dy \tag{9.14}$$

Taking M to be a constant ($M = 1$, *say*), one gets

$$\frac{1}{2}u_w^2 + v_w \int_0^\infty u^2 dy = 0 \tag{9.15}$$

This relation shows that only the case when $v_w(x) < 0$, suction through the wall is possible. Thus, for a similarity solution of this problem, (Merkin & Needham, 1986) have taken $v_w(x) = -(1/4)S$ and $u_w(x) = U_0 x^{-1/2}\varepsilon$, where U_0 is the constant jet's characteristic velocity. Eq. (9.10) has to be solved now along with the boundary conditions

$$f(0) = S,\ f'(0) = \varepsilon,\ f'(\eta) \to 0 \text{ as } \eta \to \infty \tag{9.16}$$

and the momentum condition (taking $M = 1$).

$$\int_0^\infty (f - S) f'^2 d\eta = 1 \tag{9.17}$$

with S and ε connected by the relations

$$2\varepsilon^2 = S \int_0^\infty f'^2 d\eta \tag{9.18}$$

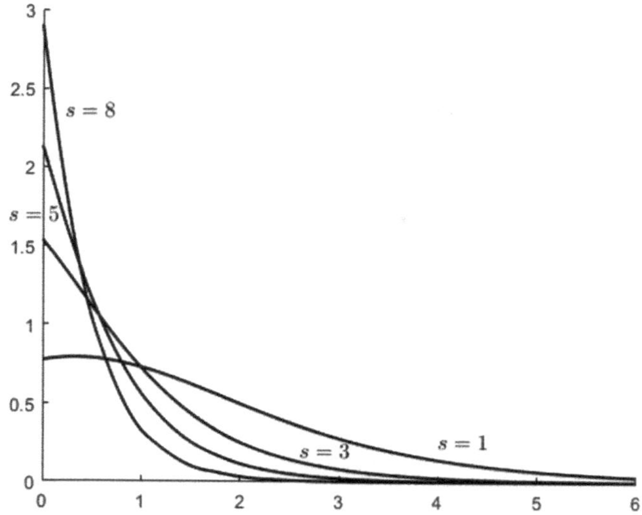

Figure 9.2 Graphs of $f'(\eta)$ for several values of *S*. *(From Merkin, J. H., & Needham, D. J. (1986). A note on the wall-jet problem.* Journal of Engineering Mathematics, *20(1), 21–26. https://doi.org/10.1007/BF00039320 25.)*

Eq. (9.10) can be integrated twice to give

$$f' = \frac{1}{6} f^{1/2}\left(\omega^3 - f^{3/2}\right) \tag{9.19}$$

where $\omega^2 = f(\infty)$ is a constant, that will be determined by solving (9.17). The velocity profiles $f'(\eta)$ for several values of S obtained by numerically solving Eq. (9.19) using appropriate values of ω given in Table 1 of the paper by Merkin and Needham (1986) are shown in Fig. 9.2.

9.2.3 Existence of the solutions

Needham and Merkin (1987) have shown that Eq. (9.10) with the boundary conditions (9.16) has no solutions when

$$i)\ S \neq 0,\ \ \varepsilon = 0;\ \ ii)\ S = 0, \varepsilon \neq 0;\ \ iii)\ S < 0 \text{ any } \varepsilon \tag{9.20}$$

while solutions are possible for some values of S and ε. These solutions can be obtained by solving the following equation:

$$f' = \frac{1}{6}\left(\omega^3 f^{1/2} - f^2\right) \equiv G(f, \omega) \tag{9.21}$$

where $\omega^2 = f(\infty)$ is determined using (9.17). An examination of this equation shows that $G(f, \omega)$ is defined only in $f(\eta) \geq 0$ where it has two branches, an upper branch, in which the positive sign is taken for the square root, and a lower branch, in which the negative sign is considered. Also, $G(f,\omega) = 0$ at $f(\eta) = 0$. Fig. 9.3 illustrates a plot of $G(f,\omega)$ as defined by (9.21) showing the upper and lower branch solutions from $f(0) = S$. A very detailed study of these solutions has been presented by Needham and Merkin (1987), and we will not present these details here.

On the other hand, (Magyari & Weidman, 2005, 2006) have solved the thermal Eq. (9.3) subject to the boundary conditions:

$$\frac{\partial T}{\partial y} = 0 \text{ at } y = 0, \quad T \rightarrow 0 \text{ as } y \rightarrow \infty \tag{9.22}$$

assuming that $T = x^{-1/3}$. Thus, Eq. (9.3) for the temperature similarity variable $\theta(\eta)$ is given by

$$(3\theta' + Pr\, f\theta)' = 0 \tag{9.23}$$

with the boundary conditions

$$\theta'(0) = 0, \quad \theta(\eta) \rightarrow 0 \text{ as } \eta \rightarrow \infty \tag{9.24}$$

The temperature profiles can be expressed as (see (Magyari & Weidman, 2005))

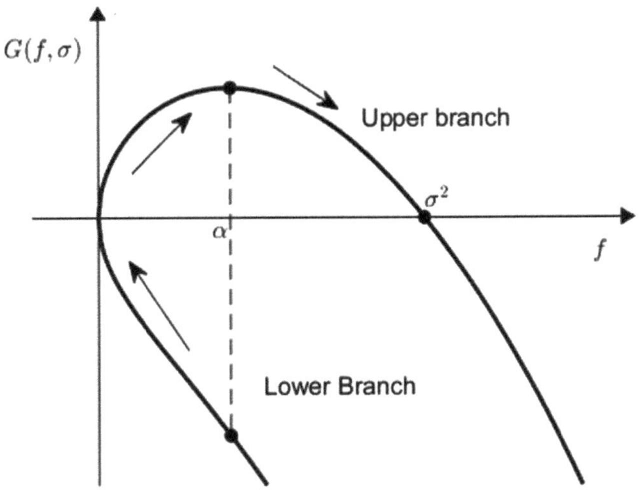

Figure 9.3 Plot of $G(f,\omega)$ as defined by (9.21) showing the upper and lower branch solutions from $f(0) = S$. *(From Needham, D., & Merkin, J. (1987). A note on the wall-jet problem II.* Journal of Engineering Mathematics, 21*(1), 17–22. https://doi.org/10.1007/ BF00127689 19.)*

$$\theta(\eta) = \theta(0)\ exp\left[-Pr\int_0^\eta f(s)ds\right] \equiv \theta(0)G(\eta) \tag{9.25}$$

where $\theta(0)$ is still undetermined and $G(\eta)$ can be written as

$$G(\eta) = \left[\frac{\sqrt{3}Ai(0) + Bi(0)}{\sqrt{3}Ai(z) + Bi(z)}\right]^{2\ Pr},\quad z = [f''(0)/2]^{1/3}\eta \tag{9.26}$$

for the Airy jet and

$$G(\eta) = \left[1 - f(\eta)^{3/2}\right]^{Pr} \tag{9.27}$$

for the classical TAG jet, where $Ai(z)$ and $Bi(z)$ are Airy functions.

The similar temperature profiles, given by Eqs. (9.26) and (9.27), are compared in Fig. 2 of the paper by (Magyari & Weidman, 2005) for $Pr = 0.1, 1$, and 10, when $f''(0) = 2/9$. It was shown that while for small values of Pr in Fig. 2a, the Airy jet temperature profile deviates substantially from that of classical wall jet, with increasing P, the deviation diminishes such that at $Pr = 10$ in Fig. 2c, the difference is almost imperceptible.

9.3 Jet profile solutions of the Falkner–Skan equation

9.3.1 Introduction

The Falkner–Skan equation subject to the no slip boundary condition is central to boundary layer theory (see (Hermann Schlichting & Gersten, 2003)). Traditionally it represents a family of similarity solutions which can be associated with local external flows varying as $U(x) \approx x^m$, where m is a constant. For accelerating flows, the solutions can specifically be associated with flow past a wedge. Solutions also exist which can be related to decelerating flows. The existence of dual solutions for negative showing forward and reverse flow, respectively, has determined (Stewartson, 1954) to discuss their relevance to the separation phenomenon. He thus, for the first time, examined the Falkner–Skan equation with respect to a zero shear boundary condition. Subject to the constraint of exponential decay, Stewartson identified new solutions displaying a velocity defect relative to the external flow.

Astin and Wilks (1996) have reported on the existence of new solutions of the Falkner–Skan equation, which in contrast displays jet-like profiles. The new solutions were presented in detail, and a general discussion indicates a variety of estimates of the boundary data which facilitate the identification of the exact numerical solutions.

9.3.2 Basic equations

The governing Eqs. (1.1) and (1.2) can be written in the Cartesian coordinates x and y, using the boundary layer approximations as

$$\frac{\partial u}{\partial x}+\frac{\partial v}{\partial y}=0 \tag{9.28}$$

$$u\frac{\partial u}{\partial x}+v\frac{\partial u}{\partial y}=U\frac{dU}{dx}+\nu\frac{\partial^2 u}{\partial y^2} \tag{9.29}$$

Astin and Wilks (1996) have looked for a similarity solution of these equations of the form

$$\psi(x,y)=\left(\frac{2\nu U}{m+1}\right)^{1/2} x^{1/2}f(\eta),\quad \eta = y\left[\frac{U(m+1)}{2\nu x}\right]^{1/2} \tag{9.30}$$

which reduces Eqs. (9.28) and (9.29) to the well-known Falkner–Skan equation

$$f''' + ff'' + \beta(1-f'^2)=0 \tag{9.31}$$

subjected to the boundary conditions

$$f''' + ff'' + \beta(1-f'^2)=0 \tag{9.32}$$

where $\beta = 2m/(m+1)$. Note that $f(\eta) = \eta$ is an exact solution of Eq. (9.29) for all β. In general, however, the solution requires the identification of $f'(0)$ for varying β. For $\beta > 0$ there are no solutions with $|f'(\eta) \leq 1|$. Stewartson's solutions covert firstly the range $-0.19884 < \beta < 0$. Here dual solutions exist with $f'(0) < 0$. The lower limit has $f'(0) = 0$ and corresponds in $(f'(0),\ f''(0),\ \beta)$ space to the separation profile of the conventional Falkner–Skan family obtained by Hartree (1937). The remainder of Stewartson's solutions relate to the range $-0.5 < \beta < -0.19884$ and are characterized by $1 > f'(0) > 0$.

Fig. 9.4 shows the representative velocity profiles $f'(\eta)$ determined by Astin and Wilks (1996) and which illustrate the wake interpretation.

Further, (Astin & Wilks, 1996) have studied the velocity jet profiles over the range $-1 < \beta < -0.5$. These profiles are illustrated in Fig. 9.5 for a

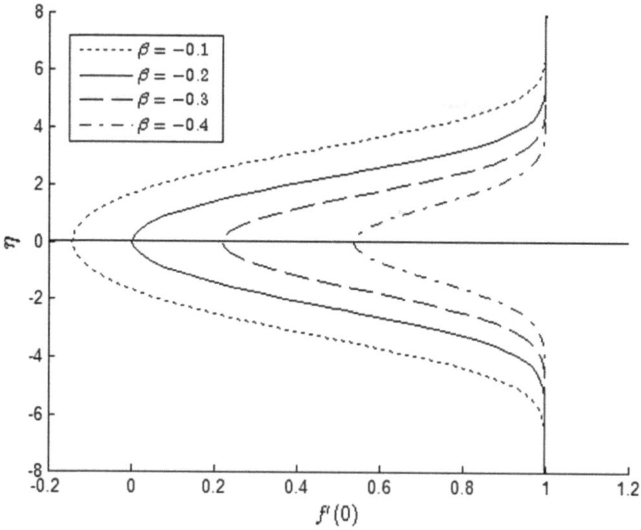

Figure 9.4 Wake profiles of the Falkner–Skan equations for $-0.5 < \beta < 0$. *(From Astin, P., & Wilks, G. (1996). Jet profile solutions of the Falkner-Skan equation.* ZAMP Zeitschrift für Angewandte Mathematik und Physik, *47(5), 790–798. https://doi.org/10.1007/BF00915275 795.)*

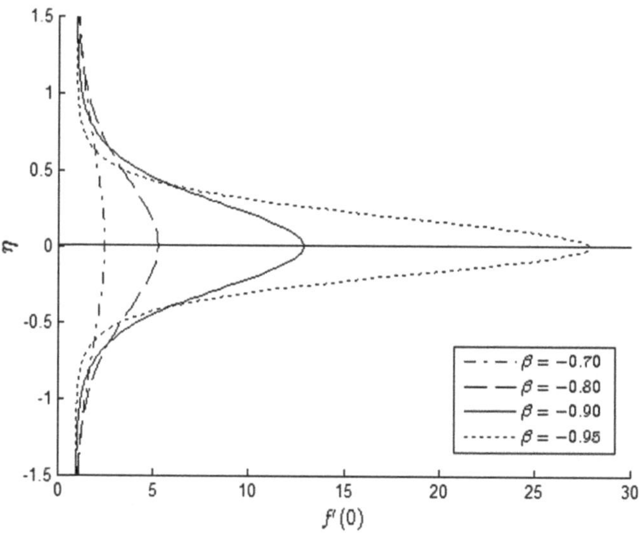

Figure 9.5 Jet velocity profiles $f'(\eta)$ of the Falkner–Skan equations for $-1 < \beta < 0.5$. *(From Astin, P., & Wilks, G. (1996). Jet profile solutions of the Falkner-Skan equation.* ZAMP Zeitschrift für Angewandte Mathematik und Physik, *47(5), 790–798. https://doi.org/10.1007/BF00915275 795.)*

selection of β values. They have also determined the following estimation for $f'(0)$

$$f'(0) = \left(\frac{(1-\sqrt{2}) + (1-2\sqrt{2})\beta)}{\beta+1}\right) \tag{9.33}$$

It transpires that (9.33) is a considerable improvement on Stewartson's original asymptotic estimates near $\beta = -1/2$. Both the profiles and the relation (9.33) highlight the singular limit of the new solutions as $\beta \rightarrow -1$.

9.4 Numerical modeling of Glauert type exponentially decaying wall jet flows of nanofluids using Tiwari and Das' nanofluid model

9.4.1 Introduction

Jafarimoghaddam and Pop (2019) have investigated very recently the laminar nanofluid flow and heat transfer of a jet spreading out over a permeable heated flat surface using the mathematical nanofluid model proposed by Tiwari and Das (2007). Similarity solutions of the boundary layer equations were been sought, according to which the forms of the velocity distribution across the jet and the heat transfer coefficient associated with the hot fluid on the surface of the plate are assumed. Explicit analytical solutions with high precision have been given for both the velocity and the temperature distributions.

9.4.2 Basic equations

The basic boundary layer equations are in dimensionless form

$$\frac{\partial u}{\partial x} + \frac{\partial v}{\partial y} = 0 \tag{9.34}$$

$$u\frac{\partial u}{\partial x} + v\frac{\partial v}{\partial y} = A_1\frac{\partial^2 u}{\partial y^2} \tag{9.35}$$

$$u\frac{\partial \theta}{\partial x} + v\frac{\partial \theta}{\partial y} = A_2\frac{\partial^2 \theta}{\partial y^2} \tag{9.36}$$

where

$$A_1 = \frac{1}{(1-\varphi)^{2.5}\left(1-\varphi+\varphi\rho_p/\rho_f\right)}, \quad A_2 = \frac{1}{Pr}\frac{k_{nf}/k_f}{1-\varphi+\varphi\left(\rho C_p\right)_p/\left(\rho C_p\right)_f} \tag{9.37}$$

the physical properties of the base fluid and nanoparticles are being defined in Table 1.1 and thermophysical properties of the nanofluids are defined in Table 1.2.

Following (Glauert, 1956), we define the similarity variables

$$\eta = x^{-3/4} y, \quad \psi = x^{1/4} f(\eta) \tag{9.38}$$

Using (9.38) in Eq. (9.35), one gets

$$A_1 f''' + \frac{1}{4} ff'' + \frac{1}{2} f'^2 = 0 \tag{9.39}$$

with the boundary conditions

$$f(0) = S, \; f'(0) = \varepsilon, \; f(\eta) \to 0 \; as \; \eta \to \infty \tag{9.40}$$

Integrating Eq. (9.39) from η to ∞, we get

$$-A_1 f'' - \frac{1}{4} ff' + \frac{1}{4} \int_\eta^\infty f'^2(s) ds = 0 \tag{9.41}$$

Multiplying this equation by $f'(\eta)$ and integrating the resulted equation from η to ∞, we obtain

$$A_1 f'^2(\eta) - \frac{1}{2} f(\eta) \int_\eta^\infty f'^2(s) ds = 0 \tag{9.42}$$

Therefore, the boundary conditions at the wall suggest

$$A_1 \varepsilon^2 - \frac{1}{2} S \int_0^\infty f'^2(s) ds = 0 \tag{9.43}$$

On the other hand, (Jafarimoghaddam & Pop, 2019) have defined the Flux of the Exterior Momentum as

$$\int_0^\infty g(\eta) f'(\eta) d\eta = 1 \tag{9.44}$$

Integrating by parts this equation, one gets

$$\int_0^\infty [f(\eta) - S] f'(\eta)^2 d\eta = 1 \tag{9.45}$$

It should be mentioned that for the pure base fluids ($A = 1$), Eq. (9.43) becomes that previously considered by Merkin and Needham (1986),

Needham and Merkin (1987), indicating that S should be positive, permeating only suction. Multiplying Eq. (9.39) $b\, f(\eta)$ and integrating once leads to

$$A_1\left(ff'' - \frac{1}{2}f'^2\right) + \frac{1}{4}f^2 f' = C_1 \tag{9.46}$$

Boundary condition $f'(\eta) \to 0$ as $\eta \to \infty$ gives $C_1 = 0$. Multiplying Eq. (9.46) by $f(\eta)^{-3/2}$ and integrating it, the following equation is obtained:

$$f'(\eta) = \frac{1}{6A_1} f^{1/2}(\eta)\ \left[\omega^3 - f^{3/2}(\eta)\right] \tag{9.47}$$

where ω^2 is given by the algebraic equation

$$\omega^8 - \frac{20}{9}\ S\omega^6 + \frac{16}{9}S^{5/2}\ \omega^3 - 5\left(8A_1 + \frac{1}{9}S^4\right) = 0 \tag{9.48}$$

Further, (Jafarimoghaddam & Pop, 2019) have determined the reduced skin friction along the jet's wall $f''(0)$ as

$$f''(0) = \frac{1}{24A_1^2}\left(w^2 - S^{3/2}\right)\left[\frac{1}{3}\left(w^3 - S^{3/2}\right) - S^{3/2}\right] \tag{9.49}$$

After using $\theta = \theta(\eta)$, Eq. (9.36) becomes

$$\theta'' + \frac{1}{4A_2} f\theta' = 0 \tag{9.50}$$

along with the boundary conditions

$$\theta(0) = 1, \quad \theta(\eta) \to 0 \ \text{as} \ \eta \to \infty \tag{9.51}$$

The analytical solution of Eqs. (9.50) and (9.51) can be expressed as

$$\theta(\eta) = 1 + \theta'(0)\int_0^\eta exp\left[-\frac{1}{4A_2}\int_0^s f(\eta)d\eta\right] \tag{9.52}$$

where

$$\theta'(0) = -\frac{1}{\int_0^\infty exp\left[-\frac{1}{4A_2}\int_0^\eta f(s)ds\right]} \tag{9.53}$$

On using Eq. (9.52), we write

$$\int_0^{\eta} f(s)ds = -4A_1 ln\left(\frac{\omega^3 - f^{3/2}}{\omega^3 - S^{3/2}}\right) \tag{9.54}$$

Hence, we have

$$\theta(\eta) = 1 + 6A_1\left(\omega^3 - S^{3/2}\right)\int_a^{\frac{A_1}{A_2}f(\eta)} \theta'(0)\left(\omega^3 - f^{3/2}\right)^{\frac{A_1}{A_2}-1} f(s)^{-1/2} df(s) \tag{9.55}$$

and

$$\theta'(0) = \frac{\left(\omega^3 - S^{3/2}\right)^{A_1/A_2}}{6A_1 \int_a^{\omega^2}\left(\omega^3 - f^{3/2}\right)^{\frac{A_1}{A_2}-1} f^{-1/2} df} \tag{9.56}$$

The quantities of practical interest are the skin friction coefficient C_f and the local Nusselt number Nu_x, which can be easily shown that are given by

$$Re_x^{5/4} C_f = \frac{2^{1/4} f''(0)}{(1-\varphi)^{2.5}}, \quad Re_x^{-1/4} Nu_x = -2^{-1/4}\frac{k_{nf}}{k_f}\theta'(0) \tag{9.57}$$

9.4.3 Comments

Figs. 9.6 and 9.7 show the variation of the reduced skin friction coefficient $f''(0)$ as a function of the nanofluid volume fraction parameter φ for various nanofluids Cu/water, Al_2O_3/water and TiO_2/water when $S = 0$ and 1. It can be deduced from these figures that the behavior of Cu/water nanofluid is noticeably different than the other types of nanofluids considered in this work. It can then be observed that the problem admit multiple (dual) solutions, an upper and lower solutions, respectively.

Also, the velocity profiles $f'(\eta)$, shown in Figs. 9.8 and 9.9, possess upper and lower branch solutions.

The same happens for the temperature profiles and reduced heat flux $-\theta'(0)$ illustrated in Figs. 20–26 in the paper by Jafarimoghaddam and Pop (2019). In order to see which of these dual solutions are stable and applicable in practice and which solutions are not stable, a stability analysis has been performed by Jafarimoghaddam and Shafizadeh (2019), but we will not present this analysis here.

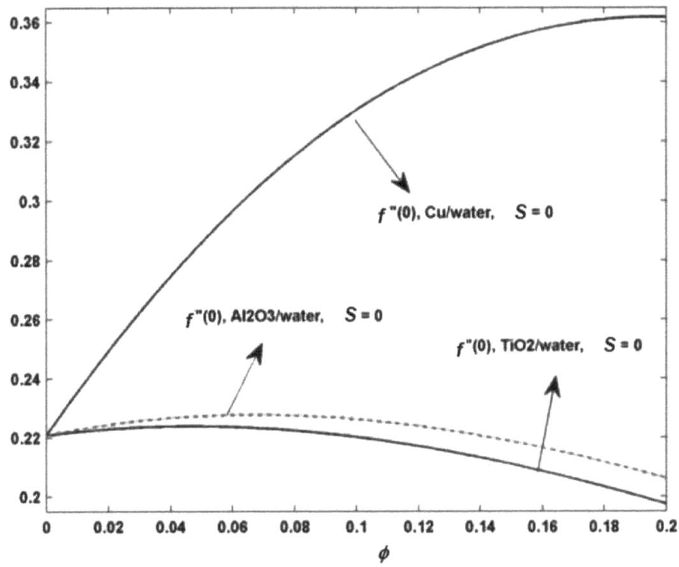

Figure 9.6 Variation of the reduced skin friction $f''(0)$ with φ for different nanofluids when $S = 0$. *(From Jafarimoghaddam, A., & Pop, I. (2019). Numerical modeling of Glauert type exponentially decaying wall jet flows of nanofluids using Tiwari and Das' nanofluid model.* International Journal of Numerical Methods for Heat & Fluid Flow, *29(3), 1010–1038. https://doi.org/10.1108/HFF-08-2018-0437 12.)*

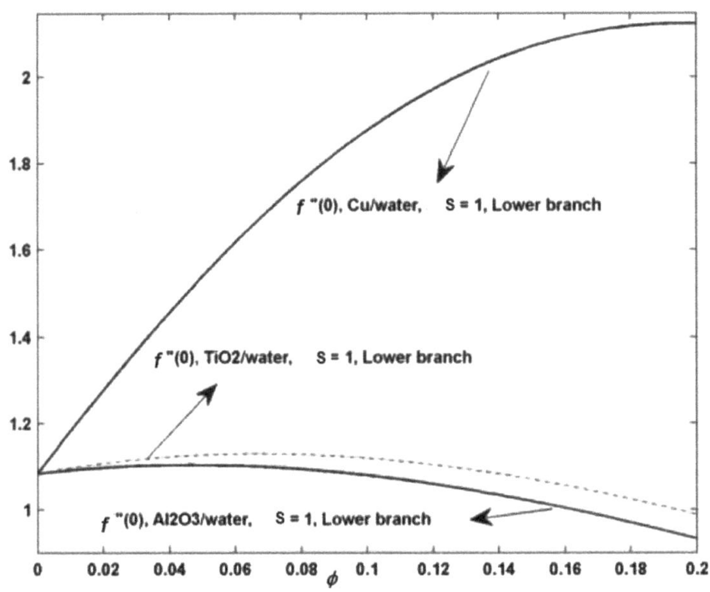

Figure 9.7 Variation of the reduced skin friction $f''(0)$ with φ for different nanofluids when $S = 1$ (lower similarity branch). *(From Jafarimoghaddam, A., & Pop, I. (2019). Numerical modeling of Glauert type exponentially decaying wall jet flows of nanofluids using Tiwari and Das' nanofluid model.* International Journal of Numerical Methods for Heat & Fluid Flow, *29(3), 1010–1038. https://doi.org/10.1108/HFF-08-2018-0437 12.)*

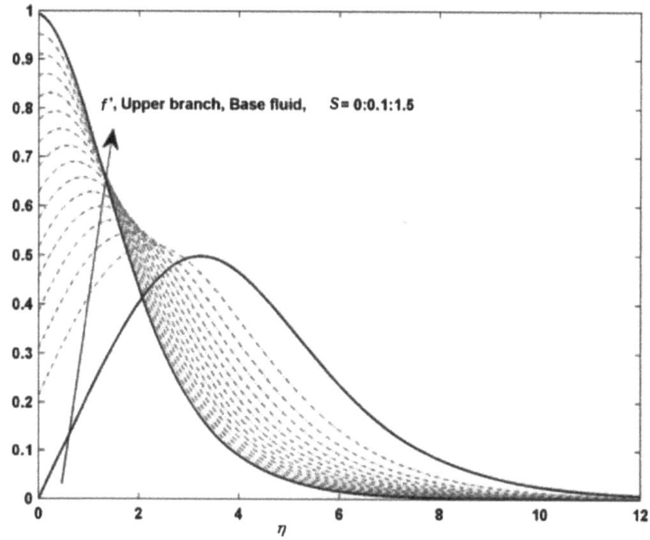

Figure 9.8 Variation of the velocity profiles $f'(\eta)$ with S for the base fluid (pure water). Upper similarity branch. *(From Jafarimoghaddam, A., & Pop, I. (2019). Numerical modeling of Glauert type exponentially decaying wall jet flows of nanofluids using Tiwari and Das' nanofluid model.* International Journal of Numerical Methods for Heat & Fluid Flow, *29(3), 1010–1038. https://doi.org/10.1108/HFF-08-2018-0437 12.)*

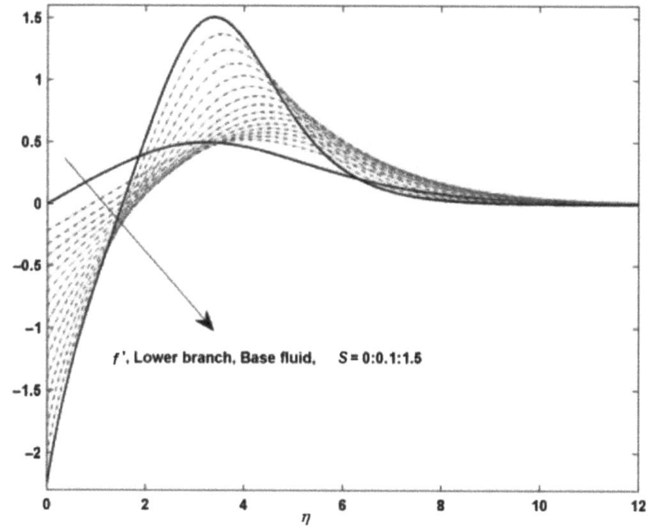

Figure 9.9 Variation of the velocity profiles $f'(\eta)$ with S for the base fluid (pure water). Lower similarity branch. *(From Jafarimoghaddam, A., & Pop, I. (2019). Numerical modeling of Glauert type exponentially decaying wall jet flows of nanofluids using Tiwari and Das' nanofluid model.* International Journal of Numerical Methods for Heat & Fluid Flow, *29(3), 1010–1038. https://doi.org/10.1108/HFF-08-2018-0437 12.)*

It should be also mentioned that (Cohen et al., 1992) presented a new family of solutions describing the mean flow of an incompressible two-dimensional wall jet subjected to boundary suction or blowing, in which the vertical velocity at the wall varies in a power law with downstream distance, described by the following equation:

$$f''' + ff'' + \Lambda f'^2 = 0 \tag{9.58}$$

subject to the boundary conditions

$$f(0) = S,\ f'(0) = 0,\ f'(\eta) \to 0 \text{ as } \eta \to \infty \tag{9.59}$$

The following three cases were considered by Cohen et al. (1992): $\Lambda > 2$, $\Lambda = 2$ Glauert solution), and $2 > \Lambda > 1$. A linear stability has been performed, and it indicates that two unstable modes, an inviscid inflectional point instability and a viscous instability, may coexist in the flow. It is demonstrated theoretically that the relative importance of each mode can be controlled by subjecting the wall jet to small amounts of blowing or suction.

Xu and Wu (2011) extended the study of (Cohen et al., 1992) to the case of steady laminar wall jet over a stretching surface in the presence of lateral suction or injection. Similarity solutions absent in previous publications are found in some new ranges of parameters in the governing equation. The accuracy and reliability of the solutions have been checked through detailed convergence study and compared with the solutions from the numerical method and analytic method, and excellent agreement has been found.

We mention also that (Raees et al., 2014) have investigated the laminar nanofluid flow and heat transfer due to a jet spreading out over a convectively heated flat surface. Using the for the nanofluid the mathematical model proposed by Tiwari and Das (2007), similarity solutions of the boundary layer equations are given, according to which the forms of the velocity distribution across the jet and the heat transfer coefficient associated with the hot fluid on the lower surface of the plate are assumed. Explicit HAM approximations, founded by Liao (2003), were given for both the velocity and the temperature distributions. Besides, the important quantities of the practical interests including the boundary layer thickness, the skin friction coefficient, the Nusselt number, as well as the overall surface heat transfer rate are computed and discussed. It is found, under others, that for the variation of the velocity profiles $f(\eta)$ with η and various values of the volume fraction parameter φ, there is a critical value η_c of η corresponding to the maximum velocity, when $\eta < \eta_c$. The velocity profile $f'(\eta)$ increases dramatically as η enlarges, while $f'(\eta)$ decreases gradually when $\eta > \eta_c$. This is due to the fluid flow near the flat surface being restrained by the resting wall, while the flow at far field is retarded by the ambient quiescent fluid.

Finally, we point out that (Merkin & Weidman, 2020) studied the free convection wall jet flow with a prescribed wall temperature, which enables the boundary layer equations to be reduced to similarity form. It is seen that there is a finite range $Pr_1 < Pr < Pr_0$ of the Prandtl number for which there are solutions. The singular nature of the solution as $Pr \rightarrow Pr_0$ and $Pr \rightarrow Pr_1$ has been examined.

In the Russian literature, we mention the paper by Akatnov (1953) and the book by Vulis and Kashkarov (1965).

Nomenclature

A_1, A_2	constants
$A_i(z), B_i(z)$	Airy functions
C_f	skin friction coefficient
(ρC_p)	heat capacitance (J/K m^3)
$f'(\eta)$	dimensionless velocity functions
$g(\eta)$	function
G	function
k	thermal conductivity (W/m K)
m	constant
M	exterior momentum flux
Nu_x	local Nusselt number
Pr	Prandtl number
Re_x	local Reynolds number
S	suction/injection parameter
T	dimensional fluid temperature (K)
$T_w(x)$	surface temperature (K)
(u,v)	velocity components (m/s)
$u_w(x)$	surface stretching velocity (m/s)
U	characteristic velocity (m/s)
U_0	external flow velocity (m/s)
$u_w(x)$	suction/injection velocity (m/s)
(x, y, z)	Cartesian coordinates (m)

Greek letters

β	constant
ε	stretching parameter
Λ	constant
η	dimensionless similarity variable
φ	volume fraction of the nanofluid
ψ	stream function (m^2/s)
ρ	density of fluid (kg/m^3)
$\theta(\eta)$	dimensionless temperature function
ω	constant

Subscripts

c	critical value
f	fluid phase
nf	nanofluid
w	condition at the wall

Superscripts

$'$	differentiation with respect to η

References

Akatnov, N. I. (1953). Development of 2D laminar jet along a solid surface. *Leningrad Politekh. Inst. Trudy, 5*, 24–31.

Aly, E. H., & Pop, I. (2020). Merkin and Needham wall jet problem for hybrid nanofluids with thermal energy. *European Journal of Mechanics - B: Fluids, 83*, 195–204. https://doi.org/10.1016/j.euromechflu.2020.05.004

Astin, P., & Wilks, G. (1996). Jet profile solutions of the Falkner-Skan equation. *ZAMP Zeitschrift für Angewandte Mathematik und Physik, 47*(5), 790–798. https://doi.org/10.1007/BF00915275

Astin, P., & Wilks, G. (1997). Laminar jet assimilation into non-uniform flows. *Proceedings of the Royal Society of London. Series A: Mathematical, Physical and Engineering Sciences, 453*(1958), 593–606. https://doi.org/10.1098/rspa.1997.0033

Bansal, J. L., & Tak, S. S. (1978). Approximate solutions of heat and momentum transfer in laminar plane wall jet. *Flow, Turbulence and Combustion, 34*(2–3), 299–311. https://doi.org/10.1007/BF00418874

Barker, J., & Wilks, G. (2002). Axisymmetric wake and jet solutions in decelerating streams. *Zeitschrift für Angewandte Mathematik und Physik, 53*, 35–47. https://doi.org/10.1007/s00033-002-8140-0

Cohen, J., Amitay, M., & Bayly, B. J. (1992). Laminar–turbulent transition of wall-jet flows subjected to blowing and suction. *Physics of Fluids A: Fluid Dynamics, 4*(2), 283–289. https://doi.org/10.1063/1.858304

Glauert, M. B. (1956). The wall jet. *Journal of Fluid Mechanics, 1*(06), 625–643. https://doi.org/10.1017/S002211205600041X

Gorla, R. S. R. (1973). Unsteady heat transfer characteristics of a two dimensional laminar wall jet. *International Journal of Engineering Science, 11*(8), 841–851. https://doi.org/10.1016/0020-7225(73)90033-5

Hartree, D. (1937). On an equation occurring in Falkner and Skan's approximate treatment of the equations of the boundary layer. *Mathematical Proceedings of the Cambridge Philosophical Society, 33*(2), 223–239. https://doi.org/10.1017/S0305004100019575

Jafarimoghaddam, A. (2017). Closed form analytic solutions to heat and mass transfer characteristics of wall jet flow of nanofluids. *Thermal Science and Engineering Progress, 4*, 175–184. https://doi.org/10.1016/j.tsep.2017.09.006

Jafarimoghaddam, A. (2018a). Two-phase modeling of magnetic nanofluids jets over a stretching/shrinking wall. *Thermal Science and Engineering Progress, 8*, 375–384. https://doi.org/10.1016/j.tsep.2018.09.008

Jafarimoghaddam, A. (2018b). Wall jet flows of Glauert type: Heat transfer characteristics and the thermal instabilities in analytic closed forms. *European Journal of Mechanics - B: Fluids, 71*, 77–91. https://doi.org/10.1016/j.euromechflu.2018.04.002

Jafarimoghaddam, A., & Pop, I. (2019). Numerical modeling of Glauert type exponentially decaying wall jet flows of nanofluids using Tiwari and Das' nanofluid model. *International Journal of Numerical Methods for Heat and Fluid Flow, 29*(3), 1010–1038. https://doi.org/10.1108/HFF-08-2018-0437

Jafarimoghaddam, A., Pop, I., & Merkin, J. H. (2019). On the propagation of the non-similar wall jet flows with suction/injection. *The European Physical Journal Plus, 134*(5), 215. https://doi.org/10.1140/epjp/i2019-12647-5

Jafarimoghaddam, A., & Shafizadeh, F. (2019). Numerical modeling and spatial stability analysis of the wall jet flow of nanofluids with thermophoresis and brownian effects. *Propulsion and Power Research, 8*(3), 210–220. https://doi.org/10.1016/j.jppr.2019.06.002

Kalteh, A. M., & Abbassi, A. (2006). Similarity solution of laminar axisymmetric jets with effect of viscous dissipation. *Journal of Heat Transfer, 128*(10), 1099–1102. https://doi.org/10.1115/1.2345435

Launder, B. E., & Rodi, W. (1979). The turbulent wall jet. *Progress in Aerospace Sciences, 19*, 81–128. https://doi.org/10.1016/0376-0421(79)90002-2

Launder, B. E., & Rodi, W. (1983). The turbulent wall jet measurements and modeling. *Annual Review of Fluid Mechanics, 15*(1), 429–459. https://doi.org/10.1146/annurev.fl.15.010183.002241

Liao, S. (2003). *Beyond perturbation. Introduction to the homotopy analysis method*. Chapman and Hall/CRC. https://doi.org/10.1201/9780203491164

Magyari, E., & Keller, B. (2001a). The free jets as boundary-layer flows induced by continuous stretching surfaces. *Heat and Mass Transfer, 38*(1–2), 111–114. https://doi.org/10.1007/s002310000170

Magyari, E., & Keller, B. (2001b). The wall jet as limiting case of a boundary-layer flow induced by a permeable stretching surface. *Zeitschrift für Angewandte Mathematik und Physik, 52*, 696–703. https://doi.org/10.1007/s00033-001-8136-1

Magyari, E., & Keller, B. (2004). The algebraically decaying wall jet. *European Journal of Mechanics - B: Fluids, 23*(4), 601–605. https://doi.org/10.1016/j.euromechflu.2003.10.003

Magyari, E., & Weidman, P. D. (2005). The preheated Airy wall jet. *Heat and Mass Transfer, 41*(11), 1014–1020. https://doi.org/10.1007/s00231-005-0617-x

Magyari, E., & Weidman, P. D. (2006). Heat transfer characteristics of the algebraically decaying Glauert jet. *Heat and Mass Transfer, 43*(2), 165–173. https://doi.org/10.1007/s00231-006-0102-1

Merkin, J. H., & Needham, D. J. (1986). A note on the wall-jet problem. *Journal of Engineering Mathematics, 20*(1), 21–26. https://doi.org/10.1007/BF00039320
Merkin, J. H., & Weidman, P. D. (2020). An anomalous self-similar solution for a laminar wall jet flowing up a heated vertical wall. *European Journal of Mechanics - B: Fluids, 80*, 42–51. https://doi.org/10.1016/j.euromechflu.2019.11.009
Needham, D. J., & Merkin, J. H. (1987). A note on the wall-jet problem II. *Journal of Engineering Mathematics, 21*(1), 17–22. https://doi.org/10.1007/BF00127689
Raees, A., Xu, H., & Raees-ul-Haq, M. (2014). Explicit solutions of wall jet flow subject to a convective boundary condition. *Boundary Value Problems, 2014*(1), 163. https://doi.org/10.1186/1687-2770-2014-163
Rahimi-Esbo, M., Ranjbar, A. A., Ramiar, A., & Rahgoshay, M. (2012). Numerical simulation of forced convection of nanofluid in a confined jet. *Heat and Mass Transfer, 48*(12), 1995–2005. https://doi.org/10.1007/s00231-012-1040-8
Rehman, M. M. U., Qu, Z. G., Fu, R. P., & Xu, H. T. (2017). Numerical study on free-surface jet impingement cooling with nanoencapsulated phase-change material slurry and nanofluid. *International Journal of Heat and Mass Transfer, 109*, 312–325. https://doi.org/10.1016/j.ijheatmasstransfer.2017.01.094
Riley, N. (1958). Effects of compressibility on a laminar wall jet. *Journal of Fluid Mechanics, 4*(6), 615–628. https://doi.org/10.1017/S0022112058000707
Schlichting, H. (1933). Laminare strahlausbreitung. *ZAMM Zeitschrift für Angewandte Mathematik und Physik, 13*(4), 260–263. https://doi.org/10.1002/zamm.19330130403
Schlichting, H., & Gersten, K. (2003). In *Boundary-layer theory*. Springer Science & Business Media. https://doi.org/10.1007/978-3-662-52919-5
Schwarz, W. H., & Caswell, B. (1961). Some heat transfer characteristics of the two-dimensional laminar incompressible wall jet. *Chemical Engineering Science, 16*(3–4), 338–351. https://doi.org/10.1016/0009-2509(61)80041-9
Stewartson, K. (1954). Further solutions of the Falkner-Skan equation. *Mathematical Proceedings of the Cambridge Philosophical Society, 50*(3), 454–465. https://doi.org/10.1017/S030500410002956X
Tetervin, N. (1948). Laminar flow of a slightly viscous incompressible fluid that issues from a slit and passes over a flat plate. *Naca Tn, 1644*, 40.
Tiwari, R. K., & Das, M. K. (2007). Heat transfer augmentation in a two-sided lid-driven differentially heated square cavity utilizing nanofluids. *International Journal of Heat and Mass Transfer, 50*(9–10), 2002–2018. https://doi.org/10.1016/j.ijheatmasstransfer.2006.09.034
Turkyilmazoglu, M. (2014). A note on the exact temperatures of free two-dimensional and circular jets. *The European Physical Journal Plus, 129*(6), 120. https://doi.org/10.1140/epjp/i2014-14120-5
Turkyilmazoglu, M. (2016). Flow of nanofluid plane wall jet and heat transfer. *European Journal of Mechanics - B: Fluids, 59*, 18–24. https://doi.org/10.1016/j.euromechflu.2016.04.007
Turkyilmazoglu, M. (2019). Free and circular jets cooled by single phase nanofluids. *European Journal of Mechanics - B: Fluids, 76*, 1–6. https://doi.org/10.1016/j.euromechflu.2019.01.009
Vulis, L. A., & Kashkarov, B. P. (1965). *Theory of viscous fluids jets*. Nauka.
Weidman, P. (2017). Hiemenz stagnation-point flow impinging on a biaxially stretching surface. *Meccanica, 53*(4–5), 833–840. https://doi.org/10.1007/s11012-017-0761-7
Xu, H., Liao, S.-J., & Wu, G.-X. (2008). A family of new solutions on the wall jet. *European Journal of Mechanics - B: Fluids, 27*(3), 322–334. https://doi.org/10.1016/j.euromechflu.2007.07.002
Xu, H., & Wu, G. (2011). The jet over a stretching wall with suction or injection. *Science China Physics, Mechanics and Astronomy, 54*(3), 502–510. https://doi.org/10.1007/s11433-010-4234-4

Index

'*Note:* Page numbers followed by "f" indicate figures and "t" indicate tables.'

Kolibri 360 GmbH
Ferdinand-Jühlke-Straße 7
99095 Erfurt, Deutschland
produktsicherheit@kolibri360.de